Microorganisms for Sustainability

Volume 39

Series Editor

Naveen Kumar Arora, Environmental Microbiology, School for Environmental Science, Babasaheb Bhimrao Ambedkar University, Lucknow, Uttar Pradesh, India

Microorganisms perform diverse roles on our planet most of which are important to make earth a habitable and sustainable ecosystem. Many properties of microorganisms are being utilized as low input biotechnology to solve various problems related to the environment, food security, nutrition, biodegradation, bioremediation, sustainable agriculture, bioenergy and biofuel, bio-based industries including microbial enzymes/ extremozymes, probiotics etc. The book series covers all the wider aspects and unravels the role of microbes towards achieving a sustainable world. It focuses on various microbial technologies related to sustenance of ecosystems and achieving targets of Sustainable Development Goals. Series brings together content on microbe based technologies for replacing harmful chemicals in agriculture, green alternatives to fossil fuels, use of microorganisms for reclamation of wastelands/ stress affected regions, bioremediation of contaminated habitats, biodegradation purposes. Volumes in the series also focus on the use of microbes for various industrial purposes including enzymes, extremophilic microbes and enzymes, effluent treatment, food products.

The book series is a peer reviewed compendium focused on bringing up contemporary themes related to microbial technology from all parts of the world, at one place for its readers, thereby ascertaining the crucial role of microbes in sustaining the ecosystems.

Natarajan Amaresan •
Dhanasekaran Dharumadurai •
Olubukola Oluranti Babalola
Editors

Agricultural Microbiology Based Entrepreneurship

Making Money from Microbes

Springer

Editors
Natarajan Amaresan
C.G. Bhakta Institute of Biotechnology
Uka Tarsadia University
Surat, Gujarat, India

Dhanasekaran Dharumadurai
Department of Microbiology
Bharathidasan University
Tiruchirappalli, Tamil Nadu, India

Olubukola Oluranti Babalola
Faculty of Natural and Agricultural Sciences
North-West University
Mmabatho, South Africa

ISSN 2512-1901 ISSN 2512-1898 (electronic)
Microorganisms for Sustainability
ISBN 978-981-19-5746-8 ISBN 978-981-19-5747-5 (eBook)
https://doi.org/10.1007/978-981-19-5747-5

This Springer imprint is published by the registered company Springer Nature Singapore Pte Ltd.
The registered company address is: 152 Beach Road, #21-01/04 Gateway East, Singapore 189721, Singapore

Foreword

Agricultural microbiology based on entrepreneurship is also referred to as science entrepreneurship or bioscience enterprise in agriculture, life science entrepreneurship, entrepreneurship in microbiology, or agriculture microbiology enterprise. It is the entrepreneurship of microorganisms which acts as a bridge of innovation connecting academia and industry. It is the business of utilizing ideas and innovations of microorganisms and products obtained from bacteria, fungi, actinomycetes, and protozoa to gain profit and serve the society. Therefore, bioentrepreneurship encloses an understanding of an idea, invents something, and brings about the sum of all activities to create an environment and infrastructure for well-trained professionals to build research-based projects and their commercialization. Bioentrepreneurship is manifested as the key driver of the new bioeconomy. For a person to be a bioentrepreneur, few skills are prerequisite like positive energy, leadership qualities, appropriate track record, technical knowledge of the field, articulating and managerial skills. Agricultural microbiology is a branch of science that studies different microorganisms that are associated with plants, soil fertility, and even animal diseases. The microorganisms associated with plants include organisms of different groups like bacteria, fungi, actinomycetes, and protozoa. Agricultural microbiology has been extensively studied for the decade as it helps to understand the importance of relevant microbial strains to agricultural applications to minimize agricultural loss and increase soil fertility and harvest. There is a symbiotic relationship between plants and the associated microorganisms where both benefit from each other. Besides, different topics within agricultural microbiology help in the diagnosis and prevention of plant diseases that might result from plant pathogens. Agricultural microbiology helps in the understanding of the specific plant requirements that include soil texture, soil nutrients, water content, and associated microorganisms. Based on the studies made in agricultural microbiology, different microorganisms can be used for different purposes, all ultimately leading to minimizing loss of plant and plant products and increasing fertility and harvest. The dominant group of microorganisms found in the soil and plants is bacteria. These bacteria are present in a symbiotic relationship with the plants to help in processes like nitrogen fixation and mineral supply. Bacteria are followed by actinobacteria, as the second most dominant group. Actinobacteria are also studied in agricultural microbiology as they produce different groups of antibiotics and also help in increasing the fertility of the soil.

Agricultural microbiology based on entrepreneurship allows for the exploration of new and advanced techniques that can be used in agricultural practices to make the practices safer and reliable business model of biofertilizers, plant growth promoters, biopesticides, and biocompost production. This book comprises 20 chapters from contributors across the world. The book content has divisions as entrepreneurship, business plan, biofertilizer, and compost production. It outlines the concept of entrepreneur and entrepreneurship, government schemes for entrepreneurship programs, skills for entrepreneurs, ethical and legal issues in microbial products in India, and intellectual property principles in microbial technologies.

The second part of the book summarizes mass multiplication, production cost analysis, and marketing of *Rhizobium*, *Azotobacter*, *Azospirillum,* cyanobacterial, VAM, and *Frankia* biofertilizers. It also describes the production cost analysis and marketing of phosphate solubilizers and *Bacillus thuringiensis, Trichoderma, Metarhizium,* and *Beauveria* bioinsecticide. A special chapter on vermicomposting is also included in this book to create a centre of attention for readers in microbiology, biotechnology, agriculture, and business management.

I appreciate and congratulate the book editors Dr. Natarajan Amaresan, Assistant Professor, C.G. Bhakta Institute of Biotechnology, Uka Tarsadia University, Surat, India, Dr. Dharumadurai Dhanasekaran, Associate Professor, Department of Microbiology, Bharathidasan University, Tiruchirappalli, India, and Prof. Olubukola Oluranti Babalola, Faculty of Natural & Agricultural Sciences, North-West University, South Africa, for bringing the book on *Agricultural Microbiology-Based Entrepreneurship: Making Money from Microbes* as immense volume to the entrepreneur to open innovative entrepreneurship.

School of Life Sciences and
Biotechnology, Shanghai Jiao Tong
University, Shanghai, China

Fengli Zhang

Preface

Entrepreneurship is currently a hot topic in policy circles all around the world. Poverty, income inequality, and unemployment are rising, and strong entrepreneurship, shown in part through the formation and expansion of small and medium businesses, is vital for economic development. Therefore, it is critical to turn research findings into money-making initiatives that the unemployed may participate in. The book focuses on developing entrepreneurial capabilities to create a source of income using microbiological products.

Agricultural Microbiology-Based Entrepreneurship: Making Money from Microbes advocates a paradigm shift in research-related investigations toward microbial product technology transfer and commercialization. The book helps prospective entrepreneurs improve their knowledge and skills to commercialize items for broad usage. The book highlights the concept of entrepreneurship, skills for entrepreneurs and intellectual property principles, and ethical and legal issues frameworks of microbial products in the agricultural sector. Biofertilizers and composting are the major microbiological products required to increase crop yield and ensure sustainable food production. In-depth knowledge about different microbial-based fertilizers, multiplication, production costs and analysis, and marketing are discussed in the book.

This book is for everyone who has an inkling of wanting to make a fortune through microbiological research products. The book encourages academics, graduate students, and the unemployed to develop entrepreneurial skills to create a source of income. They can profit from the research presented in this book.

Surat, Gujarat, India — Natarajan Amaresan
Tiruchirappalli, Tamil Nadu, India — Dhanasekaran Dharumadurai
Mmabatho, South Africa — Olubukola Oluranti Babalola

Contents

Part I Entrepreneurship and Business Plan

1 **Concept of Entrepreneur and Entrepreneurship** 3
Swapnil Chaurasia

2 **Government Schemes for Entrepreneurship Programs** 25
Ami Naik and Pooja Patel

3 **Skills for Entrepreneurs** 41
Swapnil Chaurasia

4 **Intellectual Property Principles in Microbial Technologies** 63
R. Pathirajan, S. Gomathi, V. Ambikapathy, A. Panneerselvam, G. Gayathri, and A. Anbukumaran

5 **Ethical and Legal Issues in Microbial Products in India** 81
Pooja Patel and Ami Naik

Part II Biofertilizer and Compost Production

6 **Mass Multiplication, Production Cost Analysis, and Marketing of Cyanobacterial Biofertilizers** 103
V. T. Anju, Siddhardha Busi, and Madhu Dyavaiah

7 ***Rhizobium* Biofertilizers: Mass Production Process and Cost-Benefit Ratio Analysis** 125
Komal A. Chandarana and Natarajan Amaresan

8 **Mass Multiplication, Production Cost Analysis, and Marketing of VAM Fungal Biofertilizer** 133
Komal A. Chandarana, Harshida A. Gamit, Sapna Chandwani, Hetvi Y. Naik, and Natarajan Amaresan

9 **Mass Multiplication and Cost Analysis of *Frankia* Biofertilizer** 155
Thirumagal Thirugnanam and Dhanasekaran Dharumadurai

10 Mass Multiplication, Production Cost Analysis and Marketing of *Azospirillum* 169
S. Gomathi, V. Ambikapathy, and A. Panneerselvam

11 Mass Multiplication, Production Cost Analysis, and Marketing of *Azotobacter* 185
Prittesh Patel, Sajjan Grover, and Pushpendra Singh Chauhan

12 Mass Multiplication, Production Cost Analysis, and Marketing of *Pseudomonas* 195
Komal A. Chandarana, Harshida A. Gamit, Hetvi Y. Naik, Sapna Chandwani, and Natarajan Amaresan

13 Mass Production, Formulation, and Cost-Benefit Ratio Analysis of *Bacillus thuringiensis* Bioinsecticide 209
Komal A. Chandarana, Harshida A. Gamit, Hetvi Y. Naik, Sapna Chandwani, and Natarajan Amaresan

14 Mass Multiplication, Production Cost Analysis, and Marketing of *Trichoderma* 223
Dhruti Amin

15 Mass Multiplication, Production Cost Analysis, and Marketing of *Metarhizium* 237
Sapna Chandwani, Hetvi Y. Naik, Harshida A. Gamit, Komal A. Chandarana, and Natarajan Amaresan

16 Mass Multiplication, Production Cost Analysis, and Marketing of *Beauveria* 251
Sapna Chandwani, Hetvi Y. Naik, Harshida A. Gamit, Komal A. Chandarana, and Natarajan Amaresan

17 *Chaetomium* sp.: An Insight into its Antagonistic Mechanisms, Mass Multiplication, and Production Cost Analysis 267
Parthasarathy Seethapathy, Harish Sankarasubramanian, Rajendran Lingan, and Raguchander Thiruvengadam

18 Vermicomposting: A Leading Feasible Entrepreneurship 289
P. Kavitha

19 Mass Production and Marketing of Microbial Compost 307
Swati Patel and Urjita Sheth

20 Cost-Benefit Assessment of Solid-Based Biofertilizer Production . . . 329
R. Kannan and T. Seenivasa Moorthy

Series Editor, Editors and Contributors

About the Series Editor

Naveen Kumar Arora, Fellow of International Society of Environmental Botanists (FISEB), PhD in Microbiology, is Professor and Head of the Department of Environmental Science at Babasaheb Bhimrao Ambedkar University (A Central University), Lucknow, Uttar Pradesh, India. He is a renowned researcher in the field of environmental microbiology and biotechnology. His specific area of research is plant-microbe interactions, particularly plant growth-promoting rhizobacteria. He has more than 75 research articles published in premium international journals and several articles published in magazines and dailies. He is an editor of 25 books, published by Springer. He is a member of several national and international societies, Secretary General of Society for Environmental Sustainability, serves on the editorial board of four journals, and a reviewer of several international journals. He is also the editor in chief of the journal *Environmental Sustainability* published by Springer Nature. He has delivered lectures at conferences and seminars around the globe. He has a long-standing interest in teaching at the PG level and is involved in taking courses in bacteriology, microbial physiology, environmental microbiology, agriculture microbiology, and industrial microbiology. He has been advisor to 134 postgraduate and 11 doctoral students. He has been awarded for excellence in research by several societies and national and international bodies/organizations. Although an academician and researcher by profession, he has a huge obsession for the wildlife and its conservation and has authored a book, *Splendid Wilds*. He is the President of Society for Conservation of Wildlife and has a dedicated website www.naveenarora.co.in for the cause of wildlife and environment conservation.

About the Editors

Natarajan Amaresan is an Assistant Professor at C.G. Bhakta Institute of Biotechnology, Uka Tarsadia University, Gujarat. He has received his undergraduate, postgraduate, and doctorate degrees in Microbiology. Dr. N. Amaresan has over 16 years of experience in teaching and research in various allied fields of microbiology, mainly microbial ecology, plant-microbe interactions, and others. He has been awarded young scientist awards by Association of Microbiologists of India, National Academy of Biological Sciences, and recipient of visiting scientist fellowship from National Academy of India. He has published more than 60 research articles, 13 books, and 45 book chapters of national and international repute. He has handled various projects sponsored by DBT, DST, GEMI, etc., and guided students for their doctoral and master's degrees.

Dhanasekaran Dharumadurai is working as an Associate Professor, Department of Microbiology, Bharathidasan University, Tiruchirappalli, India. He was awarded UGC-Raman Post-Doctoral Fellowship and worked in the University of New Hampshire, USA. He was awarded Fellow of Linnean Society, London, UK, and INSA Visiting Scientist Fellowship by the National Science Academy. He has graduated 12 PhDs and 76 MScs with over 106 research publications, 49 book chapters, and an H-index of 30 and citations of 2850. He filed two Indian Patents and organized 8 national level seminar, conference, and workshop programs. He has edited 11 books and authored 4 laboratory manuals. His research focus is microbiome profiling of actinorhizal root nodules, lichen, poultry gut, and cattle's reproductive system.

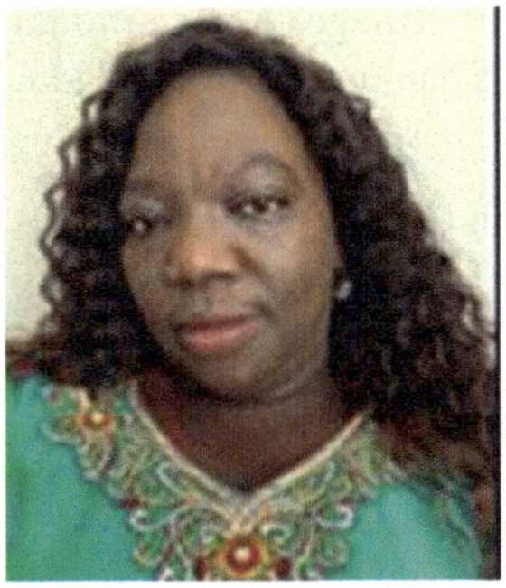

Olubukola Oluranti Babalola (Pr.Sci.Nat, MASSAF, FASLP, FTWAS, FAS), the Vice President of the Organization for Women in Science, is an NRF-rated scientist with over 20 years of experience in rhizosphere metagenomics and has an MBA. She has experience from the International Institute of Tropical Agricult^10,200 citations. Olubukola is #1 in Africa for Soil Science and Plant Nutrition. Her international experience spans the Americas, Asia, Europe, and Oceania.

Contributors

Natarajan Amaresan C.G. Bhakta Institute of Biotechnology, Uka Tarsadia University, Surat, Gujarat, India

V. Ambikapathy Department of Botany, A.V.V.M Sri Pushpam College (Autonomous) (Affiliated to Bharathidasan University, Trichy) Poondi, Thanjavur, Tamil Nadu, India

Dhruti Amin C. G. Bhakta Institute of Biotechnology, Uka Tarsadia University, Surat, Gujarat, India

A. Anbukumaran Department of Microbiology, Urumu Dhanalakshmi College, Tiruchirappalli, Tamil Nadu, India

V. T. Anju Department of Biochemistry and Molecular Biology, School of Life Sciences, Pondicherry University, Pondicherry, India

Komal A. Chandarana C.G. Bhakta Institute of Biotechnology, Uka Tarsadia University, Surat, Gujarat, India

Sapna Chandwani C.G. Bhakta Institute of Biotechnology, Uka Tarsadia University, Surat, Gujarat, India

Pushpendra Singh Chauhan Crop Life Science Ltd, Ankleshwar, Gujarat, India

Swapnil Chaurasia Shrimad Rajchandra Institute of Management and Computer Application (SRIMCA-MBA), Uka Tarsadia University, Surat, Gujarat, India

Dhanasekaran Dharumadurai Department of Microbiology, School of Life Sciences, Bharathidasan University, Tiruchirappalli, Tamil Nadu, India

Madhu Dyavaiah Department of Biochemistry and Molecular Biology, School of Life Sciences, Pondicherry University, Pondicherry, India

Harshida A. Gamit C.G. Bhakta Institute of Biotechnology, Uka Tarsadia University, Surat, Gujarat, India

G. Gayathri Department of Botany, A.V.V.M Sri Pushpam College (Autonomous) (Affiliated to Bharathidasan University, Trichy) Poondi, Thanjavur, Tamil Nadu, India

S. Gomathi Department of Botany, A.V.V.M Sri Pushpam College (Autonomous) (Affiliated to Bharathidasan University, Trichy) Poondi, Thanjavur, Tamil Nadu, India

Sajjan Grover Department of Entomology, University of Nebraska-Lincoln, Lincoln, NE, USA

R. Kannan Kalasalingam School of Agriculture and Horticulture (KSAH), Kalasalingam Academy of Research and Education (KARE), Krishnankoil, Tamil Nadu, India

P. Kavitha Department of Microbiology, Srimad Andavan Arts and Science College (Autonomous), Tiruchirappalli, Tamil Nadu, India

Rajendran Lingan Department of Plant Pathology, Centre for Plant Protection Studies, Tamil Nadu Agricultural University, Coimbatore, Tamil Nadu, India

T. Seenivasa Moorthy Kalasalingam School of Agriculture and Horticulture (KSAH), Kalasalingam Academy of Research and Education (KARE), Krishnankoil, Tamil Nadu, India

Ami Naik C.G. Bhakta Institute of Biotechnology, Uka Tarsadia University, Bardoli, Gujarat, India

Hetvi Y. Naik C.G. Bhakta Institute of Biotechnology, Uka Tarsadia University, Surat, Gujarat, India

A. Panneerselvam Department of Botany, A.V.V.M Sri Pushpam College (Autonomous) (Affiliated to Bharathidasan University, Trichy) Poondi, Thanjavur, Tamil Nadu, India

Pooja Patel Government Medical College, Surat, Gujarat, India

Prittesh Patel C.G. Bhakta Institute of Biotechnology, Uka Tarsadia University, Surat, Gujarat, India

Swati Patel ASPEE SHAKILAM Biotechnology Institute, Navsari Agricultural University, Surat, Gujarat, India

R. Pathirajan Department of Botany, A.V.V.M Sri Pushpam College (Autonomous) (Affiliated to Bharathidasan University, Trichy) Poondi, Thanjavur, Tamil Nadu, India

Harish Sankarasubramanian Department of Plant Pathology, Centre for Plant Protection Studies, Tamil Nadu Agricultural University, Coimbatore, Tamil Nadu, India

Parthasarathy Seethapathy Department of Plant Pathology, Amrita School of Agricultural Sciences, Amrita Vishwa Vidyapeetham, Coimbatore, India

Urjita Sheth Department of Biotechnology, School of Science and Technology, Vanita Vishram Women's University, Surat, Gujarat, India

Busi Siddhardha Department of Microbiology, School of Life Sciences, Pondicherry University, Pondicherry, India

Thirumagal Thirugnanam Department of Microbiology, School of Life Sciences, Bharathidasan University, Tiruchirappalli, Tamil Nadu, India

Raguchander Thiruvengadam Department of Plant Pathology, Centre for Plant Protection Studies, Tamil Nadu Agricultural University, Coimbatore, Tamil Nadu, India

Part I
Entrepreneurship and Business Plan

Chapter 1
Concept of Entrepreneur and Entrepreneurship

Swapnil Chaurasia

Abstract Through the passage of time, mankind has witnessed evolution of business structures. Right from sharp edged stones to forged metal swords and today laser sharpened sophisticated blades used for grooming purposes. All this transformation has not only been experienced through the products that we use and the businesses that we see today but also the concepts and methodology that gave rise to such transformations. One of this is entrepreneurship, a term that embraces everything that is new, everything that is useful and everything that has the potential to change the future course of action and the way we look at the world. Entrepreneurship can be considered as a natural phenomenon of every business. It is the integral nature that provides the element of an enterprise to a business moving beyond production, distribution, and exchange. The chapter aims at providing in-depth insights about various concepts pertaining to entrepreneurs and entrepreneurship through extensive review of literature covering broad dimensions of entrepreneurship. The chapter incorporates highlights from different pertinent theories related to entrepreneurship and tries to present an elaborated description of the concept and critical aspects associated with the same. As entrepreneurship is defined, understood, and percolated variously in different parts of the world, a model for entrepreneurship is also established using insights from various literatures. The last section of the chapter discusses the recent scenario of entrepreneurship in India and its accomplishments in recent times.

Keywords Entrepreneurship · Characteristics of an entrepreneur · Theories of entrepreneurship · Entrepreneurship in India

S. Chaurasia (✉)
Shrimad Rajchandra Institute of Management and Computer Application (SRIMCA-MBA), Uka Tarsadia University, Surat, Gujarat, India
e-mail: swapnil.chaurasia@utu.ac.in

N. Amaresan et al. (eds.), *Agricultural Microbiology Based Entrepreneurship*, Microorganisms for Sustainability 39,
https://doi.org/10.1007/978-981-19-5747-5_1

1.1 Introduction

Every company is driven through its vision and mission statements. Entrepreneurship is an integral element of any business life that adds to the accomplishments of the company. The people who are directly involved in company activities like the owners, management, directors, and the like are the ones who are most responsible for the realization of the vision that the company has set for itself. However, people come from a variety of backgrounds and have varied levels of education, which presents the diverse talent pool formed by entrepreneurship. This is inescapable and becomes a part of the entrepreneurial process. Furthermore, as economy and technology evolve, entrepreneurship becomes more diverse while also increasing its competitiveness. Innovation becomes the focal point from this perspective. However, overcoming the difficulties of entrepreneurship becomes a task. However, it is critical for us to recognize that not all entrepreneurship activity is beneficial to a corporate organization unless and until entrepreneurial skills and management practices are implemented within the firm to allow for change and learning. Every organization strives to understand and implement elements of entrepreneurship. Thus a close and critical look into the talents is needed. Meanwhile, a thorough understanding of the definition of entrepreneurship is required.

1.2 Defining Entrepreneurship

Since entrepreneurship is considered to be a natural phenomenon in the inception and the journey of a business that can affect the company's performance and its impact on its important stakeholders, it becomes imperative for any business enterprise or entrepreneur to first understand the definition of entrepreneurship before attempting to resolve business challenges and issues (Table 1.1).

Business historians and eminent researchers pioneered the very reservoir of information and knowledge pertaining entrepreneurship in the 1940s and 1950s. Significant institutional drivers of the research agenda in light of entrepreneurship conducted under the aegis of Harvard Business School-based Interdisciplinary Center for Study on Entrepreneurial History, through its extensive and quality-driven publications that focused on emergence and evolution of the concept of entrepreneurship. These works and areas were extensively explored by Alfred Chandler and Joseph Schumpeter and stand significant till date. However, entrepreneurship research hit certain notable methodological barriers, and attention turned to the corporation, leaving entrepreneurship research fragmented and marginalized. Thus, the world has witnessed many notions and angles through which the concept of entrepreneurship was studied and defined. The majority of the definitions of entrepreneurship emphasized on the aspects of the characteristics of an entrepreneur or the result expected out of the process of entrepreneurship (Naser et al. 2011). The

Table 1.1 Definitions of entrepreneurship

Sr. No.	Author(s)	Year	Highlights
1	Frank H. Knight	1921	Established relationship between entrepreneurship and risk taking
2	Schumpeter	1965	Exploitation of market opportunities through innovation
3	Peter Drucker	1970	Entrepreneurship is drive through challenges and risks
4	Hisrich	1990	Highlighting initial characteristics and creative mindset of an entrepreneur
5	Gartner	1990	Inclusion of innovation as one of the personality traits of an entrepreneur out of 90 attributes
6	Thomas and Mueller	2000	Stressed upon the fact that cultural value dominate the characteristics of entrepreneurship
7	Onuoha	2007	Described the concept as building and reshaping of enterprises and businesses
8	Zimmerer	2008	Entrepreneurship is about creating a new business
9	Kuratko	2009	Focused upon making a change through creation and application of ideas
10	Brooks	2009	Described entrepreneurship as the process of pursuing lucrative opportunities

following table summarizes the key highlights of different definition pertaining to entrepreneurship as it evolved over the time.

"The technique of forming new organisations or rejuvenating mature organisations, notably new enterprises often in response to identified possibilities," says Onuoha (2007). "Entrepreneurs," according to Schumpeter (1965), are "individuals who seize market opportunity through technological and/or organisational innovation." "Entrepreneurship is about taking risks," according to Frank H. Knight (1921) and Peter Drucker (1970). An entrepreneur, according to Bolton and Thompson (2000), is "a person who routinely develops and innovates to build something of recognised value around anticipated opportunities." "Someone who exhibits initiative and creative thinking, is able to organise social and economic systems to convert resources and situations into practical account, and embraces risk and failure," according to Hisrich (1990) and Ali et al. (2019). According to Thomas and Mueller (2000), entrepreneurship research should be broadened to worldwide markets in order to understand the conditions and qualities that encourage entrepreneurial activity in different countries and areas. From the above table it is evident that the definition of entrepreneurship had been shaped through various attributes expected out of an entrepreneur or the result desired from the enterprise. It is an amalgamation of various traits and characteristics of an individual taking up an entrepreneurial venture.

There have been various eras and each era highlights different sets of expected behavioral and inherent traits relished to be of critical importance for entrepreneurship (Baum and Locke 2004). Majority of such definitions of entrepreneurship focused upon the ends while some saw entrepreneurship as a well-thought process to develop a successful organization or a business. Many definitions focused upon the skillsets and the viewpoints of entrepreneurs that shape up a business enterprise (Casper 2000). The economy looks up to entrepreneurship as job generating phenomenon that results in the creation of working avenues and opportunities that would lead to economic prosperity of any country. This definition and notion of entrepreneurship formed the basis for differentiating an entrepreneur from a manager as the end results, or the desired output from both the entities is poles apart (Aulet and Murray 2013). Nonetheless, entrepreneurship is a disciplined and self-sufficient activity that has converted old behaviors into new ones. Entrepreneurship is a kind of expression (Chang and Wyszomirski 2015). According to another study, entrepreneurship "begins with action and the formation of a new organization." According to Barot (2015), entrepreneurship is a key to success, and everybody who starts a new company organization is entering a new paradigm of entrepreneurship. "Art entrepreneurship is a relatively new field of research," Chang and Wyszomirski (2015) said, "and the focus area is exploring the management process of entrepreneurship such as creativity and autonomy, adaptation, and creating artistic as well as economic and social value." Entrepreneurship necessitates dedication. Entrepreneurship is a different discipline in and of itself. Croci (2016) also characterized entrepreneurship as an autonomous and interdisciplinary discipline that may work independently.

1.3 Entrepreneurs Versus Managers

There has been a long debate since the emergence of the concept of entrepreneurship and management. Though both the terms are interrelated in more than one way, there exist distinct differences between these terms. Without a second thought there is a clear distinction between a manager and an entrepreneur, as these two entities have different roles and responsibilities and what is expected of them. The manager's foremost responsibility is to oversee the resource pooling process and effectively manage the firm's business portfolio (Baumol 1988). They play an important role when, as is often the case, businesses do not perform efficiently and fall far short of their production limits. A second, but equally important, role of the manager is to establish a reputation and foster an atmosphere of trust, which turns a conflictive system (individuals with competing goals) into a cooperative one. Managers should foster a culture of trust so that employees are less likely to engage in opportunistic behavior, even when it serves their short-term interests, while also increasing efficiency by lowering supervision and agency costs (Acemoglu et al. 2001). On the other hand entrepreneurs are resource providers. They are the ones who walk the extra mile for furnishing best of the resources that can be optimally utilized by the firm (Eisenmann and Eisenmann 2013). Entrepreneurs are not limited to the constraints of scope of functioning as with the case of managers.

The entrepreneurial function entails the identification, evaluation, and exploitation of opportunities, such as new products, services, or manufacturing processes; new strategies and organizational structures; and new markets for previously unavailable products and inputs (Shane and Venkataraman 2000). The entrepreneurial opportunity is a previously untapped and undervalued economic resource. Because various actors have different notions about the relative value of resources or when resources are transformed from inputs to outputs, entrepreneurial opportunities abound (Brown and Eisenhardt 1995). The entrepreneur's theory focuses on the diversity of opinions regarding the worth of resources. The entrepreneurial function can be defined as the identification of new economic opportunities and the subsequent production of new economic activity, frequently through the formation of a new company. As there is no market for opportunities, an entrepreneur leaves no stone unturned when it comes to utilization of available resources. Right from the stage of procurement to the last stage of conversion process the entrepreneur makes sure that the best possible resources are obtained. Thus the transformation process of input into output solely rests on the shoulders of an entrepreneur (Amit et al. 2003). The difficulty in protecting ownership rights of ideas that are not associated with patents or copyrights, as well as the different expectations held by entrepreneurs and investors on the economic value of ideas and business opportunities, and the entrepreneur's need to withhold information that may affect the project's value, are all disadvantages to the market for "opportunities" that are also viewed by many academicians as a set of probabilities with numerous determinants (Sexton and Bowman 1985). The entrepreneurial component, entrepreneurial function, entrepreneurial initiative, and entrepreneurial behavior are few terms used to describe entrepreneurship, and it is also referred to as the entrepreneurial "spirit," which is purely subjective and has been an area of study in the domain of entrepreneurship since quite some time. The entrepreneurial component is regarded to be a new factor in production that differs from the traditional conceptions of earth, work, and capital, and must be explained by the entrepreneur's remuneration in the form of money, as well as a scarcity of people with entrepreneurial qualities (Stewart and Roth 2001, 2007). Thus, entrepreneurship appropriately justifies the reason to be one of the factors of production.

The discovery and exploitation of opportunities, as well as the establishment of a business, are both considered entrepreneurial functions unlike managerial responsibilities, which are limited to administration. Entrepreneurial behavior is defined as behavior that combines creativity, risk-taking, and foresight. In other words, it includes famous theories of the innovative entrepreneur proposed by Schumpeter, the risk-taking entrepreneur who operates in an uncertain environment, and the entrepreneur with initiative and imagination who generates new chances whereas managers are acquainted with controlled atmosphere and less volatile variables pertaining to business. The importance of accurately anticipating market flaws or the ability to innovate in order to develop new and innovative combinations of strategies and resources which is highlighted by entrepreneurial "initiative." Inside or outside of an established company, entrepreneurial initiative encompasses the notions of creation, risk-taking, renewal, and innovation (Ahuja and Lampert 2001). Finally, the entrepreneurial spirit prioritizes exploration, search, and creativity over the exploitation of managerial business prospects. All of these help to explain why entrepreneurship is described in so many ways (Table 1.2).

Table 1.2 Difference between entrepreneur and manager

Entrepreneur vs. Manager	
Entrepreneur	Manager
Explores resources	Manages resources
Creates and enterprise	Administers an enterprise
Exploits opportunities	Utilizes inputs provided
Self-driven	Driven by management
Have a risk appetite	Usually risk averse
Can be intuitive and irrational	Tries to be logical and rational
Self-controlled	Requires supervision and monitoring
Opportunity driven	Data/analysis driven

Entrepreneurial functions differ from managerial functions, having said that it does not mean in any sense that an entrepreneur is not expected to perform major managerial functions like planning, organizing, directing, controlling and coordinating. The studies along with factual data analyze various behavioral and psychological variables that can be used to study entrepreneurial responses to business situations and complex problems. Recent studies along with foundation researched support the fact that various intrinsic factors act as a trigger and motivate behaviors of an entrepreneur including the need for achievement, which is posed to be greater in entrepreneurs than in managers. The tendency and nature to exercise control over opportunities and resources are also found to be greater in entrepreneurs than in managers. Thus, it can be inferred that despite the similar functions of an entrepreneur and a manager, the end outcomes and desired results of present actions differentiates the said two.

1.4 Entrepreneurial Process

The entrepreneurial process has been proposed in different ways, as the process of business development is not strictly based on science of sequence. Thus, one universal process will not suffice the need. Theories on entrepreneurship try to provide insights on broad areas that can act as a consideration for creating and nurturing an enterprise. Business processes include identifying and assessing opportunities, deciding to use or sell opportunities, procuring resources, and developing strategies and organizations for new business projects. Entrepreneurship is the "process of an individual seizing an opportunity alone or within an organization". Recently, when many business owners and businessmen have shown entrepreneurship in strategy development, it has been suggested that companies have a much brighter future than current perceptions suggest. The core activity of an entrepreneur is the establishment of a company (Mack and Pützschel 2014). The concept of entrepreneurship played a pivotal position in the emergence of any business enterprise. Since the mid-nineteenth century, various economists had critiqued the static

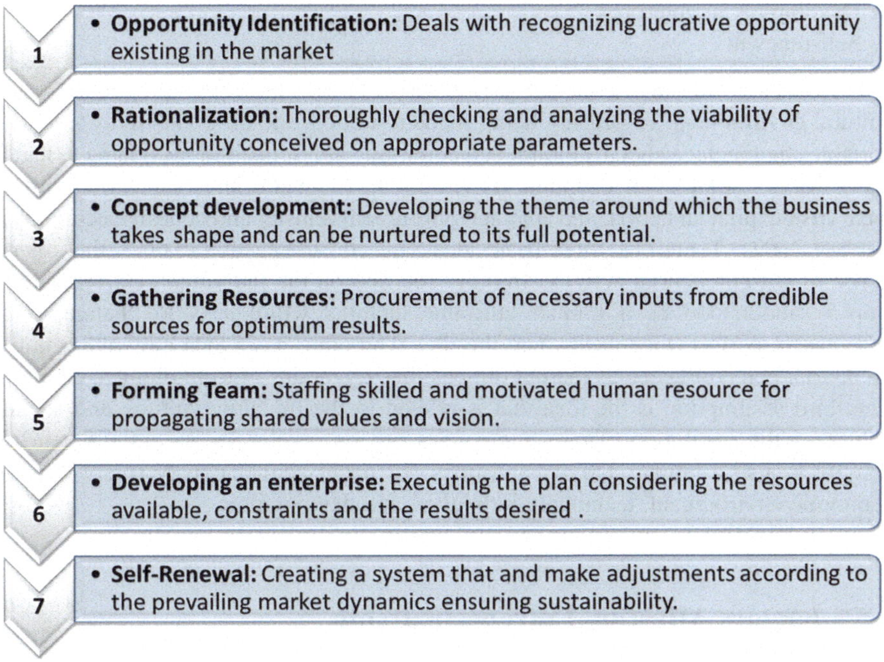

Fig. 1.1 Entrepreneurial process

theories of classical and neoclassical economic thought documenting the methods wherein the shape of economies had modified over a period of time. Many factors and determinants came into picture during the early twentieth century barring the classical notion of establishment of company and enterprise as an institution and accepting the agency of the humanistic characteristic with that of the creation nurtured under the embrace of economic conditions majorly. The changes in economic conditions led to the restructuring of the economy making the conditions conducive for entrepreneurs to incubate and nurture the business venture, thus, motivating the risk bearers in the form of entrepreneurs toward such rapid change and transformation.

Even though the entrepreneurial process for business development may take any course subjected to environmental and personal needs of an entrepreneur, a basic course of action can still be crafted so as to broadly characterize phases that would require different approaches coupled with varied skillset of an entrepreneur. The entrepreneurial process includes the following steps (Fig. 1.1).

1. Opportunity identification
2. Rationalization
3. Concept development
4. Gathering resources
5. Forming team

6. Developing an enterprise
7. Self-renewal

To develop substanc from analysis of aspects like the economic, social, and cultural environment. Efforts are being made to understand different environmental variables that include social, cultural and common values, institutions related to legal frameworks, variables of economic environment (demand) and financial environment (risk capital and cost), and spatial environment (cluster and cohesive economy) (Abbey 2002). Therefore, three basic ideas are considered that explain the emergence of entrepreneurial activity. Initially focused on the individual, entrepreneurship is understood as a human attribute such as willingness to change, risk acceptance, performance needs, and the like. The other basic idea motivates entrepreneurial behavior, such as market dimensions, dynamics of technological change. The third assumption is the idea that is related to the function, culture, and social values of the institution. These approaches are not exclusive, as entrepreneurial activity is also a human activity and does not occur spontaneously solely due to economic environment, technical, normative, and demographic.

1.5 Factors Affecting Entrepreneurship

Abundance of literature, bearing factors affecting entrepreneurship, has been made available through various credible sources realizing the criticality of aspects pertaining to entrepreneurship. These factors include national culture, education, experience, family background, need for achievement, internal locus of control, attitude, government policies, and the like (Kogut and Singh 1988). These factors act as triggers and catalysts for entrepreneurial activities and in many ways affect the entrepreneurial mindset of an individual embarking on an entrepreneurial venture.

1.5.1 National Culture

Entrepreneurs in different countries have some universal characteristics, but there may also be other characteristics that are unique to their culture (Welter et al. 2016). Entrepreneurial behavior is always associated with cultural values, which are based on the framework of the cultural aspects of Hofstede. Studies show that national culture has a great impact on entrepreneurship. This study focuses on the relationship between Turkish culture and its entrepreneurial spirit and is based on how national culture relates to the level of entrepreneurship. National culture was previously defined by many scholars. Culture is defined as a set of shared beliefs, values, and expected behaviors (Hofstede 1980a). Deeply rooted, unconscious, and even irrational shared values shape political institutions and social and technological systems, all of which simultaneously reflect and reinforce values and beliefs.

National culture has been described in various ways by scholars. Culture is described as shared beliefs, values, and predicted behaviors (Hofstede 1980b). Deeply embedded, unconscious, or even irrational shared values form political establishments in addition to social and technical systems, all of which concurrently mirror and improve values and beliefs (McCloskey and Sandberg 1971). These values form a part of many theories related to entrepreneurship and the manner in which entrepreneurship takes up a shape in different nations.

1.5.2 Education

Education is one of the important elements of human capital. It is a source of knowledge, skills, discipline, motivation, and self-confidence (Cooper et al. 1994). A significant amount of research has been done on the basis of human capital theory to investigate the impact of education on the performance of entrepreneurs and their businesses. Highly educated people are expected to run their businesses better than less educated people. Though education as an element alone does not guarantee success, it actually helps in enriching and enhancing the impact of the efforts that an entrepreneur puts into the business venture. Empirical evidence from education and entrepreneurship shows a mixed relationship (Foreman-Peck 2006; Lekoko et al. 2012). They have been positive, neutral, and, in some cases, negative correlation. They were obtained through analysis of real time data. For example, the research conducted by Stewart, Watson, and Barnir in 2003 and Neubaum in 2005 showed positive relationship between success and education of entrepreneur, whereas, the study by Bukvii in 2012 showed negative relationship.

It is not easy to discuss the importance of education to the success of an entrepreneur, as it depends on different aspects, such as the industry to which the company belongs. The role of education differs from sector to sector and from one industry to another (Fayolle and Gailly 2008). For example, education is more important to manufacturing entrepreneurs and is more obvious than entrepreneurs in other industries. A particular type of education is a more accurate prediction of a particular type of business and its performance (Jenkins and Griffith 2004). Education has different implications when measured by indicators (Hattab 2015). For example, the impact of education on entrepreneurs differs between when measured in terms of profitability and sales and when measured in terms of employment. In many countries one of the major parameters to measure the impact of education over entrepreneurial success is through the market capitalization of the venture. This assessment does neglect the qualitative aspect of the impact and measure only the quantitative impact of education. Thus, in order to measure the impact of education on entrepreneurial success both aspects should be given equal weightage.

1.5.3 Experience

Another most sought after element that affects entrepreneurship is experience. Experience provides an edge to an individual especially in critical circumstances. Through experience, people collect useful information and develop skills in different professions. So far, various aspects of previous experience have been discovered in the literature. However, the most commonly cited types of experience are entrepreneurship, management, and industry experience. Entrepreneurship experience refers to the number of previous startups and the role they play in those ventures. Industry experience refers to the experience in the industry to which the current company belongs. Management experience refers to management experience regardless of industry. As discussed earlier in the chapter, entrepreneurs have to perform all the functions of a manager to increase the sustenance of the venture. Studies on human capital confirm the importance of experience to be a successful entrepreneur. They claim that previous experience improves the company's economic performance (Judge et al. 2007). From an industrial experience perspective, researchers suggest that individuals entering areas they are familiar with are more likely to succeed in their business as they tend to make less errors and the amount of rework is reduced by a margin. They understand better than others how to meet the demands of this market and the volatility in it. As a result, these individuals are more likely to gain information that the outsiders and new comers cannot collect nor have easy access to. This experience helps build an entrepreneurial knowledge base, improve access to market information and corporate networks, and improve management skills. They may also have useful contacts and sufficient experience to understand the types of products in demand in their business context. Thus, experience acts as an equity that is reaped in critical and demanding situations.

1.5.4 Family Background

Family background acts as a backbone for entrepreneurial behavior as it shapes the responses of an individual to a great extent. It is possible to acquire the knowledge necessary to start one's own business through family background, especially when it comes to countries in which majority of the businesses in key sectors belongs to the category of family owned enterprise, like India to be precise. There are several reasons why entrepreneurial children also known as the second-generation entrepreneurs are likely to be self-employed. These factors generally derive from the mechanism of exposure and obstruction. The exposure mechanism focuses on how the parent's social position exposes the child to experiences and expectations that have a significant impact on career choices. Scientists argue that children of self-employed parents are more likely to see self-employment as a viable career choice than their peers. Independent parents can act as role models for their children. This role-modeling feature not only means that children value independence more than other forms of employment, but it can also encourage entrepreneurial behavior like risk taking ability. Another reason in support of the concept is availability of the resources in the form of privileges. Easy and instant access to capital and crucial infrastructure provide self-confidence to the individual looking to take up risky but lucrative ventures. It argues that entrepreneurial children can successfully enter self-employment because they have better access to knowledge about entrepreneurial opportunities Thus, it is a clear explanation of how the family background influences an entrepreneur and entrepreneurship as a whole.

1.5.5 Need for Achievement

Need for achievement is one the important psychological aspect that affects entrepreneurship. The need for achievement was introduced in a motivational theory by McClelland in 1961 along with two other important needs that lead to a certain kind of behavior. It was proposed that people in need of high achievement prefer to solve problems as their need for high performance is directly linked to their need for achievement. These individual set high goals, and strive to achieve those goals themselves (Kirby 2003). Certain advance theory assumes that people with high performance needs are more entrepreneurial and often perform better than those with low performance needs (Collins et al. 2004). They spend their time thinking about how to make their job better and how to achieve something that matters to them, thus enhancing their productivity through channelized efforts, which are lead to high reward achievement (McClelland 1961a, b, 1965, 1987).

1.5.6 Innovativeness

In the modern times, innovation is considered as a refined meaning of "change." Innovativeness is an integral element of entrepreneurial persona, which allows the entrepreneur foster accomplishments of their enterprise. This element emphasizes on a real time and active methodology in regard to innovation, which includes a gradual day by day attempt to enhance one's activities, work processes, and procedures. Schumpeter (1934) turned into one of the first authors to introduce the function of innovation into the entrepreneurial process. He perceived innovation as a scientific look for changes, possibilities for brand new markets, merchandise, or thoughts. Numerous articles and literature have described the potential of entrepreneurship in bringing radical as well as incremental innovation not only in markets but in the economy as a whole. As proposed by him dynamic entrepreneurship is a phenomenon in which an individual is an entrepreneur who innovates and who makes new combinations (Graham and Shuldiner 2001). These combinations of various resources, coupled with innovative capabilities of an entrepreneur, provide astonishing results in the form of new products, new markets, new methods of production, and new capabilities. Innovativeness is an aspect that provides scope and pathway for development and evolution as per the requirements of the dynamic market place.

1.5.7 Locus of Control

Locus of control is ability, a required skill, and one of the most important factors that can affect entrepreneurship in more than one ways. It refers to an individual's potential of perceiving things to persuade activities, which are of prime significance. It primarily is concerned with external and internal forces that determine the consequences and results of the behaviors and actions taken by an entrepreneur throughout his/her entrepreneurial journey. The notion that outside or external drivers determine the results forms one end of the ideology while the ones with conviction that inner and intrinsic drivers determine the ends forms the other ideology of this factor. The former is called as externals and the latter is termed as internals. Both these aspects of one's persona greatly affect the destination and the end results. The variables like randomness, chances, and the like are not appreciated by the internals and thus, they develop stronger and advanced level of need for achievement, performance, and reward.

1.5.8 Attitude

Attitude is the most important factor affecting entrepreneurship at a level, which is beyond measurement. Almost all the psychological and behavioral aspects in the form of variables some way or the other are interlinked with attitude of the entrepreneur. It is said that enterprise is shaped in the vessel of attitude. Attitude is reflected through qualitative and subjective aspects like optimism, pessimism, and realism. Entrepreneurs with optimistic approach are the ones with great intuitiveness driving their actions whereas an entrepreneur with pessimist approach is more cautious and analysis driven to avoid worst case scenarios. Realist entrepreneurs are the ones who explore all the possible opportunities to convert into profitable ventures. Whatever the mindset is, an entrepreneurial venture is impacted by the attitude an entrepreneur reflects (Fig. 1.2).

1.6 Theories of Entrepreneurship

Entrepreneurship theories and research will continue to be vital in the field's development. Scholars have proposed several hypotheses to explain the topic of entrepreneurship. Economics, psychology, sociology, anthropology, and management all have roots in these theories. Entrepreneurship theories cover a wide range of

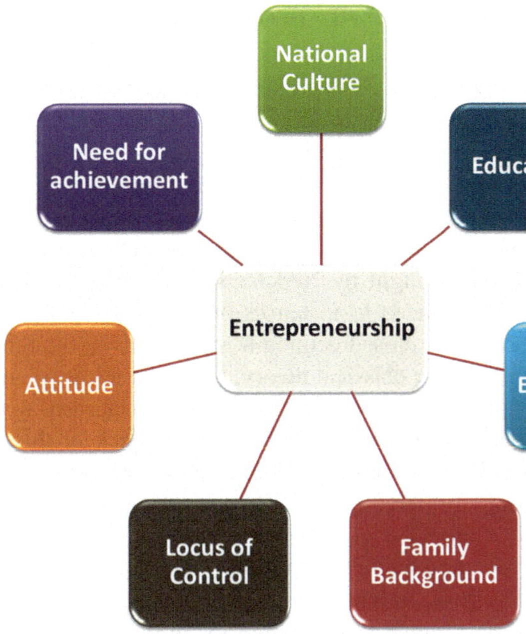

Fig. 1.2 Factors affecting entrepreneurship

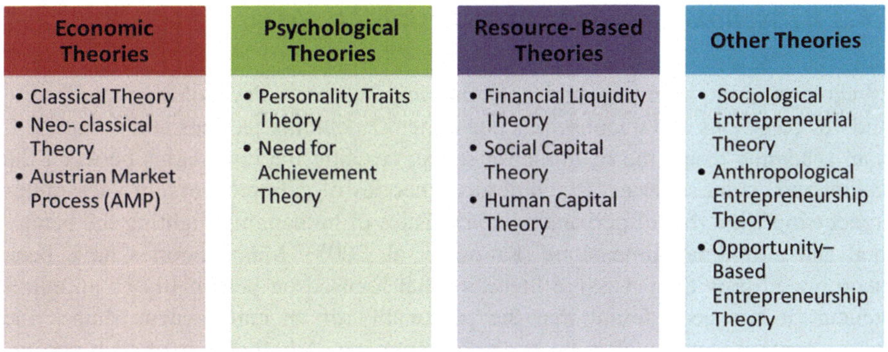

Fig. 1.3 Theories of entrepreneurship

aspects that are critically important for development of entrepreneurship and entrepreneurial traits. These theories lay emphasis on various dimensions, personality traits, situational factors, and the like, which have an effect on entrepreneurs and sometimes paves the basis for measurement of an entrepreneur's effectiveness and efficiency. These theories are broadly categorized as follows (Fig. 1.3).

1.6.1 Economic Theories of Entrepreneurship

These theories that deal with virtues of free commerce, specialization, and competition were lauded in classical theory. In a competitive marketplace, the classical theory proposed and reflected the entrepreneur's directing role in the creation and distribution of goods and services (Weber 1930, 1956). Though the hypothesis has had its own critique on the grounds that it was not able to provide explanation to the dynamic transformation brought about by the entrepreneurs of the industrial-era, the correction to this model was brought by Neoclassical theories claiming that economic phenomena could be reduced to instances of pure trade, reflect an optimal ratio, and occur in an essentially closed economic system. Even though the theory did come up as advancement over classical theory, lacuna in terms of entrepreneurship at an individual tier was still missing as the theory concentrated mainly on the macro factors like aggregate supply and demand. The setback of this theory was addressed by AMP, which is the Austrian Market Process that deliberately focused on human actions and behavior pertaining to their knowledge and its impact on an economy (Austin et al. 2006). The AMP highlighted the innovativeness of an entrepreneur in an economy and his/her capability of creating something "NEW," which in turn is regarded as the founding stone of enterprising. The stepping stone of creating something new sets the economy in a cyclical motion of production, distribution, and exchange.

1.6.2 Psychological Entrepreneurship Theories

Psychology is the scientific study of mind and behavior. Psychology includes the study of conscious and unconscious phenomena, including feelings and thoughts. It is an academic discipline of immense scope, crossing the boundaries between the natural and social sciences. Psychological theories of entrepreneurship view entrepreneurship from that of personal characteristics of human, highlighting the behavioral and attitudinal dimensions (Krauss et al. 2005). Many theories have been incorporated providing massive literature that focused on personality of an entrepreneur. It has been found that the personality of an entrepreneur shapes the entrepreneurial venture taken up by the entrepreneur. A hefty list is of such personality traits that are available in the literature, which reflects distinct behavior of an entrepreneur and supports that these traits impact the overall molding and functioning of an enterprise (Zhao et al. 2010). Traits such as risk bearing capability, taking initiatives, composure, aggressiveness, and ambitious and the like are some widely witnessed and supported traits of an entrepreneur. These traits are demonstrated by majority of entrepreneurs across the globe and thus have deep rooted establishment in the development of personality theories of entrepreneurships (Lent et al. 2000). These traits are patterns of behaviors that entrepreneurs show during their entrepreneurial journey. These traits can be inherited and are acquired too. The origin of psychological or behavioral traits is yet a debatable topic but the fact of the matter is the impact of such traits has been very strong in deciding upon the fate of the enterprise (Robinson et al. 1991). The qualities provide us with a hint or comprehension of these features or inborn potentials. In fact, understanding personality traits entails drawing conclusions based on one's actions. Entrepreneurs have traits such as being more opportunity motivated (they look for opportunities), demonstrating high levels of creativity and innovation, and demonstrating high levels of managerial abilities and business know-how. Some of the most popular psychological theories from which entrepreneurship have taken cues are Personality Trait Theory, Need Achievement Theory, Behavior Theory, Social Change Theory.

1.6.3 Resource-Based Entrepreneurship Theories

Resources are the inputs that are utilized for proper functioning of any business venture. Resources are one of the most important aspects of business. The effectiveness and efficiency of entrepreneur is also measured and assessed upon his ability to utilize these resources. Resources have been classified into various broad categories depending upon the nature, type, utilization mechanism, and sometimes the results generated out of them. These resources can be tangible, intangible, rare, adequate, common, imitable, and so on (Barringer et al. 2005). Access to these resources plays an important role in determining the success of an entrepreneur or an entrepreneurial venture as a whole. Possession of valuable resources gives an entrepreneur the opportunity to take risks, explore new avenues, try what has never been tried before, something close to trial and error and thus enhance the strength of entrepreneurial traits and behaviors being exhibited by the individual. The resource-based view of entrepreneurship theories recognizes the importance and vitality of liquid, social and human resources, and considers them as the most critical dimensions in the development of entrepreneurship. The theory highlights the capitalization of market of opportunities and development of market depends upon the possession and effective utilization of these resources. Right from the point of identification of opportunity to the point where the idea shapes up into a lucrative and viable economic possibility, these resources are in dynamic and continuous interplay.

Liquidity/Financial Capital: According to theory, entrepreneurs have individual-specific resources that help them recognize new opportunities and assemble fresh resources for their startup. According to research, some people are better at seeing and exploiting chances than others because they have more information and knowledge (Bosma et al. 2004). This information and knowledge of the market provides an entrepreneur with minute details for crafting and implementing market strategies satisfying the market dynamics in the most efficient manner. This knowledge gradually grows and develops into competitive advantage in the prevailing market and strengthens the strategic position of the business, gifting it with stability and sustainability. Social Capital or Social Network Theory treats the entrepreneurs as part of a bigger social network structure that makes up the economy. It is rightly said that "There is no 'I' in 'Team'." The social network theory supports the statement in the best suitable manner as an individual may be able to see an entrepreneurial opportunity, but he or she may lack the social contacts necessary to turn the potential into a business venture. In this era of startups it is imperative for an entrepreneur to realize the importance of team building and the astonishing results that can be generated through a well-knitted network of highly qualified and efficient manpower. Access to a bigger social network is regarded to be one solution to this problem. Social network theory provides undisputed insights over the significance of network for a business enterprise. The network is viewed as the net worth of the entrepreneur and for the entrepreneurial venture. Where the social network theory views the enterprise as a result of well-knitted network of capabilities the Human Capital Theory lays emphasis on the cell of these networks. The human capital theory emphasizes and realizes the importance of human resources in building and shaping the entrepreneurial structures. It basically takes into the consideration the limitless potential of the human resources in the development of an enterprise. Two components, namely, education and experience, underpin the human capital entrepreneurship paradigm. Knowledge earned via education and experience is a resource that is distributed differently among people, and it is therefore crucial to comprehending disparities in opportunity identification and exploitation. Human capital elements are positively associated to becoming a budding entrepreneur, according to empirical studies.

1.6.4 Sociological Entrepreneurship Theory

As the name suggests, sociological enterprise focuses on the social context of entrepreneurship. In other words, in the sociological theories the level of analysis is traditionally the society. A society is defined as a group of people coming together on the grounds of similarities of beliefs, traditions, customs, traditions, culture, work/profession, and the like (Chen et al. 2018). Mankind has evolved overtime and society has been one of the integral parts of mankind that has provided meaning to the existence in more than one way. Here, the focus is on building social relationships and bonds that promote trust and not opportunism. Establishment of a valued code of conduct is one of the core aspects of the theory (Corner and Ho 2010). A society views individuals through their deeds and thus, role of morals and ethics play a vital role. In other words, the entrepreneur should not take undue advantage of people to be successful; rather success comes as a result of keeping faith with the people (Davidsson and Wiklund 1997). The second is called the life course stage context, which involves analyzing the life situations and characteristic of individuals who have decided to become entrepreneurs. Many theories support the emergence of situational leaders as a result of circumstances. Similarly, entrepreneurs are considered the results of prevailing situations. The experiences of people could influence their thought and action; therefore they want to do something meaningful with their lives. The theory tries to present the importance of looking at the bigger picture of what entrepreneurs and entrepreneurship can make out of the environment they are a part of.

1.6.5 Anthropological Entrepreneurship Theory

Anthropological Entrepreneurship Theory takes its cues form anthropology, which is the study of a community's origins, development, practices, and beliefs; in other words, the culture of the community's residents. A culture is a complex whole of many aspects like traditions, customs, education, beliefs, values, norms, and the like. Culture has a huge influence on the behaviors and actions of individuals. These patterns are also experienced in entrepreneurial actions. According to the idea, the influence of one's culture creates new ventures. Entrepreneurial mindsets, such as creativity, are influenced by cultural practices, which in turn influence venture formation behavior. The effect of culture has already been discussed in the earlier section of the chapter. Individual ethnicity has an impact on attitude and conduct, and culture reflects individual ethnic, social, economic, ecological, and political intricacies.

1.6.6 Opportunity-Based Entrepreneurship Theory

This theory highlights the role and effect of stimulus that triggers entrepreneurial actions in the name of opportunities. The theory reflects the dominance of opportunity as a trigger that starts a series of actions and behavior that results in building and creating something new. An entrepreneur seeks satisfaction in exploration and the need is accomplished when a lucrative possibility is generated out of such exploration (Carr and Sequeira 2007). This is the result of research into the distinctions between entrepreneurial and administrative management. He finds that the "pursuit of opportunity without respect to existing controlled resources" is at the heart of entrepreneurial management. The entrepreneur is always on the lookout for change, reacts to it, and seizes the opportunity it presents. Thus, the theory emphasis on the end that acts as means for the actions of entrepreneurs.

The above mentioned theories have their relevance in explaining the phenomenon of entrepreneurship in their unique manner. However, there is no universal methodology or approach to decide upon the results and nature of entrepreneurship. Entrepreneurship is a comprehensive term that encompasses a varied range of aspects and dimensions, which are considered to be very intricate.

1.7 Entrepreneurship in India: The Startup Zone

The emergence of a nation's capital, wealth, and efficacy relies upon the competitiveness of its companies and this, in turn, is predicated basically by the competencies of its marketers and entrepreneurs (Berger et al. 2001). The essence of the cutting-edge company lies in the specialization of capabilities. "The businessmen"

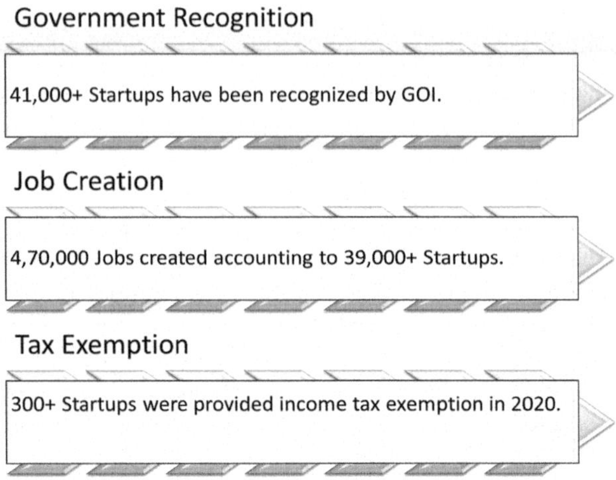

Fig. 1.4 Entrepreneurship in India—current scenario

that control monetary interest are, in the strictest sense, each managers and marketers, the latter in a double sense: the person businessman (independent) and the "company entrepreneur" who, without taking part substantially in phrases of capital, controls the company. The person entrepreneur detects or creates commercial enterprise possibilities that she or he then exploits via small- and medium-sized companies, usually taking part in investment of the capital for that company. This is not restrained to efficaciously coping with the company's belongings and coordinating and controlling its activities; in response to the modern climate, she or he has to anticipate, articulate, and control change. In other words, they have to reinvent the company on a day-to-day basis, developing new enterprise (spin-offs) and broadening the organization's network. India has witnessed a tide of startups in recent years. There has been a paradigm shift in the way entrepreneurship is viewed today. A large gamut of entrepreneurs has emerged in different sectors contributing toward the economy and market development. The Economic Survey 2021 on startup revealed the following accomplishments pertaining to entrepreneurship in India (Fig. 1.4).

Apart from the above accomplishments, India witnessed a rise in number of startups from 1.52 lakhs to 1.75 lakhs from 2019 to 2020 amounting to a 15% growth. While there has been an increase of 26% of total number of startup recognition during the same time period, India has also witnessed an addition of 12 unicorns in 2020 making a total of 38 unicorns. India has the world's third largest network of startups. Forty startups have forayed in India's Space Sector Technology (The Economic Times 2021). The sector-wise development is worth mentioning when talking about startups in India in the last year. About 15% startups contributed to enterprise technology, whereas health, education, and financial sectors with individual contribution constituted10% of the total startup composition. Startups in retail sector, consumer segment, and HR sector contributed to 7%, 6%, and 6%, respectively. Approximately 28% of the startups belonged to manufacturing, services, and technology. A total of 11% of the startups represented media and entertainment, advertising, and Supply Change Management (SCM) technology with individual breakup of 4%, 4%, and 3%, respectively (Report. distribution of startups in India 2022).

1.8 Conclusion

Business history has made vital, and sometimes overlooked, contributions to the study of entrepreneurship. Historical studies provide insights for other social scientists into how modern entrepreneurial activity might be better contextualized in time and place by integrating entrepreneurship within the broader process of historical development in industries and economies. Even though the literature has been mainly geared toward huge businesses, business history has made significant contributions to the study of entrepreneurship through its wide coverage of countries, regions, and industries. Where management research on entrepreneurship over the last two decades has been narrowly clustered in its empiricism, often drawing broad generalizations based on high-technology startup firms in a few locations, historical research had sought to understand entrepreneurship in a much wider range of settings and to establish its significance in the economic prosperity.

By reengaging core concepts and theories of entrepreneurship and drawing together the many distinct streams of study, new insights can be created through entrepreneurship research. There is much work to be done on the historical effect of culture and values on entrepreneurial activity, using more thorough approaches than before, and attempting to explain more precisely how significant culture is in relation to other variables. By examining the interaction between institutions and entrepreneurs, there are several chances to supplement previous research on the role of institutions in economic progress. Looking at the current scenario of entrepreneurship, development, and growth in entrepreneurship is evident. In India, the future of entrepreneurship looks quite promising as the domain has shown remarkable growth especially considering the market slowdown in the recent time. Sectors such as advertising and media, health, and IT solutions of various forms are expected to grow at a rapid pace. Thus, it is realized that entrepreneurship development should thrive to build a strong and developed economic infrastructure for any nation.

References

Abbey A (2002) Cross cultural comparison of motivation for entrepreneurship. J Int Bus Stud 14(1): 69–82

Acemoglu D, Johnson S, Robinson J (2001) The colonial origins of comparative development: an empirical investigation. Am Econ Rev 91:1369–1401

Ahuja G, Lampert CM (2001) Entrepreneurship in the large corporation: a longitudinal study of how established firms create breakthrough inventions. Strateg Manag J 22(6–7):521–543

Ali A, Kelley DJ, Levie J (2019) Market-driven entrepreneurship and institutions. J Bus Res 113: 117. https://doi.org/10.1016/j.jbusres.2019.03.010
Amit R, Glosten L, Muller E (2003) Challenges to theory development in entrepreneurship research. J Manage Stud 30:815–833
Aulet W, Murray FE (2013) A tale of two entrepreneurs: understanding differences in the types of entrepreneurship in the economy. Ewing Marion Kauffman Foundation
Austin J, Stevenson H, Wei-Skillern J (2006) Social and commercial entrepreneurship: same, different, or both? Entrepr Theory Pract 30(1):1–22
Barot H (2015) Entrepreneurship—a key to success. Int J Bus Manage 3(1):163–165
Barringer BR, Jones FF, Neubaum DO (2005) A quantitative content analysis of the characteristics of rapid growth firms and their founders. J Bus Ventur 20:663–687
Baum JR, Locke EA (2004) The relationship of entrepreneurial traits, skill, and motivation to subsequent venture growth. J Appl Psychol 89(4):587–598
Baumol WJ (1988) Entrepreneurship, management, and the structure of payoffs. MIT Press, Cambridge
Berger AN, Klapper LF, Udell GF (2001) The ability of banks to lend to informationally opaque small businesses. J Bank Financ 25(12):2127–2167. dec
Bolton WK, Thompson JL (2000) Entrepreneurs: talent, temperament, technique. Butterworth Heinemann, Oxford
Bosma N, Van Praag M, Thurik R, De Wit G (2004) The value of human and social capital investments for the business performance of start-ups. Small Bus Econ 23:227–236
Brown SL, Eisenhardt KM (1995) Product development: past research, present findings, and future directions. Acad Manag 20/2:343–378
Carr JC, Sequeira JM (2007) Prior family business exposure as intergenerational influence and entrepreneurial intent: a theory of planned behavior approach. J Bus Res 60:1090–1098
Casper S (2000) Institutional adaptiveness, technology policy, and the diffusion of new business models: the case of German biotechnology. Organ Stud 21(5):887–914
Chang WJ, Wyszomirski M (2015) What is arts entrepreneurship? Tracking the development of its definition in scholarly journals. J Entrepr Arts 4(2):11–31
Chen FW, Fu LW, Tsai SB, Su CH (2018) The influence of entrepreneurship and social networks on economic growth-from a sustainable innovation perspective. J Sustain 10(2510):1–19
Collins CJ et al (2004) The relationship of achievement motivation to entrepreneurial behavior: a meta-analysis. Hum Perform 17(1):95–117
Cooper AC, Gimeno-Gascon FJ, Woo CY (1994) Initial human and financial capital as predictors of new venture performance. J Bus Ventur 9:371–395
Corner P, Ho M (2010) How opportunities develop in social entrepreneurship. Entrepr Theory Pract 34(4):635–659
Croci CL (2016) Is entrepreneurship a discipline?. Honors Theses and Capstones. 296. Cited from https://scholars.unh.edu/honors/296. University of New Hamspire Scholar's Repository
Davidsson P, Wiklund J (1997) Values, beliefs and regional variations in new firm formation rates. J Econ Psychol 18:179–199
Drucker P (1970) Entrepreneurship in business. Enterprise J Bus Policy 1
Eisenmann TR, Eisenmann T (2013) Entrepreneurship: a working definition. Harv Bus Rev. https://hbr.org/2013/01/what-is-entrepreneurship
Fayolle A, Gailly B (2008) From craft to science. Teaching models and learning processes in entrepreneurship education. J Eur Indust Train 32(7):569–593
Foreman-Peck J (2006) Measuring historical entrepreneurship. In: Cassis Y, Minoglou I (eds) Entrepreneurship in theory and history. Palgrave Macmillan, New York
Graham MBW, Shuldiner AT (2001) Corning and the craft of innovation. Oxford University Press, Oxford
Hattab HW (2015) Impact of entrepreneurship education on entrepreneurial intentions of university students in Egypt. J Entrep 23(1):1–18
Hisrich RD (1990) Entrepreneurship/intrapreneurship. Am Psychol 45(2):209–222

Hofstede G (1980a) Culture's consequences: international differences in work related values. Sage, Beverly Hills
Hofstede G (1980b) Motivation leadership and organization: do American theories apply abroad? Organ Dyn 9:42–63
Jenkins M, Griffith R (2004) Using personality constructs to predict performance: narrow or broad bandwidth. J Bus Psychol 19:255–269
Judge TA, Jackson CL, Shaw JC, Scott BA, Rich BL (2007) Self-efficacy and work-related performance: the integral role of individual differences. J Appl Psychol 92:107–127
Kanwal S (2022) Distribution of start-ups in India in 2020, by sector. Statista. https://www.statista.com/statistics/1208520/india-distribution-of-start-ups-by-sector/
Kirby AD (2003) Entrepreneurship. MCGraw-Hill Education, Maidenhead
Knight FH (1921) Risk, uncertainty, and profit. Houghton Mifflin, Boston
Kogut B, Singh H (1988) The effect of national culture on the choice of entry mode. J Int Bus Stud 19:411–432
Krauss SI, Frese M, Friedrich C, Unger JM (2005) Entrepreneurial orientation: a psychological model of success among southern African small business owners. Eur J Work Organ Psychol 14:315–344
Lekoko ME et al (2012) The effectiveness of entrepreneurship education: what matters most? Afr J Bus Manag 6(51):12023–12033
Lent RW, Brown SD, Hackett G (2000) Contextual supports and barriers to career choice: a social cognitive analysis. J Couns Psychol 47(1):36–49
Mack J, Pützschel J (2014) The influence of contextual factors on the entrepreneurial process. In: Master thesis presented at Umeå School of Business and Economics
McClelland DC (1961a) Encouraging excellence. Daedalus 90:711–724
McClelland D (1961b) The achieving society. Van Nostrand, Princeton
McClelland DC (1965) Toward A theory of motive acquisition. Am Psychol 20:321–333
McClelland DC (1987) Characteristics of successful entrepreneurs. J Creat Behav 21:219–233
McCloskey DN, Sandberg LG (1971) From damnation to redemption: judgments on the late Victorian entrepreneur. Explor Econ Hist 9:89–108
Mueller SL, Thomas AS (2000) A case for comparative entrepreneurship: assessing the relevance of culture. J Int Bus Stud 31(2):287–301
Naser J, Omari AL, Ghassan. (2011) Measuring entrepreneurship characteristics of business graduate students and their effect on entrepreneurship businesses: a comparative study. J Damascus Univ Econ Law 27(4):139–168
Onuoha G (2007) Entrepreneurship. AIST Int J 10:20–32
Robinson PB et al (1991) An attitude approach to the prediction of entrepreneurship. Entrepr Theory Pract 15:13–31
Schumpeter J (1934) The theory of economic development. Harvard University Press, Cambridge
Schumpeter JA (1965) Economic theory and entrepreneurial history. In: Aitken HG (ed) Explorations in enterprise. Harvard University Press, Cambridge
Sexton D, Bowman N (1985) The entrepreneur: a capable executive and more. J Bus Ventur 1:129–140
Shane SA, Venkataraman S (2000) The promise of entrepreneurship as a field of research. Acad Manag Rev 25(1):217–226
Stewart WH, Roth PL (2001) Risk propensity differences between entrepreneurs and managers: a meta-analytic review. J Appl Psychol 86:145–153
Stewart WH, Roth PL (2007) A meta-analysis of achievement motivation differences between entrepreneurs and managers. J Small Bus Manag 45:401–421
The Economic Times (2021) What economic survey 2020–21 says about India's startup ecosystem. https://economictimes.indiatimes.com/tech/startups/what-economic-survey-2020-21-says-about-indias-startup-ecosystem/articleshow/80586774.cms

Weber M (1930) The Protestant ethic and the spirit of capitalism. Scribners, New York
Weber M (1956) The Protestant ethic and the sprit of capitalism. Allen & Unwin, London
Welter F, Baker T, Audretsch DB, Gartner WB (2016) Everyday entrepreneurship-A call for entrepreneurship research to embrace entrepreneurial diversity. Entrep Theory Pract 41(3): 311–321
Zhao H, Seibert SE, Lumpkin GT (2010) The relationship of personality to entrepreneurial intentions and performance: a meta-analytic review. J Manag 36:381–404

Chapter 2
Government Schemes for Entrepreneurship Programs

Ami Naik and Pooja Patel

Abstract Entrepreneurship introduces a critical element of dynamism into an economic system. The process of globalization and liberalization has introduced a set of changes like the introduction of dynamism into the system through the process of globalization. Micro, Small and Medium Enterprises (MSMEs) are providing a sound base of economic development and growth in India. MSMEs are taken as foundation for all industrial endeavors of almost all countries including India. They are supporting in the contribution of huge businesses worldwide. As far as India is concerned, the economic development and progress of the country has been significantly flourishing due to these MSME's. In India, various initiatives have been taken by the government from time to time for entrepreneurship development in the country. Entrepreneurship has attracted the attention of policymakers in India. A series of high-level initiatives, including Startup India, have been launched to promote private sector development. However, the role of entrepreneurship in development remains a mystery for many policy observers. The main purpose of this chapter is to study about Entrepreneurship in India and its Promotion Under "STARTUP INDIA" SCHEME. This chapter is based on secondary data collected from multiple sources of evidence, in addition to books, journals, websites, and newspapers.

Keywords Entrepreneurship · Startup India · Schemes · Policies

2.1 Introduction

Indian entrepreneurship has been assumed as a contradiction with paradoxes. Entrepreneurs are produced in India to set examples of inspirations for many as to how innovation and success could be achieved overcoming the basic constraints. The

A. Naik (✉)
C. G. Bhakta Institute of Biotechnology, Uka Tarsadia University, Surat, Gujarat, India

P. Patel
Department of Microbiology, Government Medical College, Surat, Gujarat, India

N. Amaresan et al. (eds.), *Agricultural Microbiology Based Entrepreneurship*, Microorganisms for Sustainability 39,
https://doi.org/10.1007/978-981-19-5747-5_2

family business in India and the entrepreneurial zeal has significantly boomed and transformed the industry and business interface of India (Howorth et al. 2006). Women of India have been striving hard to attain heights of leadership positions. Besides facing additional barriers of balancing responsibilities, access to entrepreneurial prospects, control over resources, and many more, women have proved themselves equitable to many areas in order to achieve economic independence and respectable positions. Many entrepreneurial development support processes have been initiated to support and uplift the entrepreneurial zeal among women in India. These support initiatives are creating and providing paths for economic independence and work opportunities to the women in India. Micro, Small, and Medium Enterprises (MSMEs) are providing a sound base of economic development and growth in India (Rajpal 2014). MSMEs are taken as foundation for all industrial endeavors in almost all countries including India. They support in contributing huge business worldwide. As far as India is concerned, the economic development and progress of the country has been significantly flourishing due to these MSME's.

2.1.1 Entrepreneurship in India

Entrepreneurship is not new to India. In fact to quote from "the Indian Industrial Commission Report (1916–1918)," at a time when the West of Europe, the birth place of modern industrial system, was inhabited by uncivilized tribes, India was famous for the wealth of her rulers and for high artistic skill of her craftsmen (Mageshwar and Jothimani 2022). And even at a much later period, when the merchant adventures from the West made their first appearance in India, the industrial development of this country was, at any rate, not inferior to that of the more advanced European nations."

In fact an earlier version of the current Make in India policy was the Swadeshi Movement launched in 1905 during the pre-Independence era to boycott British-made goods and use Indian-made goods (Sanghi and Srija 2002). The movement saw the development of the Indian textile industry, the iron and steel industry by the Tatas, publishing of vernacular newspapers, setting up of vernacular medium educational institutions, financial institutions, etc.

However, post-independence, the policy focus of increased public investment in heavy industries and setting up of PSUs did not provide an ideal environment for entrepreneurship. The main problems faced by an entrepreneur were lack of mentoring facilities, technology support, or easy availability of credit. Though different reports on employment highlighted the need for promoting entrepreneurship as means of self-employment, entrepreneurship did not scale up. To mention a few, in the S.P Gupta "Special Group Report on Targeting 10 million Employment Opportunities Per Year" (2002) recommended "appropriate programs should be launched to increase entrepreneurial capabilities and skill for self-employment (Gupta 2002)."

2.1.2 *Startup Revolution*

However, a change is being witnessed today. As quoted by Prime Minister Shri. Narendra Modi, "The convergence of technology, integration across diverse fields, distributed architecture and people willing to back an idea, have opened a new world for enterprise" (Choudhary and Vadera 2019). For startups today, there are different levels of financial support that has come to provide the initial seed capital in the form of incubators, angel funds, or venture capital funds followed by private equity and debt in that order. However, there is a danger that too many mentors/angel investors with little experience may lead to a situation of unsuccessful startups.

The government has also come a big way in promoting startups. A startup is a company that is in the first stage for meeting their productive and emergent credit needs on rate of interest and terms decided by the group. These have helped in providing livelihood opportunities to group of women to start their own business and break the shackles of poverty. Some of the best case entrepreneurial models are the SEWA, Kudumbashree, etc. (Sanghi and Srija 2002). Though entrepreneurship has been privy to India and despite various schemes being in place the country has not witnessed the natural gradation from self-employment to entrepreneurship as part of the growth process excepting a few cases.

The Government of India has undertaken several initiatives and instituted policy measures to foster a culture of innovation and entrepreneurship in the country (Baby 2021). With a significant and unique demographic advantage, India, however, has immense potential to innovate, raise entrepreneurs, and create jobs for the benefit of the nation and the world. Through the Startup India initiative, the Government of India promotes entrepreneurship by mentoring, nurturing, and facilitating startups throughout their life cycle. With a 360° approach to enable startups, the initiative provides a comprehensive 4-week free online learning program. "Fund of Funds" has been created to help startups gain access to funding.

The make in India initiative was launched in September 2014. It came as a powerful call to India's citizens and business leaders, and an invitation to potential partners and investors around the world to overhaul out-dated processes and policies, and centralize information about opportunities in India's manufacturing sector (Jayanthi 2019). This has in turn helped procure investments, foster innovation, develop skills, protect intellectual property, and build best-in-class manufacturing.

The Central Government is implementing various schemes for prospective entrepreneurs (Lalthanzuali and Devi 2022; Agarwal and Dwived 2017; Devi 2020; Alaguraja and Nedumaran 2020; Pandey et al. 2017; Singh and Sharma 2018; Mund 2020; Unnisa and Amulya 2016; Rao 2014; Chaudhari and Borkar n.d.).

Prime Minister's Employment Generation Program (PMEGP)

Description	• The scheme is implemented by Khadi and Village Industries Commission (KVIC) as the nodal agency at the national level • At the state level, the scheme is implemented through State KVIC Directorates, State Khadi and Village Industries Boards (KVIBs), and District Industries Centres (DICs) and banks • The Government subsidy under the scheme is routed by KVIC through the identified banks for eventual distribution to the beneficiaries/entrepreneurs into their bank accounts
Nature of assistance	• Bank-financed subsidy program for setting up new microenterprises in non-farm sector • Margin Money subsidy on Bank Loan ranges from 15 to 35% for projects up to Rs. 25 lakh in manufacturing and Rs. 10 lakh in the service sector • For beneficiaries belonging to special categories such as SC/ST/Women/PH/Minorities/Ex-Servicemen/NER, the margin money subsidy is 35% in rural areas and 25% in urban areas. The maximum cost of projects is Rs. 25 lakh in the manufacturing sector and Rs.10.00 lakh in the service sector
Who can apply	• Any individual, above 18 years of age can apply • Self-Help Groups, Institutions registered under Societies, Production Co-operative Societies, and Charitable Trusts
Detailed Information	• The own contribution of the beneficiary is 10% of the project cost in case of general category and 5% of the project cost in case of reserved category (SC/ST/OBC/PH/Women/Ex-Servicemen/NER) beneficiaries • If the application for loan is approved, Banks sanction and release the balance amount of 90–95 % of the total project cost suitably for setting up of the units by the beneficiaries • In order to have sustainability of the projects/units set up under the scheme, support services are also provided in the form of Backward and Forward Linkages by organizing events like awareness camps, workshops, EDP training to the beneficiaries, exhibitions, etc. • The Government of India has introduced online process flow of application and disbursement of margin money directly to financing branches • One-page online application form is mandatory for individuals and institutional beneficiaries on the e-portal • The application form/PMEGP MIS portal is mobile friendly. SMS/e-mail alerts are sent to the applicant automatically by the system or by the concerned officials at the process of each stage • Model Projects of different KVI activities have been put up on PMEGP e-portal for the benefit of potential beneficiaries • Model Village Industries projects prepared by NSIC have also been linked to the website • To increase the registration of MSMEs in the country, the Government has undertaken measures that the PMEGP units can adopt the Udyog Aadhar Memorandum (UAM) to register online
How to apply	• Apply on https://www.kviconline.gov.in/pmegpeportal/pmegphome/

Pradhan Mantri Mudra Yojana

Description	• Primary product of MUDRA will be refinance for lending to micro businesses and units • The 2015 Budget has proposed the creation of Micro Units Development Refinance Agency (MUDRA) Bank, with a corpus of Rs.20,000 crores, and credit guarantee corpus of Rs.3000 crores
Nature of assistance	• The primary product of MUDRA will be refinance for lending to micro businesses/units under the aegis of the Pradhan Mantri MUDRA Yojana • The initial products and schemes under this umbrella have already been created and the interventions have been named "Shishu," "Kishor," and "Tarun" to signify the stage of growth/development and funding needs of the beneficiary micro unit/entrepreneur as also provide a reference point for the next phase of graduation/growth for the entrepreneur to aspire for: • Shishu: covering loans up to Rs. 50,000/- • Kishor: covering loans above Rs. 50,000/- and up to Rs. 5 lakh • Tarun: covering loans above Rs. 5 lakh and up to Rs. 10 lakh
Who can apply	• Businesses/entrepreneurs/units covered would include proprietorship/partnership firms running as small manufacturing units, shopkeepers, fruits/vegetable sellers, hair cutting salons beauty parlors, transporters, truck operators, hawkers, co-operatives or body of individuals, food service units, repair shops, machine operators, small industries, artisans, food processors, self-help groups, professionals, service providers, etc., in rural and urban areas with financing requirements up to Rs.10 lakhs • The products initially being launched are as under: • Sector/activity-specific schemes, such as schemes for business activities in Land Transport, Community, Social and Personal Services, Food Product and Textile Product sectors. Schemes would similarly be added for other sectors/activities • Micro Credit Scheme (MCS) • Refinance Scheme for Regional Rural Banks (RRBs)/Scheduled Co-operative Banks • Mahila Uddyami Scheme • Business Loan for Traders and Shopkeepers • Missing Middle Credit Scheme • Equipment Finance for Micro Units MUDRA would also adopt a credit plus approach and take up interventions for development support across the entire spectrum of beneficiary segments • The highlights of such proposed interventions/initiatives are as follows: • Supporting financial literacy • Promotion and Support of Grass Root Institutions • Creation of Framework for "Small Business Finance Entities" • Synergies with National Rural Livelihoods Mission • Synergies with National Skill Development Corporation
How to apply	http://www.mudra.org.in

SAMRIDH Scheme

Description	• The main objective of the MeitY SAMRIDH scheme is to provide funding support to the startups so that they can become successful • Through this scheme, not only funding support but also skill sets will be provided to the entrepreneurs. Acceleration will be provided to startups by providing customer connect, investor connects, and international connect services • This scheme will prove to be a boon in pushing the startup ecosystem
Nature of assistance	• In order to provide various types of benefits to push *startup ecosystem* the government has launched MeitY SAMRIDH scheme • Through this scheme the government is going to provide financial support to the startups who have brilliant solution and proof of concept for their product • This scheme will inspire a lot of citizens to start their own business • Ministry of electronics and information technology will be responsible for implementing this scheme • The startups will get platform to enhance their product and secure investment in order to scale their business • Under this scheme the existing and upcoming accelerators will be provided support for extending their services • The government is going to provide funding up to Rs. 40 lakh to the startups according to their current valuation and growth stage to accelerators • The accelerators are invited to apply online to become partners with MeitY and provide a startup accelerator program of 6 months every year

Who can apply	• The Accelerator must have been in the business of incubation for more than 3 years. Accelerator must have supported at least 50 startups of which at least 10 have received non-public investment or having targeted accelerator program with an experience of running at least 3 cohorts with activities listed as desirable under Samridh
	• Accelerator must have operations in India
	• The accelerators must have necessary space and infrastructure to carry out activities of startup
	• Should have demonstration capability with regard to
	• Structure cohort for accelerating deep tech software product startup
	    • On boarded with leading business mentor
	    • Network/connect with venture capitalist/Angel investor
	    • Supporting startup for domestic and international market immersion

How to apply	

Aspire (Scheme for promotion of Innovation, Entrepreneurship, and Agro-Industry)

Description	• The main objectives of the scheme are to: (i) Create new jobs and reduce unemployment (ii) Promote entrepreneurship culture in India (iii) Grassroots economic development at district level (iv) Facilitate innovative business solution for unmet social needs, and (v) Promote innovation to further strengthen the competitiveness of MSME sector

(continued)

Nature of assistance	• 80 Livelihood business incubators (2014–2016) to be set up by NSIC, KVIC, or Coir Board or any other Institution/agency of GoI/State Govt. on its own or by any of the agency/Scheme for promotion of Innovation, Entrepreneurship and Agro-Industry organization of the M/o MSME, one-time grant of 100% of cost of Plant and Machinery other than the land and infrastructure or an amount up to Rs.100 lakhs whichever is less to be provided • In case of incubation, centers to be set up under PPP mode with NSIC, KVIC, or Coir Board or any other Institution/agency of GoI/State Govt., onetime grant of 50% of cost of Plant and Machinery other than the land and infrastructure or Rs.50.00 lakhs, whichever is less, to be provided • Assistance toward the training cost of incubates will be met out of the ATI scheme of the Ministry as far as possible for both centers. • Total budget plan is Rs.62.50 crore for 2014–2016
Who can apply	• Implement the Incubation and Commercialization of Business Ideas Program through technical/research institutes including those in the field of agro-based industry. These would be designated as Knowledge Partners and would incubate new/existing technologies for their commercialization • To provide funds for the incubator/incubation and create necessary synergy between this scheme and the Livelihood Business Incubators/Technology Business Incubators and Incubation schemes of MSME/NSIC/KVIC/COIR BOARD/Other Ministries/Departments as well as Private incubators
How to apply	Application can be sent to Aspire Scheme Steering Committee of Ministry of MSME. Scheme Steering Committee will be responsible for overall policy, coordination, and management support. The Council will be chaired by Secretary, Ministry of MSME

Dairy Entrepreneurship Development Scheme

Description	• The department of Animal Husbandry, dairying, and fisheries is implementing Dairy Entrepreneurship Development Scheme (DEDS) for generating self-employment opportunities in the dairy sector, covering activities such as enhancement of milk production, procurement, preservation, transportation, processing, and marketing of milk by providing back-ended capital subsidy for bankable projects. The scheme is being implemented by National Bank for Agriculture and Rural Development (NABARD)
Nature of assistance	• The scheme seeks to extend assistance for setting up small dairy farms and other components to bring structural changes in the dairy sector. Assistance under the scheme is extended in the form of Interest Free Loan (IFL) to individuals, SHGs, NGOs, Cooperatives, companies for selected components
Who can apply	• Farmers, individual entrepreneurs, NGOs, companies, groups of unorganized and organized sector, etc. Groups of organized sector include self-help groups, dairy cooperative societies, milk unions, milk federations, etc. • An individual will be eligible to avail assistance for all the components under the scheme but only once for each component • More than one member of a family can be assisted under the scheme provided they set up separate units with separate infrastructure at different locations • The distance between the boundaries of two such farms should be at least 500m

(continued)

How to apply	• The entrepreneurs shall apply to their banks for sanction of the project • The bank shall appraise the project as per their norms and if found eligible, sanction the total outlay excluding the margin, as the bank loan • The loan amount is then disbursed in suitable instilments depending on the progress of the unit

National SC-ST Hub

Description	• To provide professional support to Scheduled Caste and Scheduled Tribe Entrepreneurs to fulfill the obligations under the Central Government Public Procurement Policy for Micro and Small Enterprises Order 2012, adopt applicable business practices and leverage the Stand-Up India initiatives
Nature of assistance	• To achieve 4% Public Procurement target from SC-ST entrepreneurs • Facilitating SC/ST Entrepreneurs to be part of vendor development programs and mentoring support • Collection, collation, and dissemination of information regarding SC/ST enterprises and entrepreneurs • Distribution of trade-specific tool kits to trained candidates
Who can apply	• Existing and Aspiring SC/ST Entrepreneurs
How to apply	• Apply on https://www.scsthub.in/

Raw Material Assistance

Description	• The scheme aims at helping MSEs by way of financing the purchase of raw material (both indigenous amd imported). This gives an opportunity to MSEs to focus better on manufacturing quality products
Nature of assistance	• Financial assistance for procurement of raw material up to 90 days • MSEs helped to avail economics of purchases like bulk purchase, cash discount, etc. • All the procedures, documentation, and issue of letter of credit in case of imports taken care of
Who can apply	Registered MSMEs
How to apply	• Entrepreneurs can apply through the prescribed application forms along with requisite fee to regional and branch offices of NSIC • For details refer NSIC website

Credit Guarantee Scheme for Micro and Small Enterprises (CGTMSE)

Description	• To encourage first generation entrepreneurs to venture into self-employment opportunities by facilitating credit guarantee support for collateral free/third-party guarantee-free loans to the Micro and Small enterprises (MSEs), especially in the absence of collateral

(continued)

	• Any collateral/third party guarantee free credit facility (both fund as well as non-fund based) extended by eligible institutions, to new as well as existing Micro and Small Enterprises, including Service Enterprises, with a maximum credit cap of 200 lakh (rupees two hundred lakh only) are eligible for guarantee under the scheme. Recently, guarantee coverage made eligible to select NBFCs and Small Finance banks • The guarantee cover available under the scheme is to the extent of 50%/, 75%/80%, and 85% of the sanctioned amount of the credit facility. The extent of guarantee cover is 85% for micro enterprises for credit up to 5 Lakh • The extent of guarantee cover is 50% of the sanctioned amount of the credit facility for credit from 10 lakh to 100 Lakh per MSE borrower for retail trade activity • The extent of guarantee cover is 80% for (i) Micro and Small Enterprises operated and/or owned by women; and (ii) all credits/loans in the North East Region (NER) for credit facilities up to 50 Lakh. In case of default, the trust settles the claim up to 75% of the amount in default of the credit facility extended by the lending institution for credit facilities up to 200 Lakh
Nature of assistance	• Collateral free loans up to a limit of Rs.50 lakh – for individual MSEs
Who can apply	• Both existing and new enterprises are eligible under the scheme
How to apply	• Candidates meeting the eligibility criteria may approach banks/financial institutions, which are eligible under the scheme, or scheduled commercial banks and select Regional Rural Banks

Micro and Small Enterprises Cluster Development Program (MSE—CDP)

Description	• The Ministry of MSME has adopted cluster development approach as a key strategy for enhancing productivity and competitiveness as well as capacity building of MSEs and their collectives in the country • Clustering of units also enables the providers of various services to them, including banks and credit agencies, to provide their services more economically, thus reducing the costs and improving the availability of services for these enterprises • Objectives of the scheme: (i) To support sustainability and growth of MSEs by addressing common issues such as improvement of technology, skills and quality, market access, and access to capital (ii) To build the capacity of MSEs for common supportive action through the formation of self-help groups, consortia, upgradation of associations, etc. (iii) To create/upgrade infrastructural facilities in the new/existing industrial areas/clusters of MSEs (iv) To set up common facility centers (for testing, training, raw material depot, effluent treatment, complementing production processes, etc.)
Nature of assistance	• Diagnostic Study • Soft Intervention • Setting up of Common Facility Centers (CFCs) • Infrastructure Development (Upgradation/New)

(continued)

	Cost of project and Govt. of India assistance: • Diagnostic study—maximum cost Rs.2.50 lakh • Soft interventions—maximum cost of project • Rs.25.00 lakh, with GoI contribution of 75% (90% for special category States and for clusters with more than 50% women/micro/village/SC/ST units) • Hard interventions, i.e., setting up of CFCs—maximum eligible project cost of Rs.15.00 crore with GoI contribution of 70% (90% for special category States and for clusters with more than 50% women/micro/village/SC/ST units) • Infrastructure development in the new/existing industrial estates/areas; maximum eligible project cost Rs.10.00 crore, with GoI contribution of 60% (80% for special category States and for clusters with more than 50% women/ micro/SC/ST units)
Who can apply	• Industrial associations/Consortia, Clusters
How to apply	• Only online applications are considered with effect from 01-04-2012, Hard copy of applications need to be sent through State Governments or their autonomous bodies or field institutes of the Ministry of MSME, i.e., MSMEDIs. The proposals are to be approved by the Steering Committee of MSE-CDP

Scheme of Fund for Regeneration of Traditional Industries (SFURTI)

Description	The main objectives of the scheme are to: • To organize the traditional industries and artisans into clusters to make them competitive and provide support for their long-term sustainability • To provide sustained employment for traditional Industry artisans and rural entrepreneurs • To enhance marketability of products of such clusters by providing support for new products, design intervention and improved packaging and also the improvement of marketing Infrastructure • To equip traditional artisans of the associated clusters with the improved skills and capabilities through training and exposure visits • To make provision for common facilities and improved tools and equipment for artisans • To strengthen the cluster governance systems with the active participation of the stakeholders, so that they are able to gauge the emerging challenges and opportunities and respond to them in a coherent manner • To build up innovative and traditional skills, improved technologies, advanced processes, market intelligence, and new models of public–private partnerships, so as to gradually replicate similar models of cluster-based regenerated traditional Industries
Nature of assistance	• The Scheme would cover three types of interventions, namely, "soft Interventions," "hard Interventions," and "thematic interventions" • The project outlay for various clusters is as follows: • Heritage cluster (1000–2500 artisans *):Rs. 8 crore Major cluster (500–1000 artisans*): Rs. 3 crore • Mini cluster (Up to 500 artisans*): Rs. 1.5 crore. *For NER/J & K and Hill States, there will be 50% reduction in the number of artisans per cluster. Soft Interventions: Max Rs. 25.00 lakhs (100% scheme funding) • Hard Interventions: As per project requirement (75% scheme funding) Cost

(continued)

	of Technical Agency Rs. 8 % of Soft and Hard interventions (100% scheme funding). Cost of Implementing Agency/Cluster Executive: Max Rs. 20.00 lakhs (100% scheme funding)
Who can apply	• Non-Government organizations (NGOs), Institutions of the Central and State Governments and Semi-Government institutions, field functionaries of State and Central Govt., Panchayati Raj institutions (PRIs), etc., with suitable expertise to undertake cluster development
How to apply	• The above eligible agency/organization has to submit the proposal to the State Office, KVIC, and the same is to be scrutinized at State Level and Zonal Level before submitting to Scheme Steering Committee for approval

Entrepreneurial and Managerial Development of SMEs Through Incubators

Description	• The scheme endeavors to provide early stage funding for nurturing innovative business ideas (new indigenous technology, processes, products, procedures, etc.), which could be commercialized in a year • Under this scheme financial assistance is provided for setting up of business incubators
Nature of assistance	• Funding support for setting up of "Business Incubators (BI)": The cost may vary from Rs.4 to 8 lakh for each incubatee/idea, subject to overall ceiling of Rs.62.5 lakh for each BI • Items @ per BI: (a) Upgradation of infrastructure Rs. 2.50 lakh (b) Orientation/training Rs. 1.28 lakh (c) Administrative expenses Rs. 0.22 lakh Thus the total assistance per BI—Rs. 66.50 lakh
Who can apply	• Any individual or MSME having innovative ideas ready for commercialization can apply to the host institution (e.g., IITs, NITs, technical colleges, research institutes, etc.). See the list of host institutions at following web address: • http://www.dcmsme.gov.in/schemes/Institutions_Detail.pdf Any technical institution (as given in the EoI), which wants to become host institution can apply to the office of the Development Commissioner—MSME or their nearest MSME-DI for funding support
How to apply	• Application can be made by the technical institution, which wants to be host institution, once a Request for Proposal (RFP)/Expression of Interest (EoI) is released • Any individual or MSME can apply directly to their nearest host institution, a list of host institutions is given on the website: • http://www.dcmsme.gov.in/schemes/Institutions_Detail.pdf

Entrepreneurship and Skill Development Program (ESDP) Scheme

Description	• To promote new enterprises, capacity building of existing MSMEs and inculcating entrepreneurial culture in the Country Widen the base of entrepreneurship by development, achievement, motivation, and entrepreneurial skill to the different sections of the society
Nature of assistance	• Entrepreneurship/Self-employment awareness and motivation to different sections of the society including SC/ST/Women, differently abled, Ex-servicemen, and BPL persons as career options • Enterprise Facilitation for Ideation, Mentoring, and incubation, Credit facilitation, Market accessibility, Enterprise Clinic, Diagnostic studies in the event of sickness, counseling, and other facilities • Entrepreneurship and Skill Training in Agro-Based Products, Hosiery, Food and Fruit Processing Industries, Carpet Weaving, Mechanical Engineering Workshop Machine Shop, Heat Treatment, Electroplating, Basic/ Advance Welding/Fabrication/Sheet metal work, Basic/Advance Carpentry, Glass and Ceramics, etc. • Management capacity building Training to Existing Entrepreneurs and their supervisory staff in Industrial Management, Human Resource Management, Marketing Management, Export Management/Documentation and Procedures, Materials Management, Financial/Working Capital Management, Information Technology, Digital Marketing, Quality Management/QMS/ISO 9000/EMS, WTO, IPR, Supply Chain Management, Retail Management, Logistics Management, etc.
Who can apply	Aspiring and Existing Entrepreneurs
How to apply	• To be apply through MSME-DI, MSME-TC websites • The scheme link: http://dcmsme.gov.in/Enterprise&skillDevelopment.htm

Mahatma Gandhi Institute for Rural Industrialization (MGIRI)

Description	• To support, upgrade, and accelerate the process of Rural Industrialization in the country to move toward the Gandhian vision of sustainable village economy self-sufficient in employment and amenities • Science and technology Intervention and Innovations form Rural Industries • Networking and National/International Collaboration on R & D and Technology Transfer related to KVI Sector Skill Entrepreneurship Development Training for Enterprise development in KVI sector
Nature of assistance	• Any individual entrepreneur can get help in technology upgradation, product quality improvement, DPR preparation, and product testing in KVI sector and online design support for garment sector • Any individual entrepreneur can get Skill/Entrepreneurship Development Training to set up enterprise in KVI sector • Any individual entrepreneur can avail exiting portal like www.ruralhaat.com and www.Udyamisahayak.com evolved by MGIRI for obtaining information on machinery and equipment availability and product marketing support for village industries
Who can apply	• Existing and Aspiring Entrepreneurs
How to apply	• Apply on http://www.mgiri.org/

2.2 Policies and Schemes for Women Entrepreneurs in India

Regardless of gender, starting a business is a difficult prospect. Key challenges for startups in India include generating funds, limited understanding of customers, penetrating the market, hiring qualified employees, and the complex regulatory environment. For women entrepreneurs, however, there are additional barriers, which are part of a broader and more pronounced gender gap in the male-dominated Indian society (Agarwal and Lenka 2018).

Women entrepreneurship has been recognized as an important source of economic growth. Women entrepreneurs create new jobs for themselves and others and also provide society with different solutions to management, organization, and business problems. However, they still represent a minority of all entrepreneurs. Women entrepreneurs often face gender-based barriers to starting and growing their businesses, like discriminatory property, matrimonial and inheritance laws, and/or cultural practices; lack of access to formal finance mechanisms; limited mobility and access to information and networks, etc.

2.2.1 Top Government Schemes for Women Entrepreneurs

Now is the time to say that there was a time of male mainstream in society. At first, the woman had to do the housework, and the idea of the business was a dream. Time has changed; there is no shortage of women who have grown up beyond many such forms of domination. For example, Indira Nooyi, Chanda Kochhar, Ekta Kapoor, Palguni Nair, Neeru Sharma, and so on. Our current finance minister, Mrs. Nirmala Sitharaman, is also a witness to women empowerment. Some of the special schemes for women entrepreneurs implemented by the government bodies and allied institutions are provided below (Jha 2014; Gupta and Gupta 2020; Foods 2019).

2.2.2 Annapurna Scheme

This scheme is for financing women to establish a food Catering Unit for selling tiffin/food/lunch packs, etc. It offers loans up to 50,000/– rupees. The amount granted as a loan under this scheme can be used for buying utensils, other kitchen tools, and equipment. The interest rate depends upon the market rate and the concerned bank. A guarantor is required to secure the loan and it can be repaid in 3 years (36 monthly installments). Once the loan is approved, women don't have to pay EMI for the first month.

2.2.3 *Stree Shakti Package*

Women who have 50% share in the ownership of a firm or business and have taken part in the state agencies run Entrepreneurship Development Programs (EDP) are eligible for this package. The scheme also offers a discounted rate of interest by 0.5% in case the amount of loan is more than 2 lakhs. For small sector units no security is required for loans up to Rs. 5 lakhs.

2.2.4 *Orient Mahila Vikas Yojana Scheme*

This scheme, launched by Oriental Bank of Commerce, provides capital for women for starting small businesses. Women with 51% share in the business are eligible for the loan. Collateral is not needed if loans are between Rs.10 lakhs and Rs.25 lakhs for small-scale industries. Repayment is for 7 years period. It offers concession of 2% on the rate of interest.

2.2.5 *Dena Shakti Scheme*

This scheme is provided by Dena bank to those women entrepreneurs in the fields of Agriculture and allied activities, Retail Trade, Micro Credit, Education, Housing, and retail and small business enterprises. There is a concession of 0.25% on the rate of interest. The scheme offers loans up to Rs. 20.00 lakhs under retail trade, Rs. 20 lakhs under education and housing, and Rs. 50,000/—under micro credit.

2.2.6 *Udyogini Scheme*

Women entrepreneur aged between 18 and 45, who are involved in agriculture, retail, and similar small businesses are eligible for loans up to 1 lack under this scheme. Their family annual income should be 45,000 or less to avail this loan. There is exception for widowed, destitute, or disabled women. For widowed, destitute, or disabled women from SC/ST categories, a subsidy of 30% of the loan, up to Rs. 10,000, is provided and for women with general category a subsidy of 20% of the loan or Rs.7500 whichever lower is provided. Punjab and Sind Bank and KSWDC offer this loan.

2.2.7 *Cent Kalyani Scheme*

This scheme is offered by the Central Bank of India. It is for women business owners in multiple areas such as SMEs or agricultural work or retail trading. Under this scheme, loans up to Rs.1 crore are sanctioned. There is no need of collateral and guarantors. There are no processing fees and interest rate varies according to market rates.

2.2.8 *Mahila Udyam Nidhi Scheme*

This scheme is offered by Punjab National Bank and Small Industries Development Bank of India (SIDBI). This scheme provides financial assistance of up to Rs. 10 lakh to set up a new small-scale venture. It also promotes upgrading and modernization of existing projects and repayment period is 10 years. SIDBI also includes a moratorium period of maximum 5 years. The interest rate varies according to market rate.

2.2.9 *TREAD (Trade-Related Entrepreneurship Assistance and Development) Scheme*

This scheme aims for economic empowerment of women by providing credit (through NGOs), training, development, and counseling extension activities related to trades, products, services, etc. The Government grants up to 30% of the total project cost as appraised by lending institutions, which would finance the remaining 70% as loan assistance to applicant women.

References

Agarwal M, Dwived R (2017) Pradhan Mantri Mudra Yojna: a critical review. Parikalpana: KIIT. J Manag 13(2):97–106

Agarwal S, Lenka U (2018) Why research is needed in women entrepreneurship in India: a viewpoint. Int J Soc Econ 45:1042

Alaguraja M, Nedumaran DG (2020) Performance of khadi and village industries commission through micro, small, & medium enterprise. Aegaeum J 8(3)

Baby, N. (2021) Modern scenario of entrepreneurship in India Dr. B. Jothikumar. J Exclusive Manag Sci 10(3): ISSN 2320 - 866X

Chaudhari PB, Borkar R (n.d.) Mahatma gandhi institute for rural industrialization, wardha's technology is an alternative life-giving boon for farmers

Choudhary P, Vadera M (2019) Building new India through entrepreneurship: study on how entrepreneurship can drive the agenda of overall development for new India. Think India J 22(4):10972–10984

Devi, S. (2020) Importance of innovation in entrepreneurship. Int Bi-Annual Peer Rev Multidiscip Res J (Online) 2(1): ISSN: 2582-6077

Foods A (2019) Start-up marketing strategies in India. Food Proc 133(134):141

Gupta S (2002) Report of the special group on targeting ten million employment opportunities per year over the tenth plan period. Government of India, Planning Commission

Gupta P, Gupta N (2020) Women empowerment: a challenge for government as well as society. IAHRW Int J Soc Sci Rev 8(4–6):200–204

Howorth C, Rose M, Hamilton E (2006) Definitions, diversity and development: key debates in family business research. In: The Oxford handbook of entrepreneurship. Oxford University Press, Oxford

Jayanthi D (2019) A Study about entrepreneurship in india and its promotion under'startup India'Scheme. Iconic Res Eng J 2(11):118–126

Jha R (2014) Welfare schemes and social protection in India. Int J Sociol Soc Policy 34(3/4):214–231. https://doi.org/10.1108/IJSSP-10-2013-0099

Lalthanzuali S, Devi NI (2022) Prime Minister's employment generation Programme (PMEGP) initiatives in Mizoram: an overview. Int J Curr Hum Soc Sci Res 6(1):76–84

Mageshwar C, Jothimani K (2022) Entrepreneurship development in India-the focus on start-ups. Int J Eng Manage Res 12(1):144–148

Mund CS (2020) Problems of MSME finance in India and role of Credit Guarantee Fund Trust for Micro and Small Enterprises (CGTMSE). IOSR J Econ Fin 11(4):1–6

Pandey P, Jha SK, Sinha S (2017) Effectiveness of dairy based entrepreneurial developmental programmes. Int J Soc Sci Econ Res 4(2)

Rajpal NK (2014) Micro, small and medium enterprises in north east India: performance and prospects. IUP J Entrepr Dev 11(3):7

Rao V (2014) India: support to scheme of fund for regeneration of traditional industries. Retrieved from 16 Sep 2022. https://policycommons.net/artifacts/387512/india/1352029/. CID: 20.500.12592/bzp1nr

Sanghi S, Srija A (2002) Entrepreneurship development in India-the focus on start-ups. Laugh Udyog Samachar 20–27

Singh VK, Sharma P (2018) Role of micro, small & medium enterprises in Indian economy 5(4) ISSN 2349-5138

Unnisa LA, Amulya M (2016) A study on micro and small enterprise cluster development programme (MSE-CDP). Zenith Int J Bus Econ Manage Res 6(6):20–23

Chapter 3
Skills for Entrepreneurs

Swapnil Chaurasia

Abstract In the simplest words Entrepreneurship can be defined as the creation of a new business venture. Entrepreneurship may be found almost anywhere. We use a variety of products and services provided by businesses in our daily lives. Entrepreneurs are responsible for the creation of these businesses. Entrepreneurship not only helps to start new enterprises in a variety of industries, but it also has a favorable impact on economic development. Many researchers have emphasized the importance of entrepreneurship to economic development, and it is now widely acknowledged that educational and training opportunities play a critical role in cultivating future entrepreneurs and developing the abilities of existing entrepreneurs to grow their businesses to greater levels of success. Similar is the case in deciding upon traits, characteristics, and skills reflected by an individual that entitles him as an entrepreneur in the due course. Numerous literatures have pondered upon the fact that there are certain desired skills that add up to form the overall personality of an entrepreneur. These skills are integral aspects of formation of what is called an entrepreneurial persona. The chapter discusses various types of ventures and categories of entrepreneurs along with enlisting and elaborating various skills for entrepreneurs, which have to be reflected by an individual to be a successful entrepreneur. Insights have been cumulated from different literatures highlighting broad categories of entrepreneurial skills. The varied nature of entrepreneurship is presented in this chapter, which displays common outcomes as well as many points of disagreement.

Keywords Entrepreneurship · Skills for entrepreneurs · Entrepreneurial skills · Entrepreneurial capacity · Types of entrepreneur · Types of enterprise · Characteristics and traits of entrepreneurs

S. Chaurasia (✉)
Shrimad Rajchandra Institute of Management and Computer Application (SRIMCA-MBA), Uka Tarsadia University, Surat, Gujarat, India
e-mail: swapnil.chaurasia@utu.ac.in

N. Amaresan et al. (eds.), *Agricultural Microbiology Based Entrepreneurship*, Microorganisms for Sustainability 39,
https://doi.org/10.1007/978-981-19-5747-5_3

3.1 Introduction

The term "entrepreneur" can refer to a variety of things. On one hand entrepreneur is a person of exceptional ability who pioneers change being a change agent, while on the other hand, anyone who wishes to work for themselves is labeled an entrepreneur. The term "entrepreneur" comes from the French word "entreprendre," which means "to take on." It means to establish a business in a business context. An entrepreneur is considered to be the center point of any business enterprise or a venture. He is the one who organizes, manages, and takes on the risks of a business or venture. An entrepreneur is different from a manager who is responsible for administrating the resources available and is bound by the scope of working. An entrepreneur goes beyond the scope of work and takes up the responsibilities of planning, acquiring, utilizing, and disposing of the resources in the best possible manner and making sure that such utilization is sustained for the betterment of the business for the years to come. This comprehensive aspect of entrepreneurs leads to an important question, whether every entrepreneur has the same capabilities and attributes? Are they all alike? The answer is "No." While basic characteristics and skills are shared by all entrepreneurs, there is a vast spectrum of individualism among them, which is often reflected in their ideologies and methodologies. For example some athletes excel in sports not because they have better equipment or infrastructure but because they have interest in the sport, enjoy it, and have been properly trained in the sport. They have improved their abilities. Others have a lot of natural talent and don't need as much training as are said to be gifted, which helps them to succeed over others. Others, who may or may not have been instructed, just develop their own successful method of contemplating and viewing the game and connecting the mechanism to their body type and playing style to overtake their competitors. The same is true in the case of entrepreneurs' business owners. Some people undergo formal education and training in order to improve their skills. Others have an innate talent for it. Others, on the other hand, defy all odds and invent novel strategies while still succeeding. Having said that, it is not possible to point out the exact factor that triggers the entrepreneurial drive in an individual. Being entrepreneurial and cultivating an entrepreneurial culture extends beyond the fear of risk and the shame of failure, both of which have a significant impact on the entrepreneurship landscape. It is a mixed bag of talents, abilities, capabilities, and skills that together lead to entrepreneurship. Even though we can say that entrepreneurs' profiles are crucial to defining a business idea, and successfully implementing it, it is important to note that there is also the possibility of developing entrepreneurial characteristics, and this is where educational institutions come in. They should play a key role, very early in the creation of knowledge and skills related to entrepreneurship. Knowledge, skills, and abilities (KSA) are the most important aspects of self-development and are considered in many contexts where individual performance acts as the key parameter of measurement. Innovation is an example of a discipline that can be taught from the first years of school because it is a specific instrument used by entrepreneurs to explore new business prospects or a different product or service.

The world is changing with much greater pace than ever, especially with the advent of technology. Technology-driven modern businesses, in the name of start-ups, have changed the entire landscape of the business. Training has become imperative for modern business owners to cope up and survive in this dynamic market place. Knowing how to manage risk, search out new sources of innovation, employ creativity tools, and learn from the market are all talents that every entrepreneur or potential entrepreneur should possess. The effect of the above mentioned when coupled with correct training produce miraculous results. Everyday, we see new businesses, products, processes, and services emerge and die, and the purpose of entrepreneurship education is to seek out and rigorously investigate new business/ new practices that bring value to the market and simplify the economy. Day in and day out new areas are being discovered demanding unconventional approaches, which are being tackled by modern entrepreneurs through blending their skills, knowledge, abilities, talents and, most importantly, their motive to create new. In this way, entrepreneurship is founded on a variety of soft and hard talents, skills, abilities, and the like that have been extensively researched in this chapter.

3.2 Entrepreneurial Traits

Trait can be defined as a quality that forms an integral part of one's personality or character. Entrepreneurial traits thus include all the qualities that must be reflected upon by an individual to develop into an entrepreneur. A plethora of literature is available enumerating and discussing the desired traits of entrepreneurs. It is said that to be a successful entrepreneur one must have a high degree of motivation and energy and have the ability to work long shifts for long periods of time.

with a lower than average amount of rest or breaks. In short an entrepreneur must endure hectic work schedule with high degree of persistence toward any given task. The following are some of the widely accepted traits of entrepreneurs globally.

1. *Self-drive*: An entrepreneur is self-driven toward his motive. The motive acts as the fuel for this drive. The actions of entrepreneurs are directed toward achievement of their long-term goals. The stronger the goal, the greater is the thrust.
2. *Self-belief*: Belief in oneself and one's capabilities is an integral part of entrepreneur's personality. This self-belief develops self-confidence in the entrepreneurs and self-confidence triggers bold decision making. It is about the conviction that entrepreneurs show while taking up a venture. Strong conviction leads to unflinching decision making that helps entrepreneurs sail through tough times.
3. *Self-reliant*: It is necessary to be self-reliant. An entrepreneur is always the initiator while taking decisions. These decisions involve acquisition, utilization, and channelization of resources. An entrepreneur tries to self-source these resources to the maximum extent and hold himself accountable for the results or the consequences arising in the due course. An entrepreneur should not be

focused only on their personal goals and objectives; they should be considerate and aim at being self-sufficient to avoid accepting outside aid.

4. *Goal Oriented*: The capacity to define clear, attainable goals and objectives that are both challenging and practical. The goal setting is done using the SMART (Specific, Measurable, Attainable, Realistic, and Time bound) approach. This helps entrepreneurs to determine competitive goals as per the circumstances prevailing in the market.
5. *Long-term Perspective (Visionary)*: It is very important for an entrepreneur to have a long-term view/perspective toward business. Long-term perspective enables an entrepreneur to link his/her efforts in the direction of the vision of the entrepreneur. Long-term commitment to a project usually will take 6–7 years to complete or in some cases more than that. This entails a whole commitment toward aligning the strategies with the set long-term goals.
6. *Pecuniary Perspective*: Money is an important aspect of any business. The value of the venture is calculated and determined in the form of money. Money or monetary aspect forms the centerpiece of evaluation. Entrepreneurs use money as a performance indicator. Money, whether in the form of a wage, earnings, or capital gains, should be considered to gauge how well a company is performing rather than an end in itself. Thus, money provides the angle of objectivity to the business enterprise and most importantly entrepreneur's efforts.
7. *Problem-solving Ability*: An entrepreneur must be keen and determined towards solving social problems. These problems should be viewed as opportunities by the entrepreneurs. Taking the example of Redbus.com which was established with the motive to solve the commutation problems of people especially during festivals and high rush seasons.
8. *Seeks Challenges*: Entrepreneurs look out for new challenges. They experience a feeling of accomplishment when they overcome challenges. No business is risk free. Uncertainty is all pervasive and risk arising out of uncertainty is inevitable. As entrepreneurs are considered to be risk bearers, they seek for viable and feasible methods to overcome challenges of various kinds posed by the venture time to time.
9. *Calculative Risk Taker*: An entrepreneur calculates risks with precision and mostly takes moderate risks. Though an entrepreneur is known for high risk appetite, an entrepreneur also realizes that responsibilities of his team also vests on his shoulders. Thus, an entrepreneur is inclined to take moderate risks.
10. *Learner*: Learning for entrepreneurs never stops. They learn from others, their mistakes, and, many a times, they learn from their own mistakes. Recognizing one's role in a failure might help to avoid repeating the same mistakes in the future. Failure is disappointing, but it does not have to be depressing. Treating it as a minor setback is a true trait of an entrepreneur.
11. *Positive Outlook*: Entrepreneurs look at the brighter of everything. Challenges and criticisms are stepping stones to success for an entrepreneur. An entrepreneur must be able to seek and use criticism of his/her performance's style and content for the betterment of the business and the society. To take negative

feedback as constructive feedback is one of the most rewarding traits of entrepreneurs.

12. *Initiator*: Entrepreneurs take risk and this is what differentiates them from managers. They are keen observers and are aware of what is happening around them and the world to seize the best opportunity available at hand. Entrepreneurs try to take the first-mover advantage from any lucrative situation. Making the first move helps them to capture the market early and make amendments in their functioning as per the required conditions while their competitors are still planning to make a move. Making the first move has its own advantages and disadvantages. On one hand you get to capture the entire market and make a mark with your offering. One can also have first-hand the resources available. One the other hand, there is always a risk of failure and losing the market, which can be used as a counter measure for the competitors when they decide to enter into the market. In any case, waiting for others to make the first move has never been a trait of an entrepreneur.

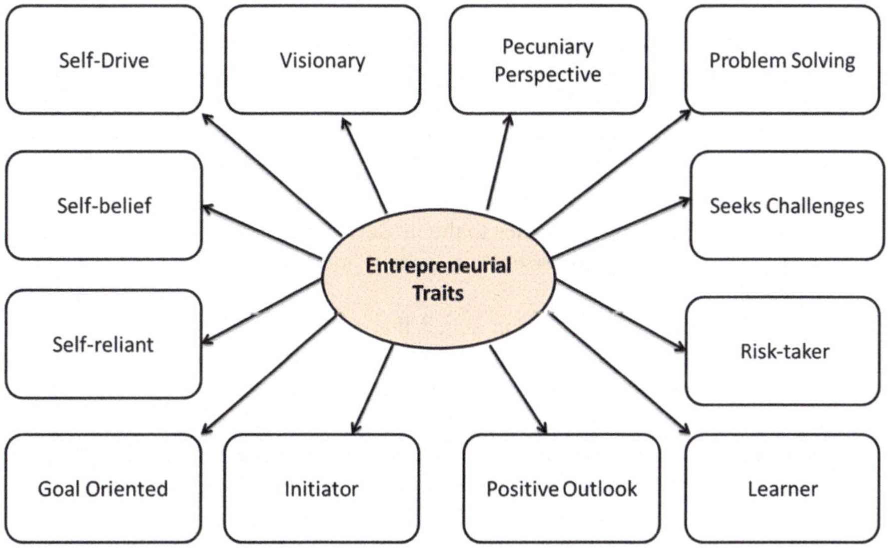

No single person possesses all of these qualities. Some qualities may dominate others depending upon the personality type of the entrepreneur. An entrepreneur must try and make efforts to strike a balance between all these qualities so as make the most of what he plans to achieve. Some qualities are reflected in the early phase while others require time and are acquired through experience. Thus, a successful entrepreneur cannot rely on few traits or selected traits but focus on acquiring and developing as many of the above mentioned as he/she can.

3.3 Entrepreneurship and Innovation

Entrepreneurship and Innovation go hand in hand. Innovation is defined as doing or creating something new. The definition is extended by including the element of creativity and profitability. That means creativity must be rewarded with profits so as to be included within the concept of innovation. Innovation plays a pivotal role in entrepreneurship as it changes the face of business dramatically. Innovation can take many forms but in the context of entrepreneurship broad areas where innovation can be made via entrepreneurs have been recognized by J. Schumpeter, which are as follows:

1. Products
2. Markets
3. Methods
4. Organizations/Enterprises

Product: Product can be defined as any offering, which can cater to the needs and wants of customers. These products can be in tangible or intangible forms. Intangible products are also known as services. Apart from products and services, an entrepreneur takes on a wide range of offerings depending upon the demand prevailing in the market like events, experiences, ideas, destinations, or places and the like. Thus, there is a wide spectrum of offerings that is capitalized by an entrepreneur for the betterment of the business and the society. Nowadays, there is a huge wave of online service in the form of applications for cellphones and other devices like laptops, television, etc. This provided lucrative opportunities to the modern entrepreneurs to take advantage of the new products and services to cater to modern needs and solve modern problems.

Markets: The traditional definition stated that a market is a place where buyers and sellers meet for purchase and sale of the product. Today, this place can be brought at the click of a button or a simple tap on the screen. The transformation of markets had been greatly supported by internet technology, which has made it easy for the entrepreneurs to reach their customers even when they are miles away. Markets are of different types on account of the nature of the transactions and parties involved like consumer markets, government markets, business markets, industrial markets, niches, and the like. New business enterprises in the form of startups have emerged and have been doing exceptionally well in the modern era. Everyday hundreds of new apps are introduced in the virtual market space addressing the needs of the customers. This is not only dominant in developed sections. Rural entrepreneurship has developed in a massive way – one of the greatest examples being e-Choupal in India. The markets are not what they used to be and will not be the same in coming future. It is sheer dedication and keen observation of entrepreneurs that direct these changing markets through upgrading their skills as per the present and future needs.

Methods: Methods are basically the set of activities undertaken for production of goods and services. These methods or set of activities, also commonly called as process, are the foundation of creation of novelties in terms of products and services. It is evident from history that mankind has evolved to a great extent and so has its processes. Right from cooking a simple meal to the extraction of oil and then using it as a fuel, all involve specific and distinct methods. Making life easier is one of the most important objectives of entrepreneurship and these methods are modified and revamped to achieve this objective. Methods/processes are divided into three broad categories, namely, manual, semiautomated, and fully automated. Manual methods are also termed as labor intensive where there is involvement of manual labor to a great extent. These methods are slowly taken over by semiautomated methods where there is some inclusion of machinery to automate certain activities, which require precision and long processing time. Fully automated methods are used in industrial manufacturing where there is involvement of heavy machinery and most of the processes are performed by specialized machines, tools, and equipment. In India, small and medium enterprises are mostly dependent of manual or semiautomated manufacturing. Thus, it can also be said that entrepreneurship in India will have to put in a lot of efforts and consideration to transform manufacturing in these segments. Also, making these segments become very lucrative for technology-driven entrepreneurs, who can tap these markets by providing something called as "Appropriate Technology". Appropriate technology is a form of technology that suits the local needs. Since MSMEs in India are found to be catering to the needs of local markets, it calls for such technology to help them out when it comes to better production with less operating costs. New enterprises and business models have come up with better methods and process for conventional production process, be it scaling, extraction, molding, packaging, and the like, paving the path for better and more advanced production line for future enterprises.

Organizations/Enterprise: An enterprise is an entity engaged in commercial, industrial, or professional activity. Various industries and sectors have different types of business organizations. These enterprises and organizations are classified according to their economic activities, namely, manufacturing, service, and trading. Enterprises are classified according to their size, such as micro-, small-, medium-, and large-scale enterprises. Though there are different categories of enterprises, there are certain elements that are inseparable to any enterprise irrespective of these categories. These include the element of trade, which is reflected by production, distribution, and exchange. Economic and social motives are the forces behind a business idea. Uncertainty and risk which an entrepreneur tries to assess, reduce and avoid in any circumstances. Some of the basic types of business enterprise include the following:

Manufacturing Enterprise: Manufacturing enterprises look for products that can be utilized as raw materials, with the goal of transforming them into finished goods. This new product might be something that can be sold directly to consumers or something that can be used throughout the value chain. These companies specialize in transforming one product into a new one. In the manufacturing process, most organizations combine raw materials, technology, labor, plant and machinery, and overheads. Manufacturing and selling shoes, garments, mobile phones, laptops, bags, metal products, and other items are examples of such firms.

Service Enterprise: As noted in the tertiary sector, a service business is concerned with the supply of intangible offerings, that is, services that do not have a physical form but create value. Businesses that provide services provide value by utilizing skills, consulting, knowledge, efficiency, and so on. Banks, law firms, chartered accountants, financial advisors, beauty salons, schools, e-commerce platforms, and event planners are just a few examples of service enterprises. In India, service sector is one of the largest and most sought out sector in terms of productivity, employability, and advancements.

Merchandise Enterprise: A merchandising business is one that involves the transfer of ownership of tangible goods. The majority of the time, things are purchased at wholesale and sold at retail in these types of enterprises. These companies make money by selling things at a higher price than they paid for them. Merchandising enterprise model is one of easiest and simplest form of enterprise as the hassle of production is eliminated from the value chain. These types of enterprises also have certain shortcomings, one of which is lack of innovation possible.

Trading Enterprise: A trading business also deals in the resale of items. The distinction between a trading business and a merchandising business is that a trader in a trading business does not have to keep inventory on hand. Trading companies deal with a variety of items and services that are sold to individuals, businesses, and government agencies. Buying products or brokering services, negotiating rates, and organizing delivery are some of the operations that trading businesses engage in. Profit margin purchase cost and selling price are how a trading business makes money. The trading industry is mostly made up of the import and export of products and services. Real estate brokers, intermediaries, importers, exporters, foreign exchange merchants, and other trading enterprises are of examples of such enterprises.

Hybrid Enterprise: Hybrid businesses are those that engage in a variety of activities such as manufacturing, service, and merchandising. These could be business groupings engaging in a variety of enterprises that don't fit well into the categories of service or manufacturing. It's possible that it won't use standard production and distribution methods. Hybrid enterprises are sometimes referred to as firms that focus on socially beneficial purposes. A typical restaurant, for example, is classed as a service business, but some are also classified as hybrid enterprises. They are engaged in manufacturing activities if they are active in combining substances to create various goods. It can be deemed merchandising if they are participating in the resale of alcoholic merchandise.

3.4 Entrepreneurship and Innovation

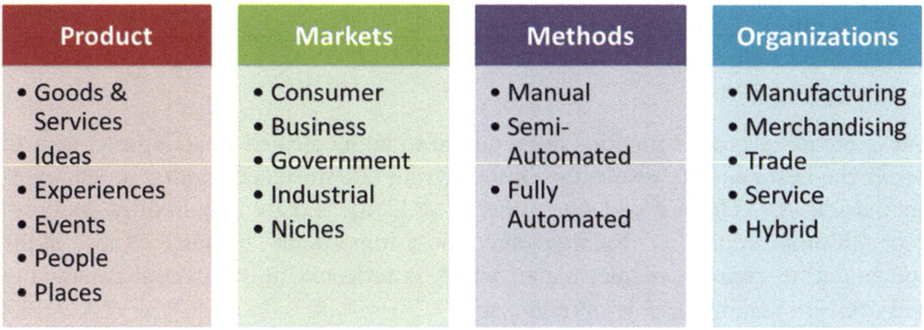

3.5 Entrepreneurial Skills

An entrepreneurial skill is one domain where most of the researches and studies have been focusing on. This domain is supported by evident facts and real-world examples of how entrepreneurship actually gets molded. Entrepreneurial skills form a wide spectrum ranging from a very simple to the most complex of skills to be acquired. Before discussing entrepreneurial skills and its varied types let us first differentiate between skills, talents, and capabilities. These terms are often interchangeably used in entrepreneurial context. Though there exist a lot of similarities between them still they minutely differ in their basic meaning and nature of use. Skills can be defined as an ability to do something well or in a better/polished way. In entrepreneurial context skills have been defined as the "ability to use one's knowledge effectively and readily in execution or performance." Thus, in simple words, skills are the abilities, which can be utilized readily for performing certain tasks in the most efficient way. It is about doing something competently. It is the power of performance, the degree to which one can excel in the similar mundane activities required for the business to grow. Talent is natural. It is a natural ability of doing something exceptionally well with ease. Talent is often athletic, creative, or artistic aptitude; it is general intelligence or mental power/ability. Capability on the other hand is the extent of someone's ability. This actually means capabilities are the upper limits of one's abilities. It is the extent to which individuals can perform at their maximum potential. Thus, capabilities can be defined as a set of tasks that can be potentially done to the maximum by one's abilities within a system. Having said that, entrepreneurs bank much more on their skills rather than anything else. It is the skills that help them sail through the tides of change, which is ever constant in the modern world. These skills are acquired, honed, and polished and modified to best suit the prevailing circumstances. There is a long list of skills researchers and academicians have come up for reflecting the true characteristic of an entrepreneur. These skills can be broadly categorized as follows.

1. Cognitive Skills
2. Social and Relational Skills
3. Technical Skills
4. Managerial Skills

These broad categories of skills are required to be inculcated in an entrepreneur to derive the best possible results out of the arising circumstances during an entrepreneurial journey. These broad categories of skills are mostly acquired by entrepreneurs through education, training, and, most importantly, practice. These skills collaborate to create a perfect blend, which is reflected in the overall personality and decision making of any entrepreneur.

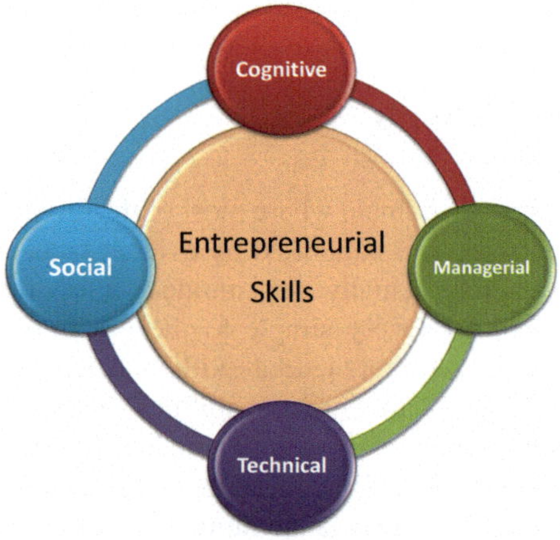

3.5.1 Cognitive Skills

Cognition can be defined as a mental process of gaining knowledge and understanding through the faculties of the senses; it also includes thinking, imagination, and perspective. The concept of learning itself is an example of cognition. It's about how the brain makes connections while remembering what it has learned. The ability to reason logically is an excellent example of cognitive ability, problem solving, and judgment. Perception includes a number of elements or processes that all have the function of describing how our knowledge is constructed and our judgments made, perceiving, learning, concept formation, reasoning, memory, problem solving, and the like. Entrepreneurial decision-making process greatly depends upon his or her cognitive ability as it lays the foundation of giving meaning and deriving the best of the outcomes from the prevailing circumstances.

Creative Thinking

Creativity is the ability to create or produce new things using skill or imagination. It can be defined as the tendency to generate or realize ideas, alternatives, or possibilities. It is useful in solving problems, communicating with others some of the most intricate issues and problems in the simplest of ways. Motivation, creative-thinking, and expertise are considered to be the three crucial dimensions of creativity. Often, creative thought involves tapping into different styles of thinking and examining information from different viewpoints to see new patterns. Anyone can foster a creative mind with some practice. Creative thinking involves tapping into different thinking styles and looking at information from different perspectives to see new patterns. Anyone can foster creativity with a little practice.

Creative thinking (along with critical thinking) is an invaluable skill for entrepreneurs. This is important because it helps them see problems and situations from a new angle. Creative thinking is a way of developing new or unorthodox solutions that are not completely dependent on past or present solutions, thus helping in the development of what is known as "Appropriate Technology." Creativity can be exercised through various methods like mind mapping, role playing, brainstorming, and the like.

Focus

Focus in the layman's language can be defined as giving all the attention to a particular thing. It not only takes into consideration the sensory receptors but also the cognition of an individual. It is the center of interest or attention; special attention that is given to somebody/something. Entrepreneurial focus or outlooks can be divided into two categories: Those who are drawn to the concept of starting a business and those who are entrepreneurs because it is their preferred method of bringing a concept to life. Since focus is the key to all aspects of thinking, including perception, memory, learning, reasoning, problem solving, and decision making, it is critical. All areas of one's ability to think will suffer if one does not have good focus. It depends on the capacity to concentrate, which is critical to entrepreneurial success. The higher the quality of work you do, the more time and concentration you can commit to it. Focus not only helps an entrepreneur to do things faster, but also ensures that the work is error-free.

Initiative

Initiative is an official action taken with an objective to solve a problem or improve a prevailing situation/condition. It is the ability to see and do what is necessary without waiting for somebody to tell you or give an approval or validation. It is an initial step, taken to try to resolve the problem. It demonstrates vigor or competence in initiating action, which is reflected through remarkable accomplishments.

The ability to take the first step in something is defined as initiative and entrepreneurs are known to be of such breed. An entrepreneur is someone who is always the first to start a new endeavor mostly taken up in the form of a new venture, a new business or a new startup. Going the additional mile and accomplishing duties before anyone tells you to is what initiative entails. Taking initiative requires foresight along with willingness and tendency to acquire new skills. Thus, it becomes a way to improve your professional life and accomplish personal goals that has been set by an entrepreneur in the long run.

Problem Solving

Problem-solving is the process of articulating a problem, locating the source of the problem, identifying, prioritizing, and selecting a potential solution for a problem, and finally executing the solution. The first step in problem-solving is to identify the issue or the problem. Problem-solving allows us to recognize and capitalize on opportunities in the environment, as well as exert (some) control over the future. Entrepreneurs and organizations alike require problem-solving abilities and the problem-solving process on a regular basis.

Entrepreneurial problem-solving is the process of bridging the gap by resolving societal, business, or technology issues through innovation and innovative solutions.

It is beneficial not just in terms of consumer acquisition, but also in the event of a problem or crisis situations. Entrepreneurs that are proactive are continually seeking for new ways to do things, connect with their employees, and enhance their product or service, inadvertently lowering the likelihood of future problems, thus securing better and a secured position for their business venture.

Perception

Perception is the process of organizing, identifying, and interpreting sensory data in order to represent and understand the information or environment that is provided. All perception is based on impulses that go through the neurological system and are triggered by physical or chemical stimulation of the sensory systems. Being aware of, perceiving, or comprehending something is the state of perception. Perception includes knowing when to try a different method with individuals to assist them learn more.

Perception has always been important in business. If properly implemented and exploited, it can be a critical component in helping a company expand and thrive in the market. It provides the company with a fresh perspective that will aid in its future growth and development. Because everyone perceives the world and addresses life challenges differently, perception is critical to understanding human behavior. Entrepreneurs keenly observe the changes in the behavior of their target market. They perceive the changes that are ought to occur in near future and alter their strategies accordingly.

Consumers will see a brand image as trustworthy, valuable, and appropriate to their own demands and self-images if perception methods are applied in commercials and promotional products. The better an entrepreneur grasps how his potential clients think as a marketer, the more effective his approaches will be.

3.5.2 Social and Relational Skills

Social skill is also an important skill that an entrepreneur can focus upon to inculcate as business is an integral part of society. Any competency that facilitates contact and communication with others, where social rules and relations are developed, transmitted, and changed, in both verbal and nonverbal ways, is referred to as a social skill. Socialization is the term for the process of gaining these skills. Social discomfort can result from a lack of such abilities. Social skills enable entrepreneurs to interact more effectively with clients, resulting in more contracts and revenues. In order for a business to flourish it has to be nurtured in the most harmonious way that can be achieved through people. Many entrepreneurial ventures focus on people as their center point. Solving their problem through innovative solution has provided many perks to these businesses. Microfinance institutions are an example of social entrepreneurship. These organizations provide banking services to unemployed or low-income individuals or groups that might otherwise be unable to obtain them.

Social skills are crucial because they enable us to engage with others in predictable ways, allowing us to better understand and be understood. Strong social skills can help people interact more effectively, which can lead to better employment results.

Impression

Entrepreneurs' behavior, attitude, and personal presentation will affect a customer's decision to choose the same entrepreneur time to time. These impressions are much more vital in digital markets especially on social media and other web platforms. Making a positive impression aids in the development of customer connections and sales. An entrepreneur never undervalues the importance of making a good first impression at the very start of a business. Customers and other stakeholders count on them to perform to their set expectations. An entrepreneur must never underestimate the significance of a positive first impression while making connections with the stakeholders who are very important for the business. After all networking plays a significant role in the growth and development of an entrepreneur.

Persuasion

Persuasion is the process of convincing someone to do or believe something. Persuasion or persuasive skills are terms used to describe influence. Persuasion can be used to try to change someone's mind about their ideas, attitudes, intentions, motives, or behaviors. Persuasion is explored in a variety of fields. A compelling entrepreneur can not only persuade customers to buy, but also build a strong network of people who want to assist their company or help their concept flourish. Persuasion is the process of persuading others to alter their minds, make a commitment, buy a product or service, or perform a certain action. In business, oral and written persuasion skills are highly valued. Persuasion is most commonly associated with sales, although it is also useful in a variety of other fields. Many entrepreneurs are aware of it, but not everyone is able to fully utilize it. Persuasion is crucial in everyday business since it can increase sales and build trust. Persuasion is used in the workplace to sell items, recruit team members, and boost efficiency. A persuasive employee can persuade others to perform effectively and achieve success.

Adaptability

Adaptability is a skill that refers to the ability to quickly pick up new skills and behaviors in response to changing situations. When hiring new employees, employers typically look for flexibility, and the talent is frequently listed in job descriptions due to its value for career advancement.

When faced with a situation, adaptability means making quick decisions. It's a way of thinking that allows entrepreneurs to manage new situations and the inevitable ups and downs that come with life and business venture mostly occurring simultaneously. Learning adaptability is exceptional at this stage through continually monitoring their personal and professional development. This proactive approach to flexibility enables them to stay ahead of the curve and successfully deal with future change and challenges.

Empathy

Empathy is a cognitive and affective process that helps people comprehend and appreciate other people's feelings, thoughts, and experiences. Empathy is defined by researchers as the ability to sense other people's emotions as well as the ability to imagine what they are thinking or feeling. Empathy, in its most basic form, is the ability to identify and comprehend others' emotions and viewpoints on a situation. If an entrepreneurs' empathy is fully developed, he/she can use it to lift someone else's spirits and encourage them through difficult times. Entrepreneurs need a good team to sail his ship across and empathy plays a crucial role in achieving this. Empathy in the workplace aids management and collaboration by allowing the entrepreneur to grasp the viewpoint of others. Putting oneself in someone else's shoes can help an entrepreneur identify a middle ground between two opposing viewpoints. Furthermore, a key competency of an entrepreneurial mindset is the ability to empathize with others.

Group Skills

The purpose of any team-building exercise is to build a stronger unit of employees. Team building has many benefits for businesses. It boosts productivity, boosts staff enthusiasm, encourages collaboration, and fosters employee trust and respect. Entrepreneurs are critical to any economy because they have the ability and initiative to anticipate requirements and bring good new ideas to market. Entrepreneurship that succeeds in taking on the risks of starting a business is rewarded with revenues, fame, and chances for continued expansion. Learning entrepreneur skills, including team building, is a part of this. Putting together a good group of people with specific abilities and a goal is what team building is all about. Any team-building activity should aim to make employees stronger as a unit. For firms, team building offers numerous advantages. It boosts productivity, staff enthusiasm, encourages collaboration, and fosters employee trust and respect.

3.5.3 Technical Skills

Technical skills are abilities or knowledge utilized to complete practical tasks in areas of science, arts, technology, engineering, and other fields. Technical abilities often necessitate the use of certain tools as well as the technologies necessary to employ those tools. Knowledge is power, and the more information an entrepreneur has, the more his company can grow. By providing such technical skills training, one is giving them assurance that they have the knowledge and competence to complete their everyday jobs to the best of their abilities.

Operations

The word "business operations" refers to a wide range of activities. In essence, it refers to what a company does on a daily basis to keep it running and profitable. As a result, those activities can vary greatly from one organization to the next. The harvesting of value from a company's assets is known as business operations. Physical and intangible assets both contribute to assessment of costs and revenues centers for an organization. The process of managing the inner workings of a company so that it functions as efficiently as possible is known as operations. The administration of business activities to achieve the best level of efficiency feasible inside an organization is known as operations management. Three dimensions of operations management include input, process, and output. Entrepreneurs are in charge of a company's productivity and efficiency. Their key responsibilities include overseeing and managing budgets and finances, managing workflow procedures and staffing requirements, managing inventories and supply chain management, and developing policies for the entire firm especially for initial years.

Design

Design opens up valuable opportunities for businesses. Its importance is often underestimated, but good design can bring some significant business benefits. The design process' research and prototype stages might help an entrepreneur come up with fresh product ideas and learn about clients' wants and preferences. Design thinking is a straightforward technique to pinpoint exactly what the issues are, often revealing a new way of thinking about them, while also giving vital insights and data for developing acceptable solutions that make a firm earn profits. Professional, clear, and appealing design components express the value of your company or product to a potential buyer right away. It gives others faith in your ability and expertise. It starts a conversation in which the audience can participate. It entices customers and leads them through an experience. Entrepreneurs bank upon design aspects a lot, as it these aspects reflect direct innovation. Not only in the case of products but also in services, design plays a vital role in attracting and engaging customers and earning the trust of stakeholders.

Research and Development

Research and development (R & D) is an important instrument for expanding and enhancing one's company. R & D is the process of studying one's market and one's customers' demands in order to produce new and better products and services to meet those needs. R & D is critical for businesses because it delivers valuable knowledge and insights, as well as enhances the existing processes that boost efficiency and save costs. It also enables companies to create new products and services in order to survive and grow in competitive markets with the help of information obtained from the markets and the intermediaries. Entrepreneurship is all founded upon innovation and newness that comes from research and development, which enables an entrepreneur to come up with better product and services, exploring and developing new markets and the like, thus, improving the quality of life especially MSME, SME, and LSMEs to reduce costs of production and become more competitive through development of all new processes as per the size and nature of business firm.

Environment Assessment

The environment offers countless opportunities, and it is critical to discover such opportunities in order to improve a company's success. Early identification allows an entrepreneur to be the first to spot opportunities rather than losing them to competition. The goal is to determine the level of risk that various environmental factors provide, as well as the commercial opportunities that these aspects present. The entrepreneurs analyze the company's strengths and weaknesses, as well as how they affect the company's ability to deal with external threats and opportunities. It is an important skill that enables and entrepreneurs to take informed decisions. It also helps an entrepreneur to counter competitor strategies and to strike a balance between the internal and the external environment through a well-equipped system.

Ergonomics

Ergonomics is the practice of adapting the work environment (equipment, furnishings, work tempo, etc.) to match the physical needs and limitations of employees rather than forcing workers to adapt to jobs that can debilitate their physical well-being over time. Ergonomics enables both industrial and service organizations to develop novel products that are more intuitive and "fit the user." Corporate culture encompasses an organization's corporate processes, rituals, and traditions. It is about developing a suitable and appropriate man–machine relationship. Organizational ergonomics, cognitive ergonomics, and physical ergonomics are three important dimensions of ergonomics that an entrepreneur can work upon for betterment of his organizational functionality. Other than these, ergonomics lead to higher productivity levels, eliminate hazards, reduce chances of errors and rework and enhance ease of work providing smooth transitions between activities and innovation processes.

3.5.4 Managerial Skills

Entrepreneurs with good management abilities can perform important tasks quickly. They must manage their professional lives alongside their personal lives, striking a balance between work and home. Every successful business starts with a fantastic idea. The entrepreneur retains ownership of the company, whereas management is the employee entity of the company. The entrepreneur will be rewarded with profit, while management will be compensated for their efforts. Whereas management does not take any risks, an entrepreneur does. Entrepreneurs must be fully versed in general management, finance, marketing, operations management, purchasing, supply chain management, human resources, and public relations. Critical managerial talents help entrepreneurs succeed and take their business to the next level.

Planning

The ability to "think about the future" or mentally predict the best approach to complete a task or achieve a specific goal is characterized as planning. It assists you in establishing clear objectives and rules for managing your company. A business plan may also be required to define staff goals, secure investment, or sell your company in the future. Entrepreneurs must be good planners as they have to decide well in advance what to do, when to do, why to do, and who is to do. Such questions have to be sorted out well in advance for smooth functioning of the business and to steer through sudden market changes. Planning also helps to determine when new employees are needed, assess financial requirements, make future predictions, etc. Thus, planning is the first and foremost step that has to be carefully taken to mold the business as per the vision of the entrepreneur.

Organizing

The capacity to use processes to get things done excellently and effectively is known as organizational skill. Creating structure and order, increasing productivity, and prioritizing chores that must be accomplished promptly versus those that may be transferred to another person or removed entirely are all examples of organization abilities. It is an acquired ability that must be cultivated and perfected through time in order to become more perfect and effective. This necessitates efficiently and successfully balancing multiple tasks performed by entrepreneurs. When a company's systems aren't well-organized, jobs build up, paperwork goes missing, and time is wasted looking for information that should be easily available. An entrepreneur's organizational abilities can save time and minimize stress. Entrepreneurs with great organizational skills are critical to guaranteeing operational efficiency and helping a business run smoothly. These abilities are needed in the enterprise to uplift the efficiency and ensure that goals of the business are accomplished on a regular basis.

Motivating

Actions or methods that evoke a desired behavior or response from others are known as motivational skills. Entrepreneurial motivation is the process that engages and inspires an entrepreneur to work harder in order to attain his or her business objectives. Three important aspects influence motivational strategies and actions; throughout, an entrepreneurial venture includes entrepreneur's personality, the target market, and the personality that the motivator wishes to influence. In the hands of managers, motivation is an efficient tool for increasing the efficiency of operations and the output of the company. When compared to other employees, motivated people perform better. When things get tough and disheartening, it gives you hope and clarity. Entrepreneurial motivation is so critical in someone's decision to start a firm. Motivation is also vital for people who work with and interact with entrepreneurs.

Marketing

Marketing is critical for entrepreneurs since without a client market, no business can establish itself and expand. At the heart of marketing is the acquisition and retention of customers. One of the smart skills for an entrepreneur is marketing skills. Both hard and soft talents are mastered by successful entrepreneurs. Hard skills such as accounting, marketing, and financial planning are necessary for running and managing a business, while soft skills such as communication, problem-solving, and decision-making aid in scaling up the business.

Without customers, a firm or a business venture would be nothing. When entrepreneurs use marketing to advertise their products, services, or companies, they increase their chances of being discovered by potential customers. A company's product or service must be known to potential customers in order for it to succeed. To raise product or service awareness, entrepreneurs must employ marketing methods. The entire act of presenting a product concept is referred to as the pitch, which is made toward prospective investors who would most like to fund the entrepreneurial idea. It is supposed to be succinct and appealing, as well as well developed to account for the interests of the intended audience. Marketing is the process of placing the appropriate product in the appropriate location, at the appropriate price, and at the appropriate time.

Finance

After all, businesses need funds to operate and survive, thus entrepreneurs must be able to efficiently handle their finances. Controlling their funds and capital, regardless of the type of business they run, will help entrepreneurs prevent losses and develop their company more swiftly and in a better manner. An entrepreneur with sound financial knowledge is more likely to have complete control over his/her company. Understanding what balance sheets and profit and loss statements mean gives an entrepreneur a clear picture of their company's financial health and allows them to make better business decisions. Financial investors and banks also bank on entrepreneurs to whom they can grant credit in order to realize profits from their activities, and the entrepreneur needs finance to fund her/his idea and innovative. Making good financial decisions about when, where, and how a firm acquires funds is critical. A company benefits the greatest when the market value of its stock rises, which not only is a symbol of growth for the company but also increases investor wealth. Financial management assists a company in determining how much money to spend, where it should be spent, and when it should be spent. It also provides an overview of the company's financial situation, assisting in the development of corporate strategy and direction, as well as contributing to the organization's goals.

3.5.5 *Skills for Entrepreneurs*

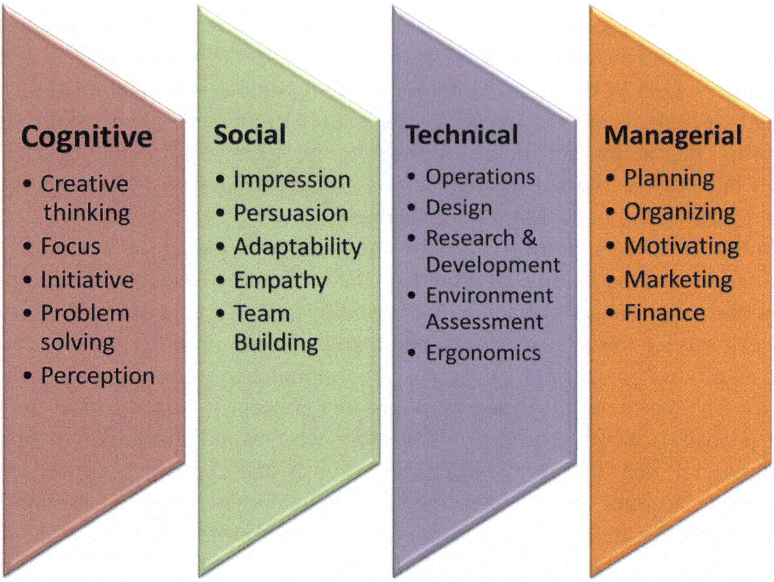

3.6 Conclusion

Skillful entrepreneurs are more likely to succeed than non-skilled entrepreneurs. An entrepreneur should inculcate the above mentioned skills like cognitive skills, social skills, technical skills and managerial skills for the betterment of the business and to provide the best to its stakeholders. Successful entrepreneurs manage to get just the right blend of these traits depending upon their business. Unlike talent, these skills are acquired and can be honed to perfection through proper training and experience. Thus, in order to excel in a business venture and to make an idea a success, an entrepreneur must exercise these entrepreneurial skills to derive the best of the results.

Bibliography

Carland JW, Hoy F, Boulton WR, Carland JA (1984) Differentiating entrepreneurs from small business owners. Acad Manage Rev 9(2):354–359
Chandler GN (1996) Business similarity as a moderator of the relationship between pre-ownership experience and venture performance. Entrepr Theory Pract 20:51–65
Chandler GN, Hanks SH (1994) Founder competence, the environment, and venture performance. Entrepr Theory Pract 18:77–89
Chatman JA (1989) Improving interactional organizational research: a model of person-organization fit. Acad manage Rev 14:333–349
Collins OF, Moore DG (1964) The enterprising man. Michigan State University, East Lancing
Cunningham JB, Lischeron J (1991) Defining entrepreneurship. J Small Bus manage 29:45–61
Mohar Y, Singh MS, Kishore K (2007) Relationship between psychological characteristics and entrepreneurial inclination: a case study of students at University Tun Abdul Razak (UNITAR). J Asia Entrep Sustain 3(2):23–131
O'Reilly CA, Chatman J, Caldwell DE (1991) People and organizational culture: a profile comparison approach to assessing person-organization fit. Acad Manage J 34:487–516
Pines AM, Dvir D, Sadeh A (2004) The making of Israeli high technology entrepreneurs: an exploratory study. J Entrep 13(1):29–52
Rauch A, Frese M (2000) Psychological approaches to entrepreneurial success: A general model and an overview of findings. In: Cooper CL, Robertson IT (eds) International review of industrial and organizational psychology. Wiley, Chichester, pp 101–142
Roure J (1986) Success and failure of high-growth technological ventures: the influence of prefunding factors. Stanford University, Stanford
Roure JB, Keeley RH (1990) Predictors of success in new technology based ventures. J Bus Ventur 5:201–220
Shenhar AJ, Dvir D (2007) Reinventing project management: the diamond approach to successful business growth and innovation. HBSP, Boston
Stewart WH, Roth PL (2004) Data quality affects meta-analytic conclusions: a response to miner and Raju, 2004, concerning entrepreneurial risk propensity. J Appl Psychol 89:14–21
Stuart R, Abetti PA (1990) Impact of entrepreneurial and management experience on early performance. J Bus Ventur 5:151–162
Wallach MA, Kogan N, Bem DJ (1964) Diffusion of responsibility and level of risk taking in groups. J Abnormal Psychol 68:263–274
Watson W, Ponthieu L, Doster J (1995) Business owner-managers' descriptions of entrepreneurship: a content analysis. J Construct Psychol 8:33–51
Wilken PH (1979) Entrepreneurship: a comparative historical study. Ablex, Norwood
Winslow EK, Solomon GT (1989) Further development of a descriptive profile of entrepreneurs. J Behav 23:149–161

Chapter 4
Intellectual Property Principles in Microbial Technologies

R. Pathirajan, S. Gomathi, V. Ambikapathy, A. Panneerselvam, G. Gayathri, and A. Anbukumaran

Abstract Intellectual property basically takes into consideration any creation or innovation which could be seen as a sort of asset or a physical property related to individual's intellect. Intellectual property rights (IPRs) provide certain exclusive rights to the inventors to reap commercial benefits from their invention. There are a variety of assets, which are labeled as intangible ones including novel ideas, invention, discoveries, unique designs, and symbols for which the owners can be granted patents under the jurisdiction of intellectual property laws and regulations. This chapter discusses about the introduction of IPR in microbial technology, Patent, copyrights, trademarks, design, international agreements on designs, microorganisms, and its patent right.

Keywords Intellectual property · Patents · Copyrights · Microorganisms

4.1 Introduction

Intellectual property basically takes into consideration any creation or innovation, which could be seen as a sort of asset or a physical property related to individual's intellect. Intellectual property rights (IPRs) provide certain exclusive rights to the inventors to reap commercial benefits from their invention. The importance of intellectual property rights was first recognized in 1883 in the Paris Convention for the Protection of Industrial Property and in 1886 in the Berne Convention for the Protection of Literary and Artistic Works. Both treaties are administered by the World Intellectual Property Organization (WIPO). As IPR has a vital role in the economy of a country, an efficient and equitable intellectual property system can

R. Pathirajan (✉) · S. Gomathi · V. Ambikapathy · A. Panneerselvam · G. Gayathri
Department of Botany, A.V.V.M Sri Pushpam College (Autonomous) (Affiliated to Bharathidasan University, Trichy) Poondi, Thanjavur, Tamil Nadu, India

A. Anbukumaran
Department of Microbiology, Urumu Dhanalakshmi College, Tiruchirappalli, Tamil Nadu, India

N. Amaresan et al. (eds.), *Agricultural Microbiology Based Entrepreneurship*, Microorganisms for Sustainability 39,
https://doi.org/10.1007/978-981-19-5747-5_4

help all countries to realize intellectual property's potential as a catalyst for economic development and social and cultural well-being. WTO and TRIPS recognized the territorial nature of intellectual property rights and drafted the agreement in very general terms maintaining appropriate local differences. The efficient and equitable intellectual property system helps strike a balance between the interests of innovators and the public interest, providing an environment in which creativity and invention can flourish, for the benefit of all. Therefore, TRIPS is the most important multilateral instrument for the globalization of intellectual property laws.

There are a variety of assets that are labeled as intangible ones including novel ideas, invention, discoveries, unique designs, and symbols for which the owners can be granted patents under the jurisdiction of intellectual property laws and regulations. These intellectual property rights (IPRs) given to the inventor further ensure protection to the author, ideas, designs, processes, products, devices, apparatus, etc. (Singh 2008).

On the other hand, when we talk about microorganisms, they can be defined as "microscopic entities" (Raghuvanshi 2017). These include viruses, bacteria, yeasts, fungi, algae, etc. Microorganisms are very well known to hold great economic value throughout the world. Despite some of the microbes causing diseases, many of them are beneficial for humans and associated animals. Microorganisms are currently being used in food, pharmaceutical, biofuel, and fermentation industries for the production of various valuable products including ethanol, antibiotics, enzymes, pigments, vaccines, and other food products (Garg et al. 2016).

As discussed earlier, the history of the utilization of microbial organisms starts in ancient civilization. In our day-to-day-life microorganisms play roles in various aspects, in therapeutics–microbial control agents or antibiotics, vaccines; metabolism-related agents, insulin, and an array of diagnostic tools; in farming system microbial strains are used in improving crop yield, conferring resistance to various stresses; food industry for different kinds of enzymes and fermenting agents (Vitorino and Bessa 2017; Kumar et al. 2018; Rai and Rai 2020). Indian patent practice and jurisprudence with respect to the patenting of biological/genetic materials are relatively new, and since 2013, guidelines to examine biotechnological inventions including microbiology made the practice and procedure settled and uniform. Four major areas under the aegis of IPR have been identified as patent, trademark, copyrights, and design.

4.2 Intellectual Property Rights (IPR)

Any creativity or invention that is considered as an individual physical property or asset has been covered under the jurisdiction of IPR. These are legal rights accorded to individuals for their creations of particular art or commercial product/ process. Under IPR laws, owners or inventors have been granted exclusive rights pertaining to intangible assets (symbols, ideas, designs, discoveries, and inventions). IPR also provides economic incentives to the creators or inventors for their original work.

Also this provides a platform to further develop and share ideas through the innovator by giving him or her temporary monopoly and rights (Sivakumaar et al. 2010).

The inventor or the innovator should be duly recognized and rewarded under the ambit of IP laws and regulations which in turn stimulates the overall techno-industrial growth and strengthens the socioeconomic fabric of the country. The advancement and modernization of life sciences with evolution of microbial biotechnology and genetic engineering have pressed the policy makers to consider the engineered microorganisms and their products to be patentable. Novel technologies have led to the creation of forms in plants and animals, which could be patented as well in the form of biopatents (Ammen and Swathi 2010).

4.3 Patents

A patent for an invention is the grant of a property right to the inventor, issued by the United States Patent and Trademark Office. Generally, the term of a new patent is 20 years from the date on which the application for the patent was filed in the USA or, in special cases, from the date an earlier related application was filed, subject to the payment of maintenance fees. US patent grants are effective only within the USA, US territories, and US possessions. Under certain circumstances, patent term extensions or adjustments may be available.

There are three types of patents:

- Utility Patents
- Design Patents
- Plant Patents

Utility Patents: It may be granted to anyone who invents or discovers any new and useful process, machine, article of manufacture, or composition of matter, or any new and useful improvement.

Design Patents: It may be granted to anyone who invents a new, original, and ornamental design for an article of manufacture.

Plant Patents: It may be granted to anyone who invents or discovers and asexually reproduces any distinct and new variety of plant.

4.4 Copyrights

General Definition of copyright "Copyright owner," with respect to any one of the exclusive rights comprised in a copyright, refers to the owner of that particular right.

Copyright is a form of protection provided by US law to the authors of "original works of authorship" fixed in any tangible medium of expression. The manner and medium of fixation are virtually unlimited. Creative expression may be captured in words, numbers, notes, sounds, pictures, or any other graphic or symbolic media.

The subject matter of copyright is extremely broad, including literary, dramatic, musical, artistic, audiovisual, and architectural works. Copyright protection is available to both published and unpublished works. Copyright protection is available for more than merely serious works of fiction or art. Marketing materials, advertising copy, and cartoons are also protectable. Copyright is available for original working protectable by copyright, such as titles, names, short phrases, or lists of ingredients. Similarly, ideas methods and processes are not protectable by copyright, although the expression of those ideas is. Copyright protection exists automatically from the time a work is created in fixed form. The owner of a copyright has the right to reproduce the work, prepare derivative works based on the original work (such as a sequel to the original), distribute copies of the work, and to perform and display the work. Violations of such rights are protectable by infringement actions. Nevertheless, some uses of copyrighted works are considered "fair use" and do not constitute infringement, such as use of an insignificant portion of a work for noncommercial purposes or parody of a copyrighted work.

4.5 Federal Registration of Copyrights

The works are protected under federal copyright law from the time of their creation in a fixed form. Registration, however, is inexpensive, requiring only a $30 (present $85) filing fee, and the process is expeditious. In most cases, the Copyright Office processes applications within 4–5 months. Copyrighted works are automatically protected from the moment of their creation for a term generally enduring for the author's life plus an additional 70 years after the author's death. The policy underlying the long period of copyright protection is that it may take several years for a painting, book, or opera to achieve its true value, and thus, authors should receive a length of protection that will enable the work to appreciate to its greatest extent.

4.6 Trademarks

Any name, word, design, symbol, device, or slogan that makes a distinction of a product or organization has been termed as trademarks. The filing or the application process (registered at a national or regional level office) usually takes between 25 and 75 weeks (www.copyrightservice.co.uk). The trademarks registered in countries like the USA, UK, Japan, etc., are country-specific and provide protection in that very country only apart from the Community Trade Mark (CTM), which is valid in all the European Union countries. Any registered trademark could be depicted as "TM" or the "®" symbol, while in the USA, there is a further differentiation between products and service marks, though protected under the trademark itself. National patent office in most of the countries has been assigned the job to administer trademarks. In simplified terms, WIPO defines a trademark as any sign that individualizes the goods of a given enterprise making them distinct from other goods in competition. For the individualization of a particular product for the consumer, TM must mention the source of that very product in order to distinguish the goods of a given enterprise from others.

4.6.1 Trademarks and the Paris Convention

The basic requirements of the Paris Convention in respect to trademarks are the same as for patents and designs, based on the two fundamental provisions preventing discrimination against nonresidents, that is, "national treatment" and the "right of priority." In order to maintain right of priority, the application for a trade mark must be lodged within 6 months of the original application in the home country. The Paris Convention includes a number of other requirements concerning trademarks and marks more generally.

Some automatic legal protection must be available for trade names without the obligation of filing or registration.

Measures must be taken in each contracting state against the use of false indications on the geographical source of the good or the identity of the producer, manufacturer, or trader.

Registration must be denied for trademarks which contain, without authorization, State emblems or official signs that have been communicated to WIPO. All member states must refuse or cancel the registration of a mark conflicting with a mark from a domestic or foreign source that is well known.

4.6.2 Trademarks and TRIPS

The TRIPS agreement is based largely on the Paris Convention in respect to trade marks, but a few extra clauses have been added. The Paris Convention requirement to prevent duplication of well-known marks has been extended to services. Moreover, the protection of well-known marks must extend to goods and services that are not similar to those in respect of which the well-known mark has been registered. Evidently, the stricter protection of well-known marks may enhance the protection of the brand names of large transnational corporations.

4.6.3 Trade Secrets

A trade secret consists of any valuable business information. The business secrets are not to be known by the competitor. There is no limit to the type of information that can be protected as trade secrets; For Example: *Recipes, Marketing plans, financial projections, and methods of conducting business can all constitute trade secrets.* There is no requirement that a trade secret be unique or complex; thus, even something as simple and nontechnical as a list of customers can qualify as a trade secret as long as it affords its owner a competitive advantage and is not common knowledge. If trade secrets were not protectable, companies would have no incentive to invest time, money, and effort in research and development that ultimately benefits the public. Trade secret law thus promotes the development of new methods and processes for doing business in the marketplace.

4.7 Design

4.7.1 Industrial Designs

The Australian Design Act grants protection to the visual appearance or design of a manufactured article, if it is new or original. In this context, design refers to the ornamental aspect of an article that is produced in quantity. This ornamental aspect may be constituted by elements that are three-dimensional (the shape of the article) or two-dimensional (lines, designs, colors) but must not be dictated solely or essentially by technical or functional considerations. Protection of an industrial design means that third parties not having the consent of the owner may not make, sell, or import articles bearing or embodying a design that is a copy, or substantially a copy, of the protected design, when such acts are undertaken for commercial purposes. The purpose of design protection is to provide an economic incentive for improving the visual appearance of manufactured products. Protection is based on a system of registrations and can last for up to 16 years. Applications for registration must be accompanied by photographs and other graphics if necessary, and describe the product on which the design is applied. Approval for registration depends on satisfying the criteria of novelty or originality (but not necessarily both).

Novelty means that the same or very similar design must be not known or previously registered in Australia.

Originality means that the design has never been applied to the particular product specified in the claim, although it may have been applied to another type of product.

The protection is only for the appearance of the article and not how it works. As for patents, detailed examination guidelines are based partly on common law (that is, previous decisions by the courts). In the absence of relevant legal precedents, the guidelines are determined by the Registrar of Designs. Design protection excludes non-visible internal parts, but does not exclude visible spare parts such as exterior panels, bumpers, or wheel trims. Removing the design protection on currently protected spare parts was recommended in the IC (1995) report on vehicle and recreational marine craft repair. The BIE (1995) report noted that there is a degree of uncertainty about the protection provided to spare parts by the current legislation. Design registration is intended to protect designs that are applied industrially, rather than a single artistic work, where copyright protection would automatically apply.

When artistic work can be applied on or to commercial articles, protection under both the Copyright Act and the Design Act may be available. However, the Copyright Act will not provide protection to a design that, when applied to an article, results in a reproduction in three-dimensions of that design (so called "corresponding design"). In most cases where dual protection is available, registered design provides a stronger legal protection than copyright, by virtue of having satisfied for registration the screening criteria regarding novelty or originality.

4.8 International Agreements on Designs

The Paris Convention's provisions concerning "national treatment" and the "right of priority," which prevent discrimination against nonresidents, also apply to industrial designs. The convention specifically states that industrial designs must be protected in each contracting state, and protection may not be forfeited on the grounds that the articles incorporating the design are not manufactured in that state. In order to retain the priority date, application for design registration in other countries must be lodged within 6 months of the original application in the home country, while in the case of patents 1 year is allowed.

TRIPS do not contain any significant amendment to the Paris Convention in respect to industrial designs. TRIPS require WTO members to grant protection for aesthetic designs, but do not require protection for designs dictated essentially by technical or functional considerations, in line with the Paris Convention. One of the articles of TRIPS contains a special provision aimed at taking into account the short life cycle and sheer number of new designs in the textile sector, requiring that the cost of relevant examinations should not unreasonably impair the opportunity to seek and obtain design protection in the textile industry.

4.9 Patents

WIPO defines patent as a document, which is issued upon once the application is filed by a regional office recognized by one or more than one country. The application is complete in a sense that it should be able to describe all about invention of a product or process. Filing a patent provides a legal protection to the invention, and only it can be exploited in the circumstances upon the approval or permission of the inventor or owner. This protection is given to the owner generally for 20 years. Different countries have their own rules and regulatory authorities, but in majority of them, inventions or creations could be protected either as short-term patent registration or in the form of a utility model. The requisite fee for such protections is also lesser than that of patents. However the duration period for such protections is shorter than that for patents. Though in terms of rights, the utility model or short-term patents are quite comparable.

"Monopolies" is another term used for patents previously giving some sort of protection to the invention. On the positive side, the patented invention is allowed to be exploited by the owner of the invention only and is prohibited to be exploited by other persons without the permission of the owner. Therefore, the owner though cannot practice his invention as his statutory right, has the right not to let others exploiting the invention commercially. In other words the inventor has the exclusive right to exclude others not to make, use, or sell the said invention. The inventor has the right to take action if anyone steals or exploits the said invention without his prior permission or agreement. These exclusive rights allow the inventor to derive material benefits as a reward for the novel work or the intellectual effort, in other words, these rights provide compensation to the inventor for the expenses incurred to the research and experimentation leading to the invention.

4.9.1 Microorganisms

If we go by the dictionary, one defines microorganisms as "microscopic organisms." These are minute or very small living microscopic entities (Raghuvanshi 2017) including viruses, bacteria, yeasts, fungi, algae, etc. Since long, microorganisms have been used as tools in food, pharmaceutical, and fermentation industries for the production of various valuable products including ethanol, antibiotics, enzymes, pigments, vaccines, and other food products. In recent years, microorganisms have been shown to produce colors, pigments, anticancer drugs, biofuel, and other metabolites (Garg et al. 2016; Yadav et al. 2016). The microorganisms and their products are in continuous demand, which is further increasing continuously. In the race of exploration of microbes for commercial purpose, intellectual property rights have become immensely important, particularly patents.

4.9.2 Microorganisms and Patent

It is a well-known fact that microbes such as yeasts, bacteria, protozoa, unicellular algae, fungi, actinomycetes, and viruses, in their original form, do not qualify for the patent application but could be patented if they are genetically altered in which case the underlying process and the product generated do qualify for the patent (Cameotra 2013). Genetically modified microbes as a result of human intervention leading to improved efficacy in comparison to the ones already existing in the nature were found to qualify for the patent (Raghuvanshi 2017). Written description about the invention has been a mandatory disclosure nowadays for the inventor especially after the establishment of the International Depository Authority under the purview of Budapest Treaty, which was not the case in earlier times. In earlier times microbial inventions were granted only product or process patents in India unlike the USA and other developed countries where these entities were allowed to be patented. India started granting patents on microorganisms since 2003 like other countries under the strict guidelines of the Budapest Treaty (Keswani et al. 2016). Currently India has two microbial repositories, i.e., Microbial Type Culture Collection (MTCC) and Microbial Culture Collection (MCC), both of them have acquired the status of International Depository Authority (IDA).

Hüttner et al. (2020) reported during the last 5 years the distribution of patent classification in the field has not changed much, with 36.8% of patent applications being classified as belonging to the domain of biotechnology, while 15.7% were classified as basic materials chemistry, and 10% as food chemistry. The biggest observable trend concerned the countries of origin of patent applications. Here, an increase in patent families originating from Asian countries (mostly China and Japan) was noticeable. The key players of the last 5 years were: Dupont (137) incl. Danisco & Genencor, Novozymes (99), DSM (57), Toray Industries (24), and Dalian Institute of Chemical Physics (23).

Patent applications related to living matter have increased rapidly due to the enormous developments in the field of biotechnology (Serina and Toledo 1999). However, over the last century, inventions which involved some kind of living matter were already protected by a patent. Initially, patent legislations were designed to protect inventions related to nonliving subjects. But recently, due to continuous developments in the technology, laws have been adapted to the peculiarities of each kind of matter including living organisms. For this reason, patent laws including the US Patent Law in 1949 were modified in such a way that they now require the deposit of the microbiological material, if the microorganism is the subject of the patent application. As a result of intensive scientific research, biotechnology has emerged as one of the most innovative and promising technologies. For this reason, the European Union has developed a new Directive (European Parliament and Council Directive on the legal protection of biotechnological inventions of July 6, 1998; Serina and Toledo 1999) to harmonize the different legal protection systems of biotechnological inventions. Some articles in this Directive, which has come in force recently, are related to microbiological inventions. The European legislation (European Patent Convention, EPC) has been modified in relation to the terminology used in this kind of inventions: the term "microorganism" has been replaced by the term "biological material," to cover all entities, which need to be deposited so as to cope with the requirement of sufficiency of disclosure (Serina and Toledo 1999).

4.9.3 Requirements for Microbiological Patent Applications

Requirements for microbiological patent applications have been discussed in detail as reported earlier by Serina and Toledo (1999). In order to disclose his or her invention, the state in exchange for the same provides certain rights and monopoly to the inventor. It is anticipated that the disclosure of the invention by the inventor should be clearly stated in the patent application. In terms of specific requirements for patenting, the most essential part is the description of the invention, which is expected to contain all the information about the properties and characteristics of the material that has been deposited and for which the patent has been applied. The date of deposition of the material to the Budapest Treaty recognized depository institution and the date of patent application should also match. The name of the depository authority to which the inventor has deposited his/her material along with the accession number is also required to be mentioned in the description. Currently, there are 30 depository authorities all across the globe, which have been duly recognized under the Budapest Treaty. The Spanish Type Culture Collection (Coleccion Espanola de Cultivos Tipo, CECT in Spain) located in the University of Valencia, at the Department of Microbiology of the Faculty of Biological Sciences, is one of the International Depository Authority under the Budapest Treaty (Serina and Toledo 1999).

4.9.4 Patentability of Microorganisms

Microbes such as bacterial, fungal, and viral entities are the major chunks of microbial kingdom that have been widely exploited by genetic engineers and biologists. The chromosomal material has been used as a raw material by these scientists, and through genetic manipulations, these microbial entities have been tailored through the use of enzymes such as restriction cutters, ligases, polymerases, etc., to generate a novel recombinant product. The recombinant DNA technology, transgenesis, and modern genetics have been known to have wide applications in many sectors of biotechnology including food biotechnology, environment biotechnology, agriculture biotechnology, and pharmaceutical biotechnology. Typical examples include use of many fungal products in bakery, wine, and antibiotic industry, bacteria/bacterial components for the manufacture of various vaccines, alteration of plants (transgenic plants) or insects' genome, etc. Living organisms were considered to be products of nature for the past 200 years or so; therefore they were not considered to be patentable. Moreover only patents were granted for the processes involved in obtaining the product majorly in the fields of chemical and mechanical engineering including some microbiological processes as well. Patenting of life forms got impetus after 1980 when it was included in the ambit of patent laws. The first patent, which was based upon the microorganisms involved in the process of fermentation of beer, was granted in the year 1873 to Louis Pasteur. Living matter is however still excluded from patentability in many countries across the globe considering them as products of nature. Under these products of nature doctrine, however, one could protect and secure his or her invention, for example, the process of microbial fermentation or the process of purification of naturally occurring compounds. More so, one could apply for a patent for microorganisms as a culture or in combination with a carrier or in the form of a consortium having synergistic effects. However the products made by the microorganisms naturally are not patentable because of the living matter. In the year 1980, however, the Supreme Court in the USA in its landmark decision granted permission for the first genetically engineered bacterium to be granted as a patent to Professor A.M. Chakraborty.

4.9.5 Types of Patentable Microbiological Inventions

Most patent legislations distinguish among three types of findings or innovations relying upon the subject area to be protected: (a) inventions that protect a product, (b) process-based inventions, which may lead to a product, and (c) applicability of the product-based inventions. However, one could claim for all of the above three types in the same invention. Typical example may include that one could stake claim for the product, process to obtain that very product, and applicability of the product in one single application (Serina and Toledo 1999).

4.10 IPR in Related Disciplines

Microbiology is considered as the base of biotechnology. All major disciplines of life, pharmaceutical, and medical sciences have many facts and principles in common. The major disciplines including microbiology, biochemistry, biotechnology, and pharmaceutical sciences are interrelated and support interfacial research. Various authors have discussed the intellectual property rights (IPRs) dimensions, facts, and issues with respect to different fields. Saha and Bhattacharya (2011) have discussed the IPR issues in pharmaceutical industry. Many investigators firmly believe that any novel idea, creation/creative expression, or discovery/invention should have the ownership to bestow the status of the property, which has been well covered under the IP rights. More so the invention or creation could be better commercialized and best protected through IPR. These legal rights are exclusively granted to the inventor or the creator or his assignee for a certain period of time, and the said invention or creation is protected for that particular time period. New and novel technologies are being explored to handle these creations or inventions providing an impetus to research and development (R & D) activities (Saha and Bhattacharya 2011). IPR definitely has many pluses in terms of protecting ones physical property with nominal investments and simultaneously saving time, money, and effort of the inventor/creator. These exclusive rights covered under IPR certainly will uplift the economic growth and development of our country as one can speculate enhanced industrial growth, output, and competitiveness in various industrial segments. Montesinos (2003) further gave insights into the patenting process, registration protocol, and how one could protect microbial pesticides from exploitation by others. The author also chalked out a plan on how to commercialize these microbial pesticides at a bigger platform/scale. There is a treaty named as Budapest Treaty, which has been signed by all countries under the aegis of the World Intellectual Property Organization (WIPO).

Under this treaty, pure culture of the microbial strains has to be deposited in a Microbial Type Culture Collection center duly recognized by WIPO countries (Montesinos 2003). Countries like the USA, the UK, and Australia were the ones where majority of the patents on biopesticides were initially deposited. All patents related to microbial pesticides have been well regulated under the Budapest Treaty. There have been vast number of patents registered in the area of microbial pesticides; only few have materialized in terms of their agriculture applications and use thereof (Montesinos 2003). For any invention or creation, therefore, the very first step was envisaged as to assure its protection, which is feasible by filing a patent before looking into commercialization aspects. Any biotechnological invention, which comprises of either a microbial product, or a process thereof, has been considered to be under the ambit of IPR, and patents, for example, on biopesticides could be applied. Many treaties at the national and international level do exist in order to provide a strong regulatory process for applying patents on microbial pesticides. Many patents related to living or attenuated bacteria or their products have been reviewed and discussed by Fialho et al. (2012) especially for their therapeutic potential having anticancer properties. When we talk about patenting of microbial products, one could apply for a patent on microbial toxins, enzymes/proteins or peptides, antimicrobial products like antibiotics, and small molecular weight proteins as well. Biffinger and Ringeisen (2008) emphasized upon the intricacies involved in the process of applying patents on microbial fuel cells (MFCs). MFCs have been widely recognized as potential alternatives to the existing standard commercial polymer electrolyte membrane (PEM) fuel cell technology. There are many potential merits associated with MFCs such as there is no need of the fuel supply to be purified. Moreover, the ambient operating temperatures are usually maintained with biologically compatible materials. More so the biological catalyst has the potential of self-regenerating. With the vast developments in the IPR, there are many advanced technologies being employed in today's context providing impetus to industrial and microbial biotechnology. Seeing the vast exploration in microbial research, it is anticipated that the pool of patents on microorganisms and their by-products and secondary metabolites will significantly increase in the near future. One may envisage further modifications and amendments in microbial patenting process in the future to make it more client-friendly according to the needs and requirements (Yadav et al. 2019).

4.11 Economic Considerations

All intellectual property involves some investment in intellectual effort, or investment in reputation in the case of trade marks. These intangible investments often can be easily copied or imitated by competitors. In many cases, without IPR protection it would be impossible to prevent "free-riding" by persons who did not contribute to the original IP investment, making it hard to recover commercially the cost of such investment. Consequently, market incentives for IP investment would be deficient. The prevention of free-riding is a key economic rationale of all IPRs.

In practice, the prevention of free-riding is not easy, and legislative considerations are further complicated by cross-border flows of goods and knowledge. The prevention of free-riding is not the sole economic objective of IPR legislation. While the prevention of free-riding is essential to ensure an adequate level of investment in IP, it is also in the interest of society to ensure the dissemination of new knowledge and ideas to the public. This is achieved by requiring patent disclosures, limiting the duration of patent protection and restricting copyright protection to expression but not ideas. There is also a noneconomic "moral" argument for IPRs, based on the "natural" right of people to the fruits of their labor. In a sense, this is a noneconomic argument because it does not address the issue of how markets operate or should operate. The focus of attention in this chapter is on economic rather than moral arguments in relation to IPR protection. Nonetheless, it must be recognized that moral arguments still play an important role in IPR policy formulation and the judicial process. Additional considerations arise in regard to technological IPRs. Given that the application of new scientific and technological knowledge by one person does not diminish its usefulness for others (non-rivalry), there can be significant positive externalities associated with research and development. In this context, positive externalities mean that social benefits exceed private benefits. Thus, there may be justification on economic grounds for public support for technological innovations beyond just the prevention of free-riding.

4.12 Geographical Indications

As far back as 1891, an agreement was worked out in Madrid for the suppression of false or deceptive indications of the geographical source of goods. Australia is not a party to this WIPO agreement, because to a large extent its provisions are already covered under the Paris Convention. According to this agreement, all goods bearing a false or deceptive indication of source by country or place of origin must be seized upon importation, or other sanctions must be applied in connection with such importation. In addition, geographical indication is one of the few IPR-related issues mentioned in the GATT, which forbade misleading labeling of geographic origin. The TRIPS agreement contains a number of clauses about geographical indications, but the issue has not yet been fully resolved. TRIPS has an in-built agenda for review and further consideration of issues that were not resolved during the original negotiating mandate. At the present, the IPR protection of new life forms is one such issue and geographical indications is the other.

Unlike the Paris Convention, which only prohibits false indication of geographical source, the TRIPS agreement prohibits the use of any mark that may mislead the public as to the geographical origin of the good. A special article on wines requires member states to have the legal means to prevent the use of inappropriate geographic indications for wines not originating in the place indicated by the geographical indication. This applies even if the geographical indication is accompanied by expressions such as "kind," "type," or "style."

However, there are possible exceptions. Members do not have to protect a geographical indication that has become a generic term for describing the product in question (manchester, china, and the like). The measures governing geographical indications shall not prejudice prior trade mark rights that have been acquired in good faith. Members availing themselves of the use of these exceptions must be willing to enter into negotiations about their continued application.

Copyrights appeared on the scene more than a hundred years after patents. The demand for legal regulation of the publishing industry arose after the introduction of printing presses in Europe in the mid-fifteenth century, which made the rewards for plagiarism much greater than before. In the beginning, copyrights were given to publishers rather than authors, and were used in much of Europe (including England) to foster the development of monopolistic printing guilds that served as instruments of censorship by church and state. The modern copyright protection of authors appeared in England in 1709, when the Act of Anne eliminated the guild monopoly on the holding of copyrights, enabling anyone to hold a copyright for a new work (David 1993). Following this change, copyright begun to assume a role in providing a commercial incentive for the expression of new forms and ideas. Later, copyrights were extended to musical compositions and the visual arts. In the 1900s, legislation had to be revised many times to cope with the emergence of new information recording and transmission technologies, such as sound recordings, films, radio, TV, and digitized information.

There are a number of International organizations and agencies that promote the use and protection of intellectual property. Although these organizations are discussed in more detail in the chapters to follow, a brief introduction may be helpful:

International Trademark Association (INTA) is a not-for-profit international association composed chiefly of trademark owners and practitioners. It is a global association. Trademark owners and professionals are dedicated in supporting trademarks and related IP in order to protect consumers and to promote fair and effective commerce. More than 5000 companies and law firms in more than 150 countries belong to INTA, together with others interested in promoting trademarks. INTA offers a wide variety of educational seminars and publications, including many worthwhile materials available at no cost on the Internet (INTA's home page at http://www.inta.org.). INTA members have collectively contributes almost US $ 12 trillion to global GDP annually. INTA undertakes advocacy [active support] work throughout the world to advance trademarks and offers educational programs and informational and legal resources of global interest. With its headquarter is in New York City, INTA also has offices in Brussels, Shanghai, and Washington DC and representatives in Geneva and Mumbai. This association was founded in 1878 by 17 merchants and manufacturers who saw a need for an organization. The INTA is formed to protect and promote the rights of trademark owners, to secure useful legislation (the process of making laws), and to give aid and encouragement to all efforts for the advancement and observance of trademark rights.

World Intellectual Property Organization (WIPO) was founded in 1883 and is a specialized agency of the United Nations whose purposes are to promote intellectual property throughout the world and to administer 23 treaties (Present 26 treaties) dealing with intellectual property. WIPO is one of the 17 specialized agencies of the United Nations. It was created in 1967, to encourage creative activity, to promote the protection of Intellectual Property throughout the world. More than 175 (Present 188) nations are members of WIPO. Its headquarters is in Geneva, Switzerland, the current Director General of WIPO is Francis Gurry who took charge on October 1, 2008. The predecessor to WIPO was the BIRPI [Bureaux for the Protection of Intellectual Property], which it was established in 1893.

4.13 Berne Convention for the Protection of Literary and Artistic Works (The Berne Convention)

An International copyright treaty called the Convention for the Protection of Literary and Artistic Works was signed at Berne, Switzerland in 1886 under the leadership of Victor Hugo to protect literary and artistic works. It has more than 145 member nations. The USA became a party to the Berne Convention in 1989. The Berne Convention is administered by WIPO and is based on the precept that each member nation must treat nationals of other member countries like its own nationals for purposes of copyright (the principle of "nation treatment"). In addition to establishing a system of equal treatment that internationalized copyright among signatories, the agreement also required member states to provide strong minimum standards for copyrights law. It was influenced by the French "right of the author."

4.14 Madrid Protocol

Its legal basis is the multilateral treaties Madrid (it is a city situated in Spain) Agreement concerning the International Registration of Marks of 1891, as well as the protocol relating to the Madrid Agreement 1989. The Madrid system provides a centrally administered system of obtaining a bundle of trademark registration in separate jurisdictions. The protocol is a filing treaty and not substantive harmonization treaty. It provides a cost-effective and efficient way for trademark holder. It came into existence in 1996. It allows trademark protection for more than 60 countries, including all 25 countries of the European Union.

4.15 Paris Convention

In 1883 March 20, France Paris convention for the protection of Industrial Property, signed, was one of the first Intellectual Property treaties, after a diplomatic conference in Paris, France, on March 20, 1883 by 11 countries. According to Articles 2 and 3 of this treaty, juristic (one who has thorough knowledge and experience of law) and natural persons are either national of or domiciled in a state party to the convention. The convention is currently still force. The substantive provisions of the convention fall into three main categories: National Treatment, Priority right, and Common Rules.

4.16 North American Free Trade Agreement (NAFTA)

It came into effect on January 1, 1994, and is adhered to by the USA, Canada, and Mexico. The NAFTA resulted in some changes to US trademark law, primarily with regard to marks that include geographical terms. The NAFTA was built on the success of the Canada-US Free Trade Agreement and provided a compliment to Canada's efforts through the WTO agreements by making deeper commitments in some key areas. This agreement has brought economic growth and rising standards of living for people in all three countries.

4.17 Conclusion

Intellectual Property Rights protecting Technology has led to increase awareness about the IP. Some individuals and companies offer only knowledge. Thus, computer consultant, advertising agencies, Internet companies, and software implementers sell only brainpower. Domain names and moving images are also protected. More than 50% of US exports now depend on some form of intellectual property protection. The rapidity with which information can be communicated through the Internet has led to increasing challenges in the field of intellectual property. The most valuable assets a company owns are its Intellectual property assets. Companies must act aggressively to protect these valuable assets from infringement (breaching, violation of law) or misuse by others. The field of intellectual property law aims to protect the value of such investments.

References

Ammen J, Swathi N (2010) Patenting life the American European and Indian way. J Intellect Prop Rights 15:55–65

Biffinger JC, Ringeisen BR (2008) Engineering microbial fuels cells: recent patents and new directions. Recent Pat Biotechnol 2:150–155

Cameotra SS (2013) Can microbes be patented? Biochem Biophys Res Commun 430:448

Fialho AM, Bernardes N, Chakrabarty AM (2012) Recent patents on live bacteria and their products as potential anticancer agents. Recent Pat Anticancer Drug Discov 7:31–55

Garg G, Sehrawat N, Yadav M (2016) Role of enzymes in food industries. In: Sharma C, Sharma AK, Aneja KR (eds) Frontiers in food biotechnology. Nova, New York

Hüttner S, Johansson A, Teixeira PG, Achterberg P, Nair RB (2020) Recent advances in the intellectual property landscape of filamentous fungi. Fungal Biol Biotechnol 7:16. https://doi.org/10.1186/s40694-020-00106-z

Keswani C, Sarma B.K and Singh H.B (2016) Synthesis of policy support, quality control and regulatory management of biopesticides in sustainable agriculture. In: Singh HB, Sarma BK, Keswani C (eds) Agriculturally important microorganisms: commercialization and regulatory requirements in Asia. Springer, Singapore, pp. 3–12

Kumar A, Singh VK, Tripathi V, Singh PP, Singh AK (2018) Plant growth-promoting rhizobacteria (PGPR): Perspective in agriculture under biotic and abiotic stress. In: Prasad R, Gill SS, Tuteja N (eds) Crop improvement through microbial biotechnology. Elsevier, Amsterdam, pp 333–342

Montesinos E (2003) Development, registration and commercialization of microbial pesticides for plant protection. Int Microbiol 6:245–252

Raghuvanshi KS (2017) Patent regime of India related to microorganisms: a critical analysis. World J Juristic Polity 1:1–10

Rai A, Rai AC (2020) Microbial endophytes and their intellectual property rights. Microbial Endophytes 349–387. https://doi.org/10.1016/B978-0-12-819654-0.00014-4

Saha CN, Bhattacharya S (2011) Intellectual property rights: an overview and implications in pharmaceutical industry. J Adv Pharm Technol Res 2:88–93

Serina I, Toledo C (1999) Microbiological patents. Int Microbiol 2:199–200

Singh KP (2008) Applied microbiology-intellectual property rights and TRIPs. C.C.S. Haryana, Agricultural University, Hisar

Sivakumaar PK, Joe MM, Sukesh K (2010) An introduction to industrial microbiology. S. Chand

Vitorino LC, Bessa LA (2017) Technological microbiology: development and applications. Front Microbiol 8:827

Yadav M, Sehrawat N, Kaur S, Singh RS (2016) Biofuel cells: concepts and perspectives for implantable devices. In: Singh RS, Pandey A, Gnansounou E (eds) Biofuels: production and future perspectives. CRC, Boca Raton

Yadav M, Meenu M, Sehrawat N, Sharma AK (2019) Intellectual property rights in microbiology. In: Singh HB et al (eds) Intellectual property issues in microbiology. CABI. https://doi.org/10.1007/978-981-13-7466-14

Chapter 5
Ethical and Legal Issues in Microbial Products in India

Pooja Patel and Ami Naik

Abstract A variety of microorganisms produce secondary metabolites, which serves as a source of natural products. These natural products have been an interest and a source of inspiration for scientists and researchers for the development and well-being of civilizations. Nowadays, modern technologies and effective scientific tools have made it possible to use such beneficiary microbes in the best possible way. Microbial-based products, especially biofertilizers and biopesticides, are known to be effective green alternatives to overcome adverse effect of chemical fertilizer and pesticides caused during green revolution. However, in India usage of such microbial-based products in agriculture is still at a very small scale. To overcome this issue, the Indian government has encouraged the use of biofertilizers and biopesticides by placing them into many of the agricultural schemes. Although, legal barriers imposed on their manufacturing, trade, import, storage, transport, and disposal restrict their production and adaptability hence, the declining interest of farmers have become a matter of concern. The main objective of this chapter is to highlight the legal provisions of the Indian laws regulating biofertilizers and biopesticides. Legal analysis stated in this chapter is based on published comparative legal research, which could enforce the manufacture process with high quality standard of biofertilizers and biopesticides present on the market and thus foster their use by farmers.

Keywords Biofertilizers · Microbial products · Legal analysis · Legal reform · Biopesticides

P. Patel (✉)
Department of Microbiology, Government Medical College, Surat, Gujarat, India

A. Naik
C. G. Bhakta Institute of Biotechnology, Uka Tarsadia University, Surat, Gujarat, India

N. Amaresan et al. (eds.), *Agricultural Microbiology Based Entrepreneurship*, Microorganisms for Sustainability 39,
https://doi.org/10.1007/978-981-19-5747-5_5

5.1 Introduction

Microorganisms play a crucial role in the biological systems of the Earth. Their existence itself serves as a major source of natural products, which originate as secondary metabolites. These chemically and structurally diverse microbial compounds and molecules have remarkable potential and hence are utilized in various industrial sectors like Medicine, Agriculture, Food and Beverages industry, and also for scientific research purpose for best possible outcome. This industry has provided valuable products that have deeply changed our lives and life spans. In the twenty-first century, these natural products are not the only area that has undergone substantial growth; advancements in research has made it possible to make advantageous gene combinations from prokaryotic and eukaryotic microbial cells, in combination with the advancement of recombinant DNA techniques. This results in an engineered microbe that can produce appreciable amounts of target compounds. As a result, large numbers of such product have been developed, approved, and marketed using different microbial expression systems; many more are currently in the development pipeline worldwide. In the near future they can be a responsible for an explosion of biologics.

The fast-paced nature of the biotechnology industry plays an important role in country's overall growth. As many industries have developed over the last two decades, its new developments often mean that legislators, regulators, and society, in general, must "catch up" in their efforts to understand the issues, the risks, and even the benefits that may result from the industry's new ways of conducting research, new products, and novel methods of product marketing and distribution. In India, agriculture is a core sector of economy and survival for mankind. Presently, it is a great challenge for agricultural scientist to increase the food productivity in order to meet the food demands of the global exceeding population, which has led to an increased reliance on chemical fertilizers and pesticides. Continuous exploitation of chemicals causes environmental pollution, loss of fertility, and, more important, economical loss. Such drawbacks have forced researchers to look forward for cost-effective and eco-friendly alternatives to get sustainable productions, more than ever before. In this regard, recent efforts have been shifted toward the use of biological organic fertilizers, which can serve as exclusive alternates to agrochemicals. Currently, the biological approach has become a subject of great attention for biosafety and sustainable agriculture. Therefore, nowadays, agriculture scientists are deriving forces to recycle the soil nutrients and, consequently, restore soil fertility with integration of plant growth-promoting microorganisms. Such a potential has prompted efforts in isolating and selecting microbial strains showing plant growth promotion capabilities through direct and/or indirect improvement of plant nutrient uptake.

Microbial-based fertilizer, pesticides, and soil amendment are known to be effective green alternatives to chemical fertilizers; however, many countries are facing major challenges with respect to the regulation of such microbial-based products, including inadequate capacity and legislation and the weak implementation of policies related to microbial-based fertilizer and pesticides. This chapter is based on the critical analysis of the laws and regulations governing the use of biofertilizers and biopesticides in India. A number of countries have amended their policies to minimize the use of chemical fertilizers and promote the use of biofertilizers and biopesticides. However, biofertilizers are still largely regulated by the system originally designed for chemical fertilizers. This situation has created market entry barriers by imposing burdensome costs on the biofertilizer industry. Other challenges include the relative immaturity of the policy network, limited resources and capabilities, and a lack of trust between regulators and producers. In India, manufacturers and importers of biofertilizers also face additional problems. For example, at the time of registration of new products, the manufacturer/trader/importer is required to generate data that are easily obtainable for chemical-based products, but which are difficult to obtain for biofertilizers. Furthermore, there are questions as to the utility of some of this data when applied to biofertilizers. The analysis in this study finds that the Indian law on fertilizers is one of the most comprehensive in the world in terms of its treatment of biofertilizers. However, challenges lie at the level of the technical or administrative personnel who deal with the registration, testing, monitoring, surveillance, inspection, and authorization tasks. Their level of understanding and their capacities are limited to chemical synthetics, and they have little or no experience with biofertilizers. Therefore, compliance and implementation of the regulations are the major challenges in India.

5.2 Analysis of the Indian Law on Biofertilizers

The various microorganisms utilized to improve plant growth have created some inconsistencies in the definition of the biofertilizers. This has also created some confusion in the market of microbial-based products for plant nutrition, which in many countries have not been regulated yet. India is probably the country with the most complete legal framework related to biofertilizers (Malusá and Vassilev 2014). After independence, the Government of India declared fertilizers an essential commodity, and began regulating the sale, price, and quality of fertilizers. In 1985, the Government of India passed the Fertilizer (Control) Order, 1985 [FCO] under Section 3 of the Essential Commodities Act, 1955. This law applies in all states and union territories of India. The Indian Ministry of Agriculture issued an order in 2006, later amended in 2009, which added biofertilizers to the Essential Commodities Act, 1955. As a result, India now has one of the world's most comprehensive legal frameworks governing biofertilizers, defined in the FCO as follows: "Biofertilizer means the product containing carrier based (solid or liquid) living microorganisms which are agriculturally useful in terms of nitrogen fixation, phosphorus solubilization or nutrient mobilization, to increase the productivity of the soil and/or crop"(Kumar and Singh 2014). The term is also covered under the broad definition of fertilizers, which "means any substance used or intended to be used as a fertilizer of the soil and/or crop." Coinciding with the insertion in the FCO of the term biofertilizer, new schedules—Schedule III, IV, and V—were also added. Schedule III relates to mixtures of biofertilizers, whereas Schedule IV concerns organic fertilizers. Similarly, Schedule V was added in 2010 to address nonedible de-oiled cake fertilizers (Malusá and Vassilev 2014), which are obtained through residue oil extraction (by expeller and/or through solvent extraction) from the crushed seeds of nonedible oilseeds (such as castor oil) for use in soil as fertilizer. Schedules III, IV, and V also introduced standards to maintain the quality of biofertilizers. In this way, a conventional law dealing with chemical fertilizers was amended to address microbial and organic products developed through novel innovations.

5.2.1 Registration of Dealers

Schedule IV of the FCO contains provisions for the registration of dealers. The registration of dealers or manufacturers is a decentralized process conducted by respective state government only; the Government of India does not perform any registration functions. Under clauses 7, 8, 9, and 10 of the FCO, a dealer (a firm, company, or organization) can register with the state Agriculture Department. This also issues licenses for their manufacture, import, trade, transport, and storage. The registration under FCO consumes half a year. In India field trial data are not required at the time of registration. Samples of biofertilizers are taken later for laboratory testing to determine properties such as Colony Forming Unit count and the sieve size of microbial agents. Thus, the manufacturing of biofertilizers in India is relatively easy and advantageous, compared to the lengthy procedure for importing. A firm or company that wishes to obtain dealership registration must apply using Form A (Appendix 1), along with Form O (Appendix 2) under clause 8 of the FCO. In order to understand the registration process, it is important to discuss these forms. In Form A, serial number 7 and 8 require the details of the fertilizer to be handled by the applicant dealer. In particular, information is required about the site and process of manufacturing required. This information is also required in further detail in Form O. The rest of the information in Form A pertains to general details about the applicant firm and the promoter of the firm. In Form O, information regarding the product source must be provided in items number 1(c) and (d) and 3. All such details must also be supported by documentary evidence. Clause 8(2) of the FCO provides for the registration by a manufacturer, importer, pool handling agency, wholesaler, or retail dealer and replaces Form A with Form A1, which is slightly different. Clause 8(4) was added in 2015 to establish the minimum qualification required for the applicant registration. They stated that an applicant for registration should possess a basic qualification in agriculture/science like BSc in Agriculture, BSc in Chemistry, Diploma in Agriculture, or a Certificate in Agri-inputs from specified institutes. Cooperatives and marketing federations are exempted from this requirement. Moreover, application forms require product's certificate of origin, test reports, microbiological performances, bioassays, toxicity reports, manufacturing process details, efficiency reports, company registration, tax-related registrations, declarations, warehouse and storage details, and several other similar documents (Arjjumend and Koutouki 2020). Registration is granted for 36 months as per clause 10 of the FCO and may be renewed as provided in clause 11.

5.2.2 Registration of Manufacturing of Biofertilizers Consortia

In part IV of the FCO, clauses 12–18 provide for the registration of manufacturing units. Special mixtures of fertilizers, biofertilizers, and organic fertilizers were included in clause 12 in 2006. Clause 13(b) states, "no person shall manufacture any biofertilizer unless such biofertilizer conforms to the standards set out in Part A of Schedule III." In different schedules, the FCO has very elaborately provided standards for all categories of mineral fertilizers, biofertilizers, and organic fertilizers. Part A of Schedule III contains specifications for all ten categories of biofertilizers: *Rhizobium*, *Azotobacter*, *Azospirillum*, Phosphate Solubilizing Bacteria, Micorrhizal Biofertilizers, Potassium Mobilizing Biofertilizers, Zinc Solubilizing Biofertilizers, Acetobacter, Carrier-Based Consortia, and Liquid Consortia. Part B of Schedule III specifies the tolerance limit of biofertilizers, while Part C explains the procedure of sampling biofertilizers. Part D comprehensively elaborates the methods of analyzing biofertilizers. The detail and guidance given in the schedules of the FCO make it user-friendly. Hence, as clause 13(b) advises, compliance with its stringent quality standards and specifications is mandatory. No manufacturing or import can be allowed until given specifications, and standards are satisfied in accordance with the provisions. Similarly, the requisite standards and specifications need to be adhered to in the case for organic fertilizers [clause 13(c) and Part A of Schedule IV]. Any firm or company that wishes to apply for registration as a manufacturer of biofertilizers or organic fertilizers must apply using Form D under clause 14(3). In accordance with clause 17 of the FCO, registration for manufacturing is issued for a period of 36 months, and renewal can be applied for under clause 18.

5.2.3 Packing and Labeling

Under clause 21(aa) of the FCO, information is given regarding packing and packaging that should not be misleading. The container should be marked with the word "biofertilizer." Other information printed on the packaging or container must be in accordance with the instructions released by the Controller (the registering authority in each state). Further, clause 21(b) prohibits tampering with packed or canned material. This clause is also applicable to imported material. However, the regulation does not address the issue of leakage and spillage from the containers, cans, or packages causing contamination, pollution, impacts of human and animal health, or occupational hazards.

5.2.4 Inspectors and Inspection

The quality and quantity of commodities available on the market for consumers are monitored and inspected by a system of inspectors. In the context of fertilizers, these inspectors are appointed by notified authorities under clause 27 of the FCO. Clause 27-B concerns the qualifications of inspectors appointed for the purpose of monitoring biofertilizers and organic fertilizers. Inspectors should have bachelor's degree in agriculture, chemistry, or microbiology, with training in the quality control of biofertilizers and organic fertilizers. Their responsibilities include the verification of information provided by manufacturers, wholesale dealers, retail dealers, importers, or pool handling agencies (clause 28a), as well as the sampling of biofertilizers (clause 28ba) in accordance with the procedure laid down in Schedule III of the FCO. In accordance with the powers vested under clauses 28c, 28d, and 28e, an inspector can inspect and examine the premises of manufacture, sale, or storage of the fertilizer or biofertilizer in addition to examining the financial accounts or documents associated with the material. An inspector can also seize or detain any fertilizer or biofertilizer. The inspection machinery having inspectors is an inevitable apparatus of the regulatory and executive wing of a state government.

5.2.5 Sampling and Analysis of Biofertilizers

In Part C of Schedule III of the FCO, the procedure for biofertilizer sampling is described in detail. Part C covers: (1) the general requirements of sampling, (2) the scale of sampling, and (3) the procedure for taking samples. Only a trained inspector can perform sampling of the biofertilizer, as the material contains microorganisms. During the handling of the samples, precautions must be taken to prevent any possible contamination and exposure to sun, dust, soot, air, or moisture. In Schedule III, the quantity of samples is also simplified. Using the Form J-1, three packets must be sampled from a consignment of up to 5000 packets. Similarly, four packets must be sampled if the consignment is of 5001–10,000 packets, and five packets must be sampled in case the consignment consists of more than 10,000 packets. The sampling method must be strictly random and instructions for the proper handling of the samples are described. One sample is to be sent to a laboratory notified by the state government under clause 29 of the FCO, to the National Centre for organic Farming (Ghaziabad), or to any of its Regional Centres of Organic Farming at Bangalore, Bhubaneswar, Hissar, Imphal, Jabalpur, or Nagpur. The inspector must send the sample to a laboratory along with the details outlined in Form K1 within 7 days from the date the samples are drawn. The FCO regulations not only guide the manner in which notified authorities and inspectors implement the legal provisions, but also provide the manufacturers, wholesale dealers, importers, pooling agents, and retail dealers with all necessary forms, reporting requirements, and technical specifications. The samples of biofertilizers received by the notified laboratory must be analyzed and tested in strict adherence to the norms given in clause 29(1-A). For each of the ten categories of biofertilizers, an elaborate method and procedure is set forth in Part D of Schedule III of the FCO. These procedures conform to the technical specifications for biofertilizers as described in Part A of Schedule III of the FCO. For example, for analyzing the Phosphate Solubilizing Bacterial (PSB) biofertilizer, a testing laboratory must have to follow these major segments: (1) Apparatus, which a testing laboratory must have before analysis, (2) Reagents required for creating medium and sterilization of plates, etc. (3) Preparation of serial dilutions for plate counts, (4) Incubation of plates, (5) Determination of soluble phosphorus using ascorbic acid, (6) Estimation of total viable propagules, (7) Estimation of infectivity potential, (8) Maintenance and preparation of culture and quality control at broth stage. After sample analysis, the laboratory must send the test report to the notified authority within 30 days from the date of receipt of the sample. This sample analysis report is completed in Form L2.

5.2.6 Quality Control

The participants stated that quality monitoring is performed during manufacturing and before the release of the products into the market. Quality control issues pertaining to biofertilizers have received special focus in the stringent standards and specifications set in the FCO. Having stringent high standards checks the crop failure and crop loss that occurs due to the ineffectiveness of biofertilizers, which will help to avoid economic and agronomic costs to farmers. The FCO's Schedule III includes standards and specifications for all ten categories of biofertilizers identified and recommended by the Advisory Committee, which functions according to the provisions of clause 38 of the FCO. These standards set out seven quality parameters for biofertilizer samples: the physical form, the minimum count of viable cells, the contamination level, pH, the particle size in case of carrier-based materials, the maximum moisture percent by weight of carrier-based products, and the efficiency character. For example, in the case of bacteria, the minimum count of viable cells is 5×10^7 cells/g of solid carrier, or 1×10^8 cells/mL of liquid carrier (Part A of Schedule III of the FCO). For products containing mycorrhizal fungi, at least 100 viable propagules must be present per gram of finished product. In addition, the efficiency of nitrogen fixation must be shown with different tests: Rhizobia shall show effective nodulation, *Azotobacter* strains shall be capable of fixing at least 10 mg N/g of sucrose consumed, and *Azospirillum* strains must be able to form a white pellicle in semisolid N-free bromothymol blue media. The activity of phosphate solubilizing bacteria (PSB) can be assessed spectrophotometrically (30% P solubilization) or by the formation of a solubilization zone of at least 5 mm in a media having at least 3 mm thickness. Similarly, products with mycorrhizal fungi shall be able to provide 80 infection points in roots per gram of inoculum used (Malusá and Vassilev 2014). Sample analysis in laboratories is described in Part D of Schedule III of the FCO.

5.2.7 *Ecological and Health Safety Issues*

Having been amended in 2006 and 2009 to insert provisions for biofertilizers and organic fertilizers, the FCO is the most progressive regulation in the world concerning fertilizers. Yet while the FCO has addressed quality control issues indirectly in several ways, other quality-related aspects, ecological risks, human and animal health safety, spillage, contamination, and occupational hazards are not considered anywhere in the regulation. Though the chemical fertilizers have higher risks to ecosystems, human and animal health, and occupational safety, microbial biofertilizers may also pose risks to the ecosystems and human health if not handled carefully. Amended legislation, i.e., the FCO, does not have direct instructions to regulate ecological hazards and health safety issues that may emanate from biofertilizers.

5.3 Analysis of the Indian Law on Biopesticides

In India, the concept of biocontrol of plant diseases has been in practice for a very long time. The Neem tree (*Azadirachta indica* A. Juss) and its derivatives, i.e., leaf extract, oil, and seed cake have been used as fertilizers and also for minimizing the risk of post-harvest loss in stored cereals (Brahmachari 2004). It was realized that biocontrol is the only means that can be utilized as a safe, cost-effective, and eco-friendly method to control the widespread resistance of chemical insecticides toward pest insects. Biopesticides are classified under three categories (1) Microbial-based biopesticides (2) Botanical biopesticides, and (3) Semiochemicals. This chapter highlights legislation process of only microbial-based biopesticides, which is mainly secondary metabolites produced by the different microbes, i.e., bacteria, fungi, algae, virus, nematodes, and protozoa, which have reported remarkable insecticide effect on crop (Van Lenteren 2012). One of the major obstacles in promoting biopesticides as an alternative to chemical pesticides is the lack of appropriate recognition of biopesticides, reflecting the weakness of the underlying policy framework in India (Kumar and Singh 2015). The relative immaturity of the policy framework, limited resources and capabilities, and a lack of trust between regulators and producers are also serious problems. Investment risks involved in opting for biopesticides on farmers' fields, and farmers' confidence in the quality and performance of the products, continue to be debated in India. The Government of

India introduced a Pesticides Management Bill (pending in parliament for many years), which is intended to replace the existing Insecticides Act, 1968. However, this bill still fails to differentiate biopesticides from conventional chemical pesticides. The analysis of Indian legislation points to the need to create a separate and distinct legal framework for biopesticides. The regulatory options governing biopesticides should also be in line with novel microbial technologies. These changes would ultimately contribute to achieving the SDGs and support the flow of goods and services for organic agriculture and horticulture in India. When considering legislation governing biopesticides, two crucial concerns should be taken into account. First, regulations must be formulated to ensure human and environmental safety and to consistently and reliably characterize the quality of biopesticide products. Second, registration and regulatory agencies require a biopesticide data portfolio, a concept originating from the framework governing chemical pesticides. Such data includes information about the mode of action, toxicological and eco-toxicological evaluations, and host range testing (Chandler et al. 2011). There is also a need to critically analyze the existing Indian legal framework to understand the gaps and weaknesses hindering the overall trade, manufacture, and use of biopesticides in the country.

5.3.1 The Insecticides Act, 1968: Scope of the Law

The Insecticide Act (1968) (amended in 2000) is the only legislation under the Indian Government that governs the import, manufacture, sale, transport, distribution, and use of all types of insecticides, including biopesticides with a view to prevent risk to human beings and animals, as well as all connected matters. In exercise of the powers conferred by Section 36 of the Insecticides Act, 1968 (46 of 1968), the Central Government, after consultation with the Central Insecticidal Board (CIB) made insecticides Rules, 1971, which governs the manufacture, grant of a license, expiry of the license, product labeling, packaging and sale, and use of insecticides. The Registration Committee grants registrations, only after the data is provided on the efficacy and safety of products to human beings and animals. The rule also assures that the samples of pesticides should be regularly checked for quality purposes. In the case of biopesticides, shelf-life, cross-contamination, moisture content, and packaging is considered. In 2015, the government also passed a bill known as the Insecticides Bill, 2015. The Bill added a modification to Section 9 of the Insecticide Act (1968), after sub-Section (3C); the sub-sections of nanotechnology-based pesticides were inserted. Based on the guidelines of the Organization for Economic Co-operation and Development (OECD), the CIB has not only streamlined the guidelines and data requirements for registration but also mentioned minimum infrastructural facilities required for the production of biopesticides. Guidelines/data requirements for minimum infrastructure facilities and the same for the registration of biocontrol products under Sections 9 and 9 (B) are being governed by Registration Committee of CIB. Although the guidelines and recommendations of the Committee promise the development of high-quality biocontrol products, for the manufacturers, these are problematic. In the case of bacterial and fungal biopesticides, the guidelines ask for bio-efficacy data that need to be generated from ICAR, SAUs, Council of Scientific and Industrial Research (CSIR) or the Indian Council of Medical Research (ICMR) institutes. Similarly, the data required for toxicity generation also requires a vigorous workout. For claiming shelf life, the registrants should provide data of two different agro-climatic locations at ambient temperature along with the meteorological data. However, the requirement of such agro-climatic and meteorological data creates an extra burden on manufacturers and discourages them in expanding their business. For example, microbes isolated from a particular agro-climatic region showing efficient biocontrol activity may or may not show the same results in a different agro-climatic zone. Hence, CIBC should also take into account such and other issues that are directly affecting the manufacturing process of biopesticides (Mishra et al. 2020).

The Insecticides Act, 1968 has registered 739 chemical pesticides in India and banned a number of toxic pesticides. Noticeably, 970 companies have registered biopesticides as of 9 April 2021. Through its Gazette Notification no. 147, dated 26 March 1999, the Government of India included the following categories of biopesticides (Arjjumend and Koutouki 2021):

Antagonistic Fungi and Bacteria • *Bacillus subtilis* • *Gliocladium* species • *Pseudomonas* species • *Trichoderma* species
Entomogenous Fungi • *Beauveria bassiana* • *Metarrhizium anisopliae* • *Nomuraea rileyi* • *Verticilium lecanii*
Grannulosis viruses (GV)
Nuclear polyhedrosis viruses (NPV)
Bacillus species • *Bacillus thuringiensis* var. *israelensis*, *Bacillus thuringiensis* var. *kurstaki* • *Bacillus thuringiensis var. galleriae*, and *Bacillus sphaericus*
Recently, the following 25 genera, species, or strains have also been added into the schedule: • *Nomuraea rileyi* • *Hirsutella* spp. • *Verticillium chlamydosporium* • *Streptomyces griseoviridis* • *Streptomyces lydicus* • *Ampelomyces quisqualis* • *Candida oleophila* • *Fusarium oxysporum* (non-pathogenic) • *Burkholderia cepacia* • *Coniothyrium minitans* • *Agrobactarium radiobacter* strain 84 • *Agrobacterium tumefaciens*

(continued)

- *Pythium oligandrum*
- *Erwinia amylovora* (hairpin protein)
- *Phlebia gigantean*
- *Paecilomyces lilacinus*
- *Penicillium islanidicum* (for groundnut)
- *Alcaligenes* spp.
- *Chaetomium globosum*
- *Aspergillus niger* strain AN27
- VAM (fungus)
- *Myrothecium verrucaria*
- *Photorhabdus luminescences* akhurustii strain K-1
- *Serratia marcescens* GPS 5
- *Piriformospora indica*

In addition to the above microbial biopesticides, the following "plant origin biopesticides" are also the part of schedule:

- Pyrethrins (pyrethrum)
- Neem products
- Karanjin
- Extracts of *Cymbopogon* spp.
- Oxymatrine
- Reduced Azadirachtin(s)
- Tripterygium of wilfordii Hook GTW (plant extract)
- Bitterbarkomycin
- Squamocin
- Eucalyptus leaf extract

5.3.2 Central Insecticides Board and Registration Committee (CIBRC)

Constituted under Sections 4 and 5, the Central Insecticides Board and Registration Committee (CIBRC) regulate pesticides in India along with the Food Safety and Standards Authority of India (FSSAI). CIBRC is responsible for advising central and state governments on technical issues related to the manufacture, use, and safety of pesticides. The functions of CIBRC are twofold as specified in the Act: (a) advising on the risk to human beings or animals involved in the use of insecticides and the safety measures to prevent such risks (Section 2(2a)); and (b) advising on the manufacture, sale, storage, transport, and distribution of insecticides with a view to ensure safety to human beings or animals (Section 2(2b)). These functions of CIB are further expanded under Rule 3 of the *Insecticide Rules, 1971*, and include:

(a) Advising the Central Government on the manufacture of insecticides under the Industries (Development and Regulation) Act, 1951.
(b) Specifying the uses of each class of insecticides based on their toxicity and whether or not they are suitable for aerial application.
(c) Advising tolerance limits for insecticide residues and establishing minimum intervals between the application of insecticides and harvest in respect of various commodities.

(d) Specifying the shelf life of insecticides.
(e) Suggesting colorization, including coloring matter, which may be mixed with concentrates of insecticides, particularly those of highly toxic nature.
(f) Carrying out such other functions as are supplemental, incidental, or consequential to any of the functions conferred by the *Act* or these rules.

The Department of Biotechnology within the Indian Ministry of Science and Technology is the technical agency that evaluates effectiveness, quality and safety issues during the approval process. Before authorization and registration, it must be determined that the microorganism and its metabolites pose no concerns relating to pathogenicity or toxicity to mammals and other nontarget organisms that will likely be exposed to the microbial product; that the microorganism does not produce a known genotoxin; and that all additives in the microbial manufacturing product and in the end-use formulations are of low toxicity and have little potential to harm human health or the environment.

Section 5a(i) of the Insecticides Act, 1968 speaks about the constitution of the Registration Committee, which is given the tasks of registering insecticides and pesticides (including biopesticides) after scrutinizing their formula and verifying claims made by the importer or the manufacturer regarding their efficacy and safety to human beings and animals. The Registration Committee, under Rule 4 of the Insecticide Rules, 1971, has been given the following tasks: (a) specify the precautions to be taken against poisoning through the use or handling of insecticides; (b) carry out such other incidental or consequential matters necessary for carrying out the functions assigned to it under the Act or Rules. While the Registration Committee is expected to emphasize toxicological and ecosystem safety issues, the majority of these concerns apply to toxic organo-chemicals. Yet most biocontrol agents are ecologically safe and nontoxic. Thus, as far as biosafety is concerned, a separate legal framework is required to provide regulatory guidance for different categories of biopesticides in a systematic and comprehensive manner. Treating all categories of pesticides under one regulatory framework harms the economic viability of biocontrol agents (affecting manufacture, trade, supply, etc.).

5.3.3 Registration Process

Applications for registering the manufacture and import of a new pesticide or biopesticide can be made in the prescribed Form I (Rule 6) to the Secretary, Registration Committee, Directorate of Plant Protection, Quarantine and Storage, Ministry of Agriculture, NH-IV, Faridabad—121,001 Haryana. The application process requires a tremendous amount of data. Form I does not contain elaborate guidelines specifying the information that is required at the time of application submission. Several relevant parameters on which data are required for registration of biopesticides are mentioned in Appendix 3. The requisite tests and analysis of each biopesticide to be registered are relatively exhaustive and specific. In addition, recently, highly technical tests such as DNA barcoding and fingerprinting are made mandatory for new registrations of microbial biopesticides. Under Section 16 of the Act, the Central Insecticide Laboratories are notified from time to time. Under Rule 5 of the Insecticide Rules, 1971, the functions of the laboratory are described. Under Section 9(3) of the Insecticides Act, 1968, the period for registration of an imported or manufactured biopesticide is 12 months from the date of application. This period may be further extended by 6 months if the Registration Committee is unable to arrive at a decision within the said period on the basis of the materials before it. This lengthy registration period is impractical from a business perspective. It is also unsuitable for biopesticides, as the shelf life of biocontrol agents is very short. Often, laboratory tests take such a long time that the effective shelf life of the particular strain contained in the biopesticide expires before registration is granted. Due to delay in testing and short shelf life, the sampled strain does not fit the standards set for that particular category of the biopesticide. Therefore, the length of time required for the registration of biopesticide must be shortened in accordance with the shelf life of the various biopesticide strains.

Section 9(3) of the Insecticides Act, 1968 also requires the Registration Committee to investigate claimed safety precautions for humans and animals, including wildlife. In cases where the precautions claimed are insufficient or, notwithstanding the observance of such precautions, the use of the insecticides involves serious risk to human beings or animals, the Committee may refuse the registration. Similarly, according to Sections 9(3B) and 9(3C), the Registration Committee must take precautionary measures when the insecticide is being introduced and registered for the first time in India. Such provisions are also applicable to biopesticides. However, unless there is a serious biosafety issue involved, biopesticides should be treated differently from chemical pesticides, with due care to the ecological and public health effects of biopesticides.

As per the provisions of Section 9 of the *Act*, applications for licenses to manufacture registered pesticides (also under Rule 9) and to obtain licenses for sale, etc., of registered pesticides (Rule 12) are lodged. Sections 9 (application) and Section 13 (license granting) of the Insecticides Act, 1968 are somewhat inconsistent with each other. The registration of biopesticides is carried out at a federal level, whereas the license is granted by state governments after the registration is done by the Central Government. Contrary to actual practice, there is no mention in Sections 9 and 13 that licenses would be issued for insecticides (or biopesticides for that matter) only after their registration by the Central Government.

5.3.4 Packing and Labeling

Under Rules 16–20 of the Insecticides Rules, 1971, the importance, manners, procedures, and prohibitive actions are described for packing, packaging, and labeling of pesticides or biopesticides. Without the prescribed proper packing and labeling, no pesticide or biopesticide is allowed to be exhibited or sold. According to Rule 17(1), "every package containing the insecticide shall be of a type approved by the Registration Committee." Rule 17(2) makes it more explicit by stating, "before putting any insecticide into the primary package, every batch thereof shall be analyzed as per the relevant specifications of the manufacture thereof, in accordance with the approved methods of analysis and the result of such an analysis shall be recorded in the register maintained for the purpose." Each package needs to contain a leaflet having necessary information about the pesticide or biopesticide. Important disclosure is to be made about the particulars regarding chemicals harmful to human beings, animals, and wildlife. Warning and cautionary statements must be included regarding the symptoms of poisoning, suitable and adequate safety measures and emergency first aid treatment, and decontamination or safe disposal of used containers. Under Rule 19, the procedure of labeling is elaborately explained. Toxicity caused by the pesticide in case of leakage, spillage, usage, or accidental contamination is addressed in all these instructions. However, there is no word mentioned about biopesticides in any section of the Act or any rule, despite the fact that various notifications and amendments are adopted both in the Act as well as rules.

5.3.5 Inspection, Sampling, and Analysis of Biopesticides

Under Section 21 of the Act, inspectors are given powers to inspect and collect samples of pesticides. Their duties are also fixed under Section 22 of the Act. However, the training of these inspectors relates only to toxic chemicals; they lack the proper training and knowledge to handle biopesticides. This lack of training may have grave implications for the trade and free use of biopesticides. In the Insecticides Act, 1968, there is no specific instruction given to inspectors on how to sample or handle the biopesticides. Unlike the Fertilizer Control Order, the Insecticides Act, 1968 has not prescribed any sampling method. Under Section 21(1)(e) of the Act and Rule 24(2) of the Insecticides Rules, 1971, the procedure prescribed for the analysis of insecticides (or for biopesticides) is the same as that prescribed by the Indian Standards Institution16 (ISI). Analysis of samples is conducted only by the Central Insecticides Laboratory constituted under the provisions of Section 16 of the Act. In accordance with the Rule 21 of the Insecticide Rules, 1971, an analyst should be qualified in Agriculture, Science, or Chemistry apart from training in analyzing insecticides.

5.3.6 Disposal of Pesticides

Rule 44 of the Insecticide Rules, 1971 emphasizes environmental and safety aspects. The manufacturers, formulators, and operators must dispose of packages or surplus materials and employ safe washing methods in order to avoid environmental or water pollution. The packages must also be broken and buried away from human habitation.

Appendix 1

PRESCRIBED FORMS UNDER FERTILIZERS CONTROL ORDER 1957/1985
FORM A
(See Clause 8)
Form of Application to obtain Dealer's
(Wholesale or Retail or Industrial)
Certificate of Registration

To
The Registering Authority / Controller,
Delhi

1. Full Name and address of the applicant:
   (a) Name of the concern and postal address:
   (b) Place of business (Please give exact address)
      (i) for Sale
      (ii) for Storage

2. Is it a proprietary / partnership/limited Company / Hindu Undivided family concern? Give the name(s) and address(es) of the proprietor partners/manager karta.

3. In what capacity is this application field:
   (i) Proprietor
   (ii) Partner
   (iii) Manager
   (iv) Karta

4. Whether the application is for wholesale or retail or industrial dealership?

5. Have you ever had a fertilizer dealership registration certificate in the past? If so give the following details:
   (i) Registration Number:
   (ii) Place for which granted
   (iii) Whether wholesale or retail or industrial dealership.
   (iv) Date of grant of registration certificate.
   (v) Whether the registration certificate is still valid?
   (vi) If not when expired
   (vii) Reasons for not removal
   (viii) If suspended / cancelled and if so when?
   (ix) Quantity of fertilizers handled during last year?
   (x) Names of products handled
   (xi) Names of source of supply of fertilizers.

6.Was the applicant ever convicted under the Essential Commodities Act 1955 or any order issued thereunder, including the Fertilizers (Control) Order, 1957, during the last three years proceeding the date of Application, if so give details:

7. Give the details of the fertilizer to be handled:

S.No.	Name of Fertilizer	Source of Supply

8. Please attach certificate (s) of source from the supplier(s) indicated under column 3 of sl. No. 7.

9. I have deposited of the registration fee of Rs.......................... vide Challan no................ dated .... in treasury / Bank .... or enclose the Demand Draft No. .... Dated .... for Rs.................. Drawn on .... Bank, in favour of .... Payable at .... Towards registration fee (Please strike out which ever in not applicable).

10. Declaration:

(a) I/We declare that the information given above is true to the best of my/our knowledge and belief and no part there is of false.
(b) I/We have carefully read the terms and conditions of the Certificate of Registration
given in Form 'B' appended to the Fertiliser (Control) Order, 1985 and agree to abide by them.
(c) I/We declare that I/We do not possess a certificate of registration of Industrial dealer and that I/We shall not sell fertilizers for industrial use (Applicable in case a person intends to obtain a wholesale dealer or retail dealer certificate of registration, excepting a state Government, a manufacturer or a pool handling agency).
(d) I/We declare that I/We do not passes a certificate of registration for wholesale dealer or retail dealer and that I/We shall not sell fertilizers for agricultural use. (Applicable in case a person intends to obtain a industrial dealer certificate of registration, excepting a State Government a manufacturer or a pool handling agency).

Dated .... Signature of the Applicant(s)

Note:
1. Where the business of selling fertilizers is intended to be carried on at more then one place a separate application should be made for registration in respect of each such place.
2. Where a person intends to carry on the business of selling fertilizers both in retail and wholesale business should be made
3. Where a person represent intends to represent more than one State Government, Commodity Board manufacturer of Wholesale dealer, separate certificate of source from each such source should be enclosed.

For use in the office of Registering Authority/Controller.

Date of receipt. Name & Designation of
Office receiving the application

Appendix 2

FORM'O'
[See Clauses 8 and 11]
Certificate of Source for Carrying on the Business of Selling Fertilizers in Wholesale/Retail for Industrial Use
No.001.Date of Issue 2018-06-14

1.Particulars of the concern issuing the certificate of source:

(a)Name and full address:

(b)Status:

(i) State Government
(ii) Manufacturer
(iii) Pool handling agency
(iv) Wholesale dealer

(c) If manufacturer of mixture of fertilizers, the details of certificate of manufacturing of mixture of fertilizers being possessed:

(i) Number
(ii) Date of Issue
(iii) Date of expiry
(iv) Grades of mixtures of fertilizers allowed to be manufactured
(v) Authority by whom issued

(d) Details of certificate of recognition:

(i) Number:
(ii) Date of issue:
(iii) Date of expiry:
(iv) Authority by whom issued:

2.Particulars of the person to whom the certificate of source is being issued:

(a) Name and full address:

(b)Status:

(i) Wholesale dealer
(ii) Retail dealer
(iii) Industrial dealer

(c)If holds a valid certificate of registration, the details thereof:

(i) Number:
(ii) Date of issue:
(iii) Date of expiry:
(iv) Authority by whom issued:

(d)Purpose of obtaining the certificate of source:

(i) For obtaining a fresh certificate of registration
(ii) For renewal of the certificate of registration

3.Details of the Fertilizers to be supplied:

Sl.No.	Name of Fertilizers	Trade Mark/ Brand Name
1	2	3
1		
2		
3		

4. Declaration: Declared that the fertilizers mentioned above will be supplied conforming to the standards laid down under the Fertilizer (Control) Order, 1985, and as the case may be, grades/formations (of mixtures of fertilizers) notified by the Central/State Government and packed and marked in container as provided under Clause 21 of the Fertilizer (Control) Order, 1985.

Signature with stamp of the
Authorized Officer

Appendix 3: Technical Data Required for Registering a Biopesticide in India

1. Strain specification
   - Genus and species.
   - Rhizosphere competence.
   - Biological control capability.
   - Growth promotion capability.
   - Wide range of growth parameters like pH and temperature.
2. CFU count
   - The data required for claiming 1 year shelf life of the product is for 15 months for talc-based formulation, i.e., the microbe should remain viable for 15 months with a colony forming units (CFU) count not less than 2 × 106 spores/mL or g on selective media (SM).
   - Pathogenic contaminants such as *Salmonella, Shigella,* and *Vibrio* should not be present. Other microbial contaminants not to exceed 1 × 104 counts/mL or g.
3. Target fungi
4. Moisture content
   - Maximum moisture content of the product should not exceed more than 8% for dry formulation of fungi and 12% for bacteria.

5. Chemistry

   - Systematic name: Genus and species.
   - Common name, if any.
   - Natural occurrence: morphological description.
   - Manufacturing process: solid or liquid state fermentation.
   - Qualitative analysis.
   - CFU on selective medium.
   - Absence of Gram– bacterial contaminants (*Salmonella, Shigella,* and *Vibrio*).
   - Moisture content.
   - Shelf life claim: two different locations along with meteorological data.

6. Technical bulletin/Product profile.
   Bioefficiency:

   - Lab bioefficacy test: The product should be tested against target pathogen/pest at one of the laboratories of ICAR/SAUs/CSIR/ICMR system.
   - Field bioefficacy test: The intended product should be tested for field bioefficacy under Indian conditions.
   - Field bioefficacy guidelines have been recently revised/enhanced w.e.f. 01.01.2011.
     Field Bioefficiency:
   - 9(3B): Provisional Registration.
   - One crop: two seasons (rabi and kharif)/year, two agro-climatic conditions (four bioefficacy trial reports).
   - 9(3): Permanent Registration.
   - One crop: two seasons (rabi and kharif)/year, three agro-climatic conditions (six bioefficacy trial reports).
   - Safety data on nontarget organisms.

7. Toxicity

   - Toxicological studies may be conducted by recognized institutes viz. IITR Lucknow, IIBAT Chennai, JR Foundation Vapi, INTOX, Pune.

     (a) (For formulated products to be directly manufactured).

     Single dose oral—Rat (21 days) (Toxicity/Infectivity/Pathogenicity).
     Single dose oral—Mouse (21 days) (Toxicity/Infectivity/Pathogenicity).
     Single dose Pulmonary—Rat (14 days) (Toxicity/Infectivity/Pathogenicity).
     Single dose Dermal—Rabbit (21 days) (Toxicity/Infectivity/Pathogenicity).
     Single dose Dermal—Intraperitoneal (21 days) (Toxicity/Infectivity/Pathogenicity).
     Primary skin irritation.
     Eye irritation.

(b) Human Safety Records.

Environmental Toxicological Studies (For formulation only) On Nontarget Vertebrates.

Toxicity to chicken, pigeon, freshwater fish.

Dossier preparation for 9(3b) and 9(3) registration.

References

Arjjumend H, Koutouki K (2020) Legal barriers and quality compliance in the business of biofertilizers and biopesticides in India. J Legal Stud 26(40):81–101

Arjjumend H, Koutouki K (2021) Analysis of Indian and Canadian laws regulating the biopesticides: a comparison. J Environ Law Policy 1:26

Brahmachari G (2004) Neem—an omnipotent plant: a retrospection. Chembiochem 5(4):408–421

Chandler D, Bailey AS, Tatchell GM, Davidson G, Greaves J, Grant WP (2011) The development, regulation and use of biopesticides for integrated pest management. Philos Trans R Soc Lond B Biol Sci 366(1573):1987–1998

Kumar S, Singh A (2014) Biopesticides for integrated crop management: environmental and regulatory aspects. J Biofertil Biopestici 5:e121

Kumar S, Singh A (2015) Biopesticides: present status and the future prospects. J Fertil Pestic 6(2): 100–129

Malusá E, Vassilev N (2014) A contribution to set a legal framework for biofertilisers. Appl Microbiol Biotechnol 98(15):6599–6607

Mishra J, Dutta V, Arora NK (2020) Biopesticides in India: technology and sustainability linkages. 3 Biotech 10(5):1–12

Van Lenteren JC (2012) The state of commercial augmentative biological control: plenty of natural enemies, but a frustrating lack of uptake. BioControl 57(1):1–20

Part II
Biofertilizer and Compost Production

Chapter 6
Mass Multiplication, Production Cost Analysis, and Marketing of Cyanobacterial Biofertilizers

V. T. Anju, Siddhardha Busi, and Madhu Dyavaiah

Abstract Cyanobacteria are the precious bio-source for potential application in sustainable agriculture. Cyanobacteria or blue-green algae hold biological nitrogen fixation properties through which they fix atmospheric nitrogen to reduced forms. These algae fix nitrogen either as a free-living form or in symbiotic associations. More than 20 kg of N ha^{-1} and organic matter available in soil are fixed by cyanobacteria. The easily available and less expensive eco-friendly cyanobacterial fertilizers control the nitrogen deficiency in soil, aeration capacity, and water holding capacity of the soil. The high cost of chemical nitrogen fertilizers replaces with the cost-effective cyanobacterial fertilizers. The economically feasible methods of cultivation enabled the successful large-scale production of cyanobacterial fertilizer. The initial production methods of cyanobacterial fertilizers were easy and inexpensive and widely adopted by the farmers in their fields. There are open and closed system cultivation methods using sunlight and artificial light. The effective application of cyanobacterial biofertilizers has reduced the burden of economically weak farmers to substitute expensive chemical nitrogen fertilizers. Nitrogen biofertilizers are the most putative fertilizers in the Asia-pacific biofertilizer market and received huge attention from consumers. Recently, there has been an enormous demand for cyanobacterial biofertilizers in Asia, especially the Indian subcontinent, for extensive application in the paddy fields. The deteriorated soil health in India accelerated the translation of cyanobacterial biofertilizer technology from the laboratory to small-scale, large-scale, and commercial sectors. This chapter focuses on the mass multiplication, production cost analysis, and marketing of cyanobacterial biofertilizers.

Keywords Cyanobacteria · Biofertilizer · Large-scale production · Cost · Marketing

V. T. Anju · M. Dyavaiah
Department of Biochemistry and Molecular Biology, School of Life Sciences, Pondicherry University, Pondicherry, India

S. Busi (✉)
Department of Microbiology, School of Life Sciences, Pondicherry University, Pondicherry, India

N. Amaresan et al. (eds.), *Agricultural Microbiology Based Entrepreneurship*, Microorganisms for Sustainability 39,
https://doi.org/10.1007/978-981-19-5747-5_6

6.1 Introduction

Cyanobacteria are Gram-negative prokaryotic microbes involved in photosynthesis through the pigment, chlorophyll a, and many accessory pigments. They entrap sunlight and participate in photosynthesis. Cyanobacteria are generally found in terrestrial and aquatic environments such as streams, rivers, lakes, etc., where they are involved in the nitrogen, oxygen, and carbon cycle. They are also named as blue green algae, which comprises around 2000 species and 150 genera with broad range of shapes and cell structures (Vincent 2009). The occurrence of cyanobacteria is dependent on the season such as they are dominant in eutrophic lakes during summers and present in some extreme environments (e.g.: oil polluted areas and high saline regions) throughout the year. For instance, they are found to be the chief producers in cryoecosystems (Quesada and Vincent 2012).

In terms of their shape and structure, they have close resemblance to bacteria through their cellular structure. They lack defined nucleus and other membrane bound organelles. Their cell wall is made of peptidoglycan and bounded by mucilaginous sheath. Their cellular size is 1 mm for unicellular organisms whereas 30 mm for multicellular organisms that is generally larger than bacteria in size. Though, the four most frequent shapes observed in cyanobacteria are of sphere, rod, filamentous, and spiral shapes. Interestingly, cyanobacteria are able to change their morphology to adapt to developmental or environmental signals (Singh and Montgomery 2011). Some filamentous cyanobacteria exist with heterocyst or without them. Heterocysts are specific cells that are involved in the nitrogen fixation and substantiate their survival in nitrogen-deprived conditions. The major genera involved in nitrogen fixation are *Anabaena*, *Aphanizomenon*, and *Gloeotrichia* (Kumar et al. 2010).

The contamination of drinking water by toxin producers of cyanobacteria may cause acute illness like gastroenteritis. The major producers of cyanobacterial toxins are *Aphanizomenon flosaquae, Anabaena flosaquae*, *Nodularia* species, and *Microcystis aeruginosa*. They produce different toxins such as lipopolysaccharide endotoxins, neurotoxins, and hepatotoxins. Some toxins are produced by the prolific growth of cyanobacteria as algal blooms in eutrophic lakes and other water reservoirs. Algal blooms are toxic and harmful to the ecosystem including the animal and human health (Percival and Williams 2014). Other than the detrimental effects of cyanobacteria, they play a vital role in the agriculture, ecosystem, and environment sustainability. The specific features such as photosynthesis, the ability to produce high biomass, ability to grow on contaminated and polluted water bodies, production of bio-fuels, improved soil fertility properties by acting as biocontrol agents, etc., enable them to be excellent bio-resource agent in sustainable agriculture and environment (Singh et al. 2016).

Cyanobacterium is one of the best alternatives for the chemical fertilizers and other pesticides in agriculture and farming. They enhance the fertility of soil by enriching the nutrient requirement and improve the growth and yield of crops. The chemical features of soil are improved by cyanobacteria as it can release several phytohormones and involve in the nutrient transport to plants. This aids in the soil agglomeration for improved soil quality. The plant hormones produced by cyanobacteria are auxins, gibberellic acid, and cytokines and other bioactive compounds such as vitamins, amino acids, and peptides. Farmers apply cyanobacterial fertilizers as the best and cheapest nitrogen-fixing fertilizer in the fields. The commonly used cyanobacterial fertilizers are either unicellular or filamentous such as *Anabaena*, *Nostoc*, *Aulosira*, *Tolypothrix*, *Cylindrospermum*, and *Stigonema* (Rodríguez et al. 2006; Joshi et al. 2020). The functional roles of cyanobacteria in agriculture and environment differ and it needs to be cultivated in large quantities to meet the requirement. In agriculture, cyanobacteria are most commonly used for nitrogen fixation as biofertilizer, to improve soil fertility, reclamation of wasteland, as biocontrol agents, and to enhance crop production rate. Cyanobacteria are best utilized for the production of biofuels, bioremediation, and as food supplements for the sustainable environment. The different method of mass cultivation of biofertilizers include pit method, trough method, field method, and nursery cum algal cultivation method based on the requirement of cyanobacterial fertilizers (Joshi et al. 2020). The different methods of large-scale production and its marketing strategies are discussed further in the chapter.

6.2 What Are Cyanobacteria

On Earth, cyanobacteria are considered to be the first among the living organisms. The first fossils obtained for cyanobacteria are from 3.8 billion years ago. The significance of cyanobacteria in the construction of aerobic atmosphere owing to the emergence of numerous cyanobacterial species is well noticed. They are involved in oxygenic photosynthesis. The oxygen released during the photosynthesis and the emergence of organisms due to photosynthesis caused the emergence of aerobic species and oxygenic conditions on Earth (Kulasooriya 2012). They are Gram-negative prokaryotes, also known by different names such as cyanophytes, cyanoprokaryotes, and blue–green bacteria. The different names are a result of the pigment (c-phycocyanin) production by them that participate in photosynthesis. They exhibit prokaryote like cellular organization and still display complex systems of respiratory and photosynthetic mechanisms found in eukaryotes. Cyanobacterial morphology are found in unicellular, filamentous, and colony forms. Colonies of cyanobacteria are bounded by a type of gelatinous sheath based on circumstances. Nostoc is filamentous in morphology and can produce spherical colonies of 3–4 cm diameter. Three different types of cells are observed in some filamentous organisms with specific functions like vegetative, climate resistant akinetes, and heterocysts (Singh et al. 2011b).

These organisms carry a nucleus and organelles such as mitochondria, endoplasmic reticulum, Golgi apparatus, and chloroplasts devoid of membrane. The significant functions which carried out by the eukaryotic cell organelles are performed by the cyanobacterial cell membrane (Encyclopedia Britannica n.d.). Other than their plasma membrane, they have a thick layer of peptidoglycan. Peptidoglycan layer is found between the plasma membrane and an outer membrane as a sandwich layer. A gelatinous sheath of glycocalyx present outside the cell wall helps them to withstand desiccation and protect from predators and phages (Kumar et al. 2010). Cyanobacteria are found in broad range of habitats on Earth and they are genetically diverse organisms. They inhabit fresh water, marine, terrestrial, air, and extreme environments (deserts, freezing ecosystems, hot springs, and hypersaline environments). Sometimes, they exist as free living whereas some species are observed as the symbiosis of plants and animals. They can also live by interacting with other organisms through the formation of microbial mats, biofilms, and/or as benthic communities (Kulasooriya 2012; Thajuddin and Subramanian 2005).

6.2.1 Applications of Cyanobacteria

Cyanobacteria hold several functional applications in the field of health care sector, pharma companies, waste water treatment plants, and agriculture as fertilizers, etc. The availability of most potent bioactive compounds enables the application of cyanobacteria in different fields. Most of them are cultivated in large quantities to extract compounds of interest with biological properties such as anticancer and antitumor activities and as nano vectors, etc. (Qamar et al. 2021). The cyanobacterial secondary metabolites are vast in number with a multitude of biological properties. Some cyanobacteria interact with the surrounding organisms and produce signaling molecules. These signals help the partner organisms in competition (Niedermeyer 2015). There are some cyanobacterial species involved in the production of growth supplements (Plavšić et al. 2002). Apart from this, the biotechnological aspects describe that cyanobacteria are rich sources of inhibitory bioactive metabolites, nutritional supplements, pigments, dyes, chromophores, biofuels, food supplements, biofertilizers, etc. (Mazard et al. 2016). The novel application of cyanobacteria includes their ability to participate in the energy-efficient designs. Green buildings and landscape designs made of algae architecture for their ability to recycle CO_2 and waste water into high value products is one of most accepted design with high success rate (Llinares-Millán et al. 2014).

Cyanobacteria are yet to be explored for unfolding their rich bioactive compounds diversity through genetic engineering and synthetic biology approaches. Recent research and approaches focused on the several metabolic pathways leading to the production of biologically active metabolic products. These bacteria are feasible for large-scale production and to explore their products owing to their small and simple genome. Hence, the role and widespread applications of cyanobacteria remain crucial in the field of health care sector, energy, food, pharmaceutical, and biotechnology industries (Fig. 6.1).

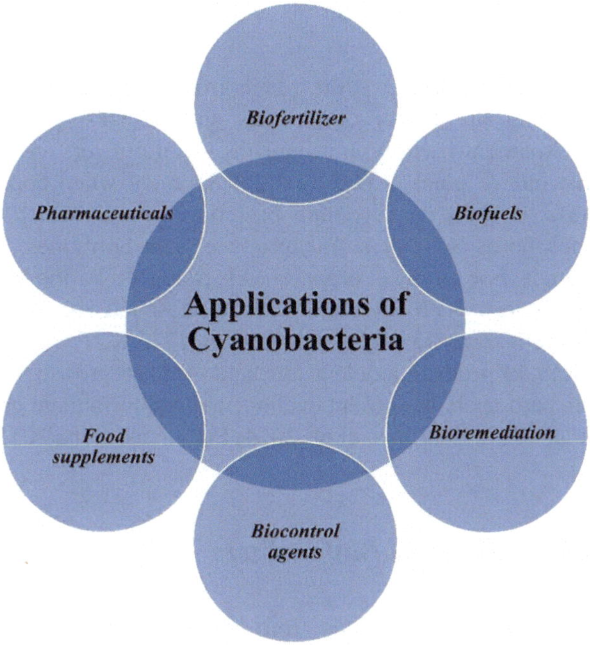

Fig. 6.1 Applications of cyanobacteria in different fields. Cyanobacteria are best utilized for the production of biofuels, biofertilizers, food supplements, pharmaceuticals, and biocontrol agents and effectively used in bioremediation

6.3 Cyanobacterial Biofertilizers

The report released by UN Food and Agriculture Organization (FAO) concluded that more food production (more than 70%) is required to meet the global demands by 2050. Several approaches and implements came into effect to meet the criteria. Thers is an emergence of global health threat owing to the continuous and simultaneous application of chemical fertilizer for enhanced crop yield rate. The indiscriminate use of chemical and toxic fertilizers creates pollution problems as well as cause adverse health effects on animals and humans who consume it (Pirttilä et al. 2021). Traces of chemical fertilizer reach aqueous bodies through rain or other way when there is a low or little uptake of them by plants, which cause detrimental effects on water organisms and other animals. The cost of chemical fertilizers is very high and it may create adverse effects on health of the ecosystem. The effects of fertilizers on soil health and quality are numerous. They interfere with the water holding capacity of soil, soil fertility, and cause disproportion of soil nutrients (Sabier Sae et al. 2015). Thus, the demand for alternative, effective, and environment-friendly fertilizers is the need of hour. Biofertilizers made of microorganisms, namely, bacteria and fungi, are capable of improving plant growth and health with enhanced production yield; thus they are best alternatives or replacements for toxic chemical fertilizers (Mahanty et al. 2017).

6.3.1 Biofertilizers

Biofertilizers improve the plant growth and production rate without negatively affecting the health of animals and humans. In addition, biofertilizers adopt environment-friendly mechanisms. Biofertilizers contain living microorganisms that induce plant growth and development when applied on soil, seeds, or plant surface. These may contain bacteria or fungi, which can fix nitrogen, solubilize phosphorus, oxidize sulfur, produce plant hormones, or decompose organic compounds. For instance, organic acids produced by the biofertilizer strain, *Pseudomonas fluorescens* K-34, are capable of phosphate solubilization. Other bacteria such as *P. fluorescens* K–34, *P. fluorescens* 1773/K, *P. trivalis* BIHB 745, and *Bacillus circulans* produce indole acetic acid for plant growth. Generally, biofertilizer strains can participate in nutrient cycling, and results in plant growth and crop yield (Verma et al. 2019; Bhardwaj et al. 2014; Parani and Saha 2012).

6.3.2 Types of Biofertilizers

There are several strains from bacteria, fungi, and algae, which are effectively employed as biofertilizer in agriculture farms and fields. Among them, the most commonly used and effective biofertilizers are plant growth promoting rhizobacteria (PGPRs), endo and ecto mycorrhizal fungi, and cyanobacteria. They are involved in the enhanced nutrient uptake, plant growth and development and, resistance to environmental stress (Singh et al. 2011a). These biofertilizers used in organic farming enrich the soil with micro and macro nutrients by fixing nitrogen, phosphate solubilization, and the production of plant growth hormones and factors. The mode of application of biofertilizer is either as seeds or as inoculants in soil that multiply and involve in biogeochemical cycle and increase crop production rate. In most of the cases, approximately, 60–90% of fertilizers applied is lost in different ways, only 10–40% is absorbed by the plants (Sinha et al. 2010; Adesemoye and Kloepper 2009).

Some of the examples of PGPRs found in soil are *Azospirillum* sp., *Azotobacter* sp., *Rhizobium* sp., phosphorus and potassium solubilizing microbes, cyanobacteria, and mycorrhizal fungi. For instance, strains of *Azospirillum*, *Azotobacter*, *Phosphobacter*, and *Rhizobacter* fix nitrogen for the growth of *Helianthus annus* and improve the plant height, seed dry weight, and the number of leaves. Likewise, the physiology and root morphology of rice plants improved when *Azotobacter*, *Azospirillum*, and *Rhizobium* strains were applied in the rice field (Dhandapani and Nadu 2012; Choudhury and Kennedy 2004).

The rhizosphere region of plants includes the soil next to the plant roots that contain high beneficial microbial diversity and nutrient content. The factors that control the rhizosphere region and microbial load are biotic and abiotic factors such as nutrients, overall root morphology, and other biotic populations and plants. The most commonly found rhizosphere microbes are PGPRs and mycorrhizal fungi (Hu et al. 2018; Enagbonma and Babalola 2019). PGPRs live in rhizosphere region and exhibits several functions enhancing the plant growth such as nitrogen fixation, phosphate mineralization, and production of siderophores, phytohormones, volatile organic compounds, and Aminocyclopropane-1-carboxylate deaminase. They also interfere with the formation of biofilms, quorum-sensing pathways, and toxin production by pathogens and favor beneficial symbiosis with other microorganisms (Bhattacharyya and Jha 2012). There are two types of PGPRs, namely, intracellular and extracellular depending on their site of attachment to the plant rhizosphere region. Examples of PGPR genera include *Azotobacter*, *Arthrobacter*, *Bacillus*, *Chromobacterium*, *Pseudomonas*, *Agrobacterium*, *Azospirillum*, *Serratia*, *Frankia*, the endophytes, *Bradyrhizobium*, *Allorhizobium*, *Bradyrhizobium*, and *Mesorhizobium* (Verma et al. 2010) (Bhattacharyya and Jha 2012). The marketed PGPRs are *Pseudomonas*, *Enterobacter*, *Bacillus*, *Azotobacter*, *Klebsiella*, *Azosprillum*, *Variovorax*, and *Serratia*. However, irrespective of the benefits of PGPRs to the crops, their application in agriculture and farming is very limited. The varying properties of inoculated PGPRs in the field affected the crop production and yield, which reduced the use of PGPRs in global agriculture practice (Vejan et al. 2016).

Among the fungi, arbuscular mycorrhizal (AM) fungi are called as natural fertilizers due to their connection between plants and their soil nutrients. These symbionts interact with most of terrestrial plants and several crops and participate in nutrient and water exchange and provide pathogen protection (Berruti et al. 2016). Particularly, the symbiotic relationship of AM fungi with plants available in phosphorus deficient soil enhances the plant growth by increasing the phosphorus content in soil (Igiehon and Babalola 2017). AM fungi are ubiquitous and their abundance varies among different ecosystems that are natural as well as manmade. AM fungi functions along with PGPRs for better and effective biofertilizer properties. Thus, their role in crop field management, along with other biofertilizers, is well established and reported. AM fungi are highly recommended for agriculture practice as they are ecosystem friendly and pollution free (Sadhana 2014).

Cyanobacteria or blue green algae (BGA) are the prokaryotes found to be effective biofertilizer for many crops particularly for rice plants. They are involved in the biological nitrogen fixation and help to fix nitrogen in N deficient soil for enhanced productivity. One of the advantages of using BGA fertilizer in paddy field is its low cost technology (Pimratch et al. 2015). The major roles of cyanobacteria, which make them the best biofertilizers are: (1) they can make soil more porous and secrete several adhesive particles, (2) they produce plant hormones such as gibberellins, auxins, amino acids, and vitamins, (3) they enhance the water holding property of soil as it hold jelly like properties, (4) they eliminate the growth of weeds, and (5) they possess phosphate solubilization and bioremediation properties (Roger and Reynaud 1982; Rodríguez et al. 2006; Saadatnia and Riahi 2009; Ibraheem 2007).

BGA fertilizer in wetland rice cultivation system offers a potential alternative to the traditional chemical fertilizers (Mishra and Pabbi 2004). Along with the nitrogen fixation ability, some of the secondary metabolites produced by BGA rescue rice plants from major diseases caused by *Rhizoctonia solani* and *Xanthomonas oryzae*. Thus, the BGA exhibited a major increase in the total rice production throughout the globe owing to their dual roles as a biofertilizer and biopesticides (Bao et al. 2021). The most commonly used BGA fertilizer such as *Anabaena* and *Nostoc* species found on rocks and soil particles can fix N by 20–25 kg/ha. Individually, *Anabaena* are reported to fix N by 60 kg/ha in a season and augments soil with organic matter for the growth of crops (Moore 1969). *Anabaena* requires a host, *Azolla* for symbiotic nitrogen fixation. The *Azolla-Anabaena* combination has been widely applied for rice, oats, tomato, barley, cotton, chilly, radish, lettuce, and sugarcane (Malliga et al. 1996; Thajuddin and Subramanian 2005).

6.4 Mass Production of Cyanobacterial Fertilizers

BGA plays a vital role in the production of rice and other crops, which meet the global requirement of food production. Global and national initiatives promoted the mass production and application of BGA fertilizer. BGA biofertilizer is found to be an effective alternative for chemical fertilizer as it is easily available, as a cheap resource for cultivation and production, has ease in application, and non-toxic and economically feasible (Mazid and Khan 2014). Thus, the demand for biofertilizer production has increased in the past few years. The major abiotic parameters that determine the successful cultivation of BGA in large scale are light, temperature, pH, nutrients (C, P, N, S, Fe, and K), water and CO_2. Some of the large-scale cultivated BGA and microalgae by economically feasible methods are *Chlorella*, *Arthrospira*, *Dunaliella*, and *Haematococcus* (Flynn et al. 2010; Rosenberg et al. 2008).

There are several methods that enable the large-scale production of cyanobacterial biofertilizers. The first step is focused on the isolation of potent cyanobacteria capable of being used as biofertilizer followed by mass or large-scale production. Initial production of BGA fertilizer is performed in Fogg's medium. Top soil (5 g) obtained from rice field is inoculated into Fogg's medium (100 mL) and incubated at room temperature after proper mixing. Growth of algal culture is improved by providing an illumination of 1500 lux for 16 h followed by dark incubation for 1 week (8 h). The culture is added to 10 mL of water upon finding a confluent algal growth. Algal filaments are obtained as separate strands by mixing the culture thoroughly and serially diluted to grow them in Fogg's medium in larger flasks. The process is repeated until pure culture is obtained and pure cultures of BGA in Fogg's media is preserved as agar slant starter culture for mass production (Atnoorkar 2021).

6.4.1 Multiplication and Production Methods

After the initial mother culture preparation, inoculants can be stored at 15 °C for short time period whereas for the storage of more than 6 months it needs to be stored at 33 °C (Chittora et al. 2020). Large-scale production of BGA fertilizer in laboratories is performed in photobioreactors. The second step involves the mixing of culture with appropriate sterilized carrier materials such as charcoal, perlite, peat, etc. Later on, the mixture containing carrier along with broth culture is packed into thin polythene bags. The carrier-based fertilizer should contain 108 live cells/g of carrier material. The storage conditions are 25–30 °C and pH of 6–7.5 (Chittora et al. 2020).

There are several methods of large-scale production such as in trays or trough, pits, field, and nursery cum algal production. The methods of cultivation in trough or tray and pit are for individual farmers (Fig. 6.2). The field and nursery cum algal methods are majorly used for commercial scale mass production (Jhala et al. 2017). The most commonly employed BGA in biofertilizer cultivation are *Anabaena* sp., *Aulosira fertilissima*, *Nostoc muscorum*, and *Tolypothrix tenuis* (Atnoorkar 2021).

Pereira et al. (2009) developed biofertilizer containing filamentous nitrogen-fixing cyanobacteria to apply for rice crops. The mass cultivation of algae was initiated by homogenizing the sample with the help of a porcelain Potter. Watanabe agar media was inoculated using homogenized culture and incubated at 25 °C under the supply of constant light. To produce large quantities, the broth was inoculated from the plates and incubated at similar culture conditions. The above process was able to produce 0.425 g/L in a 1 month time period (Pereira et al. 2009). Other than the above methods, the parameters such as light and surface area for production can be optimized depending on specific methods to enhance the algal production. These

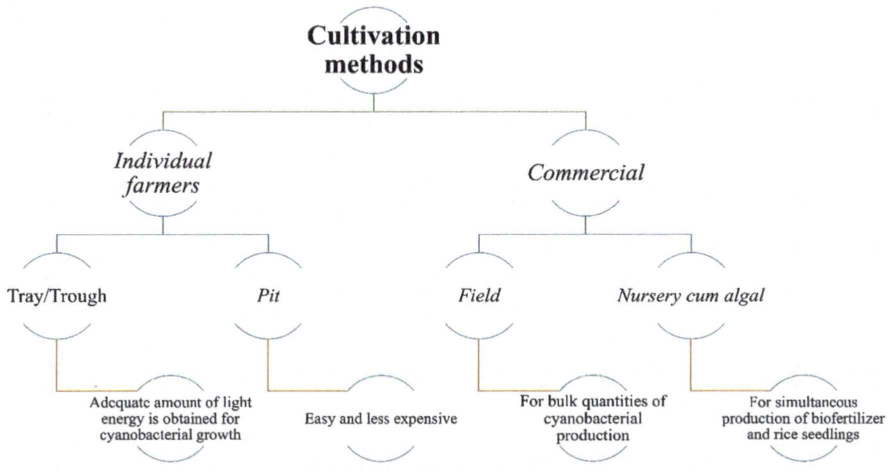

Fig. 6.2 Different mass cultivation methods of cyanobacteria used by farmers and industries

include cultivation in open and closed systems using natural (sunlight) and artificial light.

In trough or tray method of production, shallow permanent tanks of size 2 m × 1 m × 23 cm are used. The size of the tray can be optimized according to the quantity of biomass required. Here, approximately 4–5 kg of river soil is mixed with phosphate (100 g) and molybdate (2 g) and water is added to the tray of 5–15 cm depth. The ingredients are mixed thoroughly and allowed to settle for the next 8–10 h. Afterward, the surface of tray water is inoculated with the 250 g of starter mother culture of BGA. The pH of tray soil should be kept as neutral until algal growth appears. BGA will be grown and observed as hard flakes in a time period of 10–15 days. The algal flaks have to be dried and separated from soil. Usually, around half a ton of algal growth is obtained from tray/pit till 1 m^2. Further, the tray soil is preserved and algae will be allowed to grow again. In this way, dried fertilizer can be collected 2–3 times without adding starter culture (Kuraganti et al. 2020).

Pit method of cultivation is different from tray method, where tank pits are made in ground. Then, the pits are arranged by different layers of thick polyethene sheets to absorb water. Otherwise, the pits can be plastered with the help of cement. Pits are cheap and simple methods suitable for farmers than tray methods (Baweja et al. 2019). In the field method, an economical scale production of BGA through a scaled-up trough method is followed. First, the area of size 40 m^2 for algal cultivation is maintained. The soil in field after the crop harvest is prepared by maintaining the water content with a depth of 2.5 cm followed by addition of enough phosphate (12 kg). Then, carbofuran is applied to manage insect and pest attack in the field. Algal culture (5 kg) is added to the land only if previous exposure to algal culture is not done. The algal growth occurs according to the type and nutrient content of soil. In loamy soil, it grows within 4 days whereas in clay soil it happens in 14 days. Once a confluent growth takes place, floating algal mats are sun dried and stored in bags. The average yield of algal growth is estimated as 16–30 kg during summer months. The field can be reflooded for harvesting algal mats continuously (Kuraganti et al. 2020).

Nursery cum algal production method involves the cultivation of algae along with their nursery plants. Here, farmers can allot an extra land of 40 m^2 for algal cultivation along with their 320 m^2 land of nursery plants. This will reduce the dependency of farmers on chemical fertilizers for their nursery plants too. Algalization is the process through which soil is inoculated with a definite mixture of BGA species (Mishra and Pabbi 2004). The soil containing starter culture can be best utilized by farmers for the production of biofertilizers with less little inputs. This can also be an added advantage if the inoculation is performed 3–4 times continuously. The repeated algalization will help to establish the algal growth in the land permanently. The nursery cum algal production method involves the mixing of 500 g fertilizer with dried farm soil (4 kg). The algae should be multiplied in the presence of standing water for next 3–6 days. The nursery land should be water logged for around 10–12 days to ensure the maximum algal growth. In addition, supplementing small quantities of phosphate fertilizer to the BGA fertilizer inoculated in land enables the faster and effective multiplication of algae (Mishra and Pabbi 2004; Ugwu et al. 2008).

6.4.2 Open System Cultivation Using Sunlight

In this, large raceways or open circular and shallow ponds are used for the mass cultivation where solar light is the source of energy. The major advantage of using this system is the availability of cost-free source of energy. Another factor is that the system allows only the growth of specific organisms, which demands particular growth parameters and factors (Ugwu et al. 2008; Cañedo and Lizárraga 2016). Open systems support only few of the BGM species such as *Spirulina* as the method adopts specific parameters such as pH or high salinity (Iwaki et al. 2006; Zahra et al. 2020) (Fig. 6.3).

6.4.3 Closed System Cultivation Using Sunlight

Closed transparent systems made of either plastic or glass materials are used for this cultivation. The vessels are placed outside to get natural solar light source for efficient illumination (Khatoon and Pal 2015). The contamination of algal growth by competitors and predators are eliminated owing to the transparent vessels. Also other advantage of using closed systems over open systems is the high surface to volume ratio, which supports more algal growth for more biofertilizer production (Ugwu et al. 2008; De Morais and Costa 2007). For successful large-scale

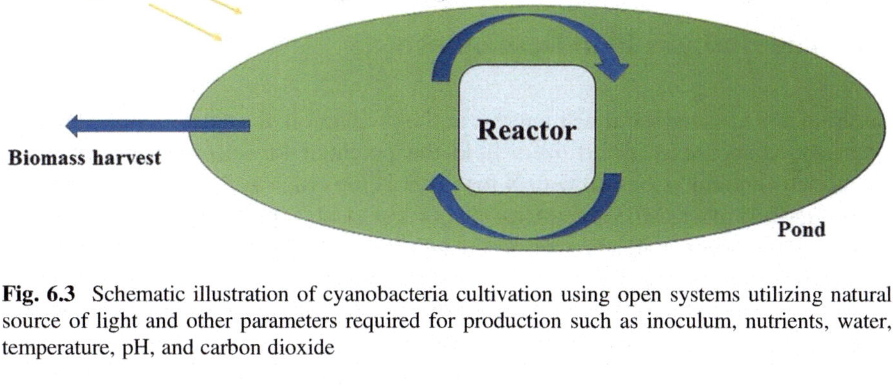

Fig. 6.3 Schematic illustration of cyanobacteria cultivation using open systems utilizing natural source of light and other parameters required for production such as inoculum, nutrients, water, temperature, pH, and carbon dioxide

production, optimum temperature and oxygen concentration should be checked thoroughly in order to prevent the failure of culturing (Khatoon and Pal 2015).

6.4.4 Closed System Cultivation Using Artificial Light

The indoor cultivation systems of this type utilize artificial light as a source of energy. The vessels are more or less similar to a traditional fermenter without using an organic carbon source. These vessels are called photobioreactors that carry out the processes and optimize the culture and growth parameters in real time with the help of software. The use of plastic or glass material for producing the vessels makes the whole system expensive compared to other systems although, this type of closed system yields high quality and quantity biomass (Pathak et al. 2018; Apt and Behrens 1999).

6.5 Method of Application of Fertilizer on Field

For field application in 1 acre of land, around 500 g of fertilizer is thoroughly mixed with 4 kg of dried and sieved soil. The fertilizer is applied to rice plants after transplantation to the soil. The reproduction and algal establishment will accelerate once large quantities of biofertilizer are provided. One should be cautious to maintain water logged condition in the land for 10–12 days after adding the fertilizer for best algal growth. Cyanobacterial fertilizers can be mixed and applied along with one third dose of effective nitrogen fertilizers for improved activity. In addition, the mechanism of cyanobacteria does not interfere with normal pest control management program. The biofertilizer should be applied on farm during four successive seasons. The cyanobacteria can regrow and establish by themselves on farm with full potential with favorable conditions, limiting the continuous application on land (Jhala et al. 2017).

6.6 The Business Plan Development

Biofertilizers has attained much interest in the production of organic foods globally. Specially, cyanobacterial fertilizers hold the potential for sustainable agriculture practices by enriching the soil with N for several crops such as rice, soybean, tomato, cotton, radish, maize, and sugarcane (Kuraganti et al. 2020). The large-scale production and commercialization of products have been always on focus to improve the global production and its usage for field applications. The major objectives of current entrepreneurship development include promoting commercially valuable products and their technologies, encouraging public partnerships for strong collaborations, and human resource development for business promotion. The steps involved in the commercialization include selecting promising technologies, awareness technologies, checking the feasibility of business plan, identifying strong entrepreneurs, evaluating entrepreneurs, fixing business incubation support, and educating entrepreneur for successful marketing (Rao 2011). In India, the Department of Science and Technology introduced cyanobacterial fertilizer technology among farmers to increase yield. In the 1960s, the Indian Agricultural Research Institute (IARI), New Delhi developed cheap and simple algal biofertilizer technology for rice plants especially for marginal farmers (Venkataraman 1972). Algalization technology containing soil-based cyanobacterial fertilizer was adopted by two Indian states: Tamil Nadu and Uttar Pradesh. The technology was simple and economically feasible and found to be useful to farmers. Later on, several projects were started and found to be effective for farmers to improve the productivity rate in an economically feasible manner. In order to popularize the cyanobacterial technology, several steps were taken by the authorities. Those were popularization and dissemination of technology, demonstration and creation of awareness, evaluation of farmers' feedback, and dissemination of technology through public–private partnerships (PPP) model.

Intellectual Property Management and Technology Transfer/ Commercialization guidelines were proposed to foster commercialization. The Indian Council of Agricultural Research established a three-step mechanism (Agro-Technology Management Center, Zonal Technology Management Centers and Institute Technology Management Units) to support intellectual property-based commercialization. Along with this, the Business Planning and Development unit of IARI developed effective and innovative management and fast-tracking transfer of technology, proprietary, and new technologies. This supported agripreneurs and farmers to boost their business. Also, the new approach created stronger industrial conditions that provided quality products at reasonable price to customers (Bhooshan et al. 2020).

The reports from Indian sources show the journey of cyanobacterial biofertilizer technology from rural to commercial enterprise in 2010–2011 and 2016–2017. Five Indian industries received license for the technology and three among them launched their product in market. The statistics stated that around 49,000 tons of biofertilizer was produced by them and distributed to several fields (27,000 ha) in Indian states (Panjab, Bihar, Andhra Pradesh, Uttarakhand, and Uttar Pradesh). The PPP model by the government was found to be very significant for the success of cyanobacterial technology (Bhooshan et al. 2020).

6.7 Production Cost and Market Value of Cyanobacterial Fertilizer

The production cost of BGA varies among different types of multiplication methods. The production cost also varies with the small-scale and large-scale production methods with the difference in the type of media used and equipment employed for production. For instance, the cost is cheap in the case of open system ponds. Though, the cost is cheap, it supports only few species of algae and often results in culture failure owing to the sudden climate changes and external contaminations. In the case of indoor systems, the cost is expensive but provides good algal growth of significant species by proper culture conditions. The cost of production for algae is estimated to be US $4 to 300/dry weight of algae. In order to reduce the production cost below US $100, large quantities of BGA should be produced through these methods (Lavens and Sorgeloos 1996). In a comparative analysis, production cost for traditional nitrogen fertilizers is very high whereas it is cheap for organic cyanobacterial nitrogen-fixing fertilizers. Cyanobacterial nitrogen fertilizer can fix nitrogen 10–30 kg/ha of N yearly, which clearly indicates its significance in sustainable agriculture (Hegde et al. 1999; Chittora et al. 2020). Table 6.1 provides information on different cyanobacteria and its beneficial role in crop yield and growth. In a survey conducted from 2015 to 2016, the implementation of BGA technology among paddy field reduced the use of chemical fertilizers and cost of production and increased the yield and income. The cost of production was reduced from 9717.0 to 9531.0 INR/acre. Also, an increased income of 3.9% was observed by farmers (Bhooshan et al. 2018).

The value of biofertilizers in agriculture market is continuously broadening along with the increasing economy and populations. In 2012, the global market value of all biofertilizers was found to be $440 million and continues to grow more than 10% per annum (Owen et al. 2015). There are several stakeholders (production industries and governmental regulatory networks) and companies who ensure the safe multiplication, production, and distribution of biofertilizers to meet the demands of growing populations. Yet, African and some Asian countries lack the access to safe and effective agricultural practices. Thus, biofertilizers are recommended for the above issues to enable improved productivity in agriculture and farming (Nosheen et al. 2021). The global BGA fertilizer market trend during the time period 2021–2026 is predicted. BGA market highlights are segmented based on the end user, crop type, and growth drivers. Still, during the forecast period, we may observe lack of awareness among farmers, which restrict the complete adoption of BGA in their fields (Intelligence M 2019). Figure 6.4 illustrates the global BGA fertilizer market trend during the forecast period 2021–26.

As the demand for more food production increased, government authorities focused on financial as well as marketing strategies to promote more biofertilizer production. The Indian government released around 20 lakhs to support the rising need of emerging population. This is to support the biofertilizer production through laboratory aid and to enhance the production by 150 t/year. The grant was intended

Table 6.1 Role of different cyanobacterial inoculum on the growth of plants and nitrogen fixation with benefits

Organisms	Nitrogen fixation	Benefits	Crop	References
Paddy Cyanobacteria	20–30 kg/ha	Improved crop yield of 10–15%	Rice	Mazid and Khan (2014)
Free-living cyanobacteria	<10 kg of N $ha^{-1}y^{-1}$	High yielding varieties	Rice	Pathak et al. (2018)
Dense mats of cyanobacteria	~10–30 kg of N ha^{-1}	High yielding varieties	Rice	Pathak et al. (2018)
Cyanobacterial mats on soil surface	12–16 kg/ha	More nitrogen fixation	Rice	Kaushik (2014)
Cyanobacterial species found on paddy field	13.8 to 44.4 kg/ha	Improved crop production	Rice	Kaushik (2014)
Tolypothrix tenuis	20 kg/ha	Assimilating atmospheric nitrogen	Rice	Kaushik (2014)
Nostoc	15–51 kg N/ ha/ year	—	Natural and agricultural habitats	Kaushik (2014)
Anabaena oryzae, *Nostoc ellipsosporum* and *Synechococcus* sp.	—	Shoot length increased by 120–242%	*Sorghum durra* and *helianthus annuus*	Essa et al. (2015)
Commercial inoculant of cyanobacteria	Improved availability of nitrogen	Improved the dry weight, total nitrogen, and pigment content	Wheat	Abd-Alla et al. (1994)
Filamentous nitrogen-fixing cyanobacteria	50 kg/ha	Increased nitrogen use, grain in quality and yield	Rice	Pereira et al. (2009)
Calothrix elenkinii	More N fixation	Elevated plant growth, nitrogenase activity, production of auxins and defense enzymes	Rice	Priya et al. (2015)

for use of the modern equipment in the production of fermenter or photobioreactor, laminar hood, and automated sand filling and bagging machine.

In India, the production of biofertilizers increased greatly from 2004 onward. The maximum production of biofertilizers was found to be by Agro Industries Corporation, followed by state agriculture sectors, national biofertilizer development area, state agricultural university, and, lastly, by different private companies. The production of BGA improved during the period 2006–2007. Blue green algal technology has been recommended by the Indian government for more rice yield. According to the Government of Tamil Nādu, which is one of the states with the highest amount of production, 267.72 thousand tons of BGA and 20.38 thousand tons of *Azolla* are

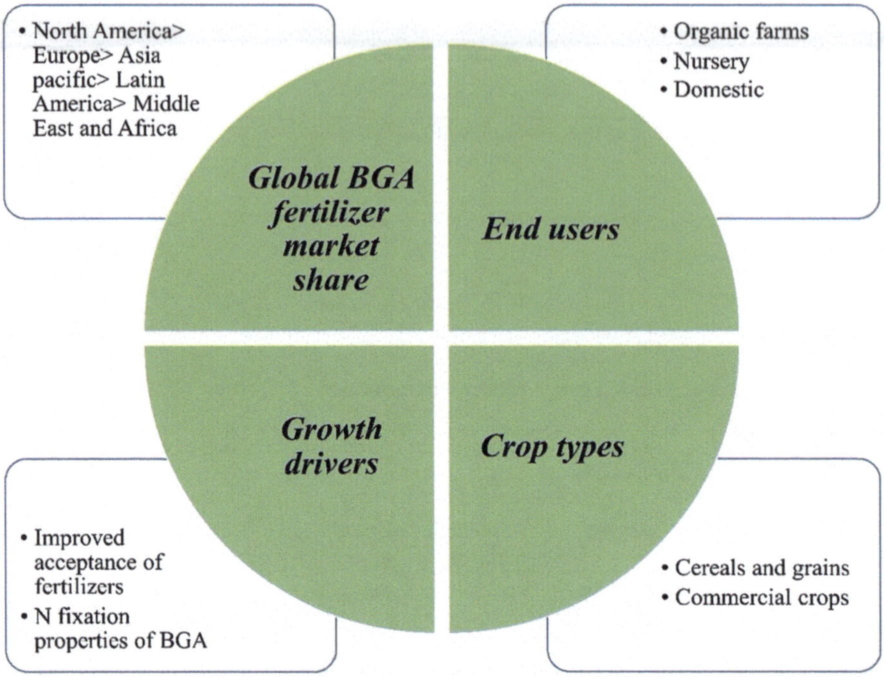

Fig. 6.4 The highlights of global BGA fertilizer market during the forecast period 2021–2026, describing the market share, end-users list, crop types, and its growth driving factors

required. As the production and installation costs of biofertilizer over chemical fertilizer are simple and cheap, the companies that have well-established marketing networks will go through this. The biofertilizer units that lacks well-organized establishments in both urban and rural areas may not be able to produce biofertilizers in an economically feasible manner (Singh et al. 2014). The outline of production describes the major production requirements such as factors affecting the production, medium used for effective and maximum production, and equipment. The equipment used in both small and large scale is very important as it represents the major portion of the infrastructure. From the Indian perspective, the overall production cost and profit for small-scale and large-scale production of BGA fertilizer indicated the profitability of using mass cultivation methods rather than the laboratory methods (Zhu et al. 2018; Venkataraman 1981). The cost difference between in the two systems can be observed in Table 6.2. Table 6.2 outlines the various parameters of small-scale and large-scale BGA production.

The value of global biofertilizer market is calculated to be US $2.6 billion in the year 2021 whereas it is predicted to hike up to US $4.5 by 2026. In addition, in the Asia-pacific region, the production and use of carrier based biofertilizer is expected to elevate from 2021 to 2026. During the forecast period, it is expected to enrich the soil properties and soil microbiome by biofertilizers such as algae. After the Asian

Table 6.2 Outline of the cyanobacterial biofertilizer production from an Indian perspective with the information on requirements such as abiotic factors, media, laboratory equipment, production cost, expenditure, and net gain or profit

Requirements	Small-scale production	Large-scale production
Abiotic factors	Light, pH, temperature, carbon dioxide, and nutrient supplements (C, N, P, S, K, Fe)	Light, pH, temperature, carbon dioxide, and nutrient supplements (C, N, P, S, K, Fe)
Media	Agar and liquid medium	Liquid medium
Raw materials		Blue-green algal culture, superphosphate, sawdust, carbofuran
Laboratory equipment	Conical flasks, rotatory shakers, autoclave, laminar air flow chamber, hot air oven, pH meter, refrigerator, and incubators	Autoclave, laminar air flow chamber, hot air oven, pH meter, fermenters, automatic filling machine
An Indian perspective (Zhu et al. 2018; Venkataraman 1981)		
Cost of production	70.00 lakh for 150 metric tonnes/annum	US $0.022/kg
Infrastructure	12 lakh for building, 2.50 lakh for labor, and 41 lakh for equipment and apparatus	US $8.98 for land preparation and labor
Recurring and non-recurring expenditure	41.00 lakh without marketing expenditure	US $60 for 2 tonnes per 2000 m^2 in 25–30 days
Net gain or profit	14.870 lakh with 100% production	US $251 profit to the farmers with a sale price of produce US $311

market, American global biofertilizer market is expected to nurture during the above time period (Research and Markets 2021). Among all, the fast-growing biofertilizer market is from North America in the year 2020 and largest market is in Europe. The competitive landscape in global biofertilizer market includes several small and large companies such as Kiwa Bio-Tech Products Group Corporation, Lallemand Inc., Camson Biotechnologies Limited, Agrinos AS and Novozymes A/S. They focus mainly on the biofertilizer production launches, different partnerships and international collaborations for expanding to reach throughout the world (Intelligence M 2019).

6.8 Challenges and Future Perspectives

The major constraints in the production and marketing of biofertilizers include technical, marketing biological, and field level and financial limitations. These include the competition of biofertilizer strains with the natural microbial flora of soil, deprived soil features, occurrence of pollutants, absence of proper bioinoculants, lack of appropriate carries, inaccessibility of skilled staffs at units, lack of funds, transportation failure, lack of proper attention from farmers, lack of suitable, commercialization parameters such as inaccessibility of proper inoculum at appropriate time and place, absence of established marketing networks, and proper regulations and standards regarding marketing (Nosheen et al. 2021). Latest research is focusing on technical and marketing improvements for the expansion of cyanobacterial fertilizer production and other related products. As consumer demand has been increasing from the past few years, establishment of marketing networks in several areas is required. Irrespective of the marketing strategy, the potential of cyanobacteria in biotechnology is not well explored. This requires elaborate and well-organized research facilities for the isolation and large-scale production of functional products from cyanobacteria.

6.9 Conclusions

The continuous application of chemical or traditional fertilizer to meet the emerging demands of human population and to increase the food supply has negatively affected the health of humans and environment. The cost associated with the fertilizer production was also a concern among the farmers and industries. Thus, the need for alternative and ecosystem friendly as well as economically feasible fertilizers arose among farmers. Thus, biofertilizers containing natural microorganism with plant growth-promoting properties were introduced by researchers and government authorities. Cyanobacteria or blue green algae lead the list of natural fertilizers with increased plant growth and productivity rate. Recent studies indicate that BGA fertilizer use and production has been increased to meet the growing population demand. They are known to be one of the best and alternative nitrogen-fixing fertilizers over traditional fertilizers. They work by utilizing carbon dioxide, water, natural or artificial light, and nutrients to convert or fix atmospheric nitrogen. BGA biofertilizers are cost effective and eco-friendly with several other advantages such as their ability to improve soil properties, control soil pathogens, synthesize plant growth-promoting substances, and mineralize phosphates. To date, they are one of the efficient fertilizers for sustainable agriculture and environment. Though, farmers were unaware of the vast potential of cyanobacteria in agriculture productivity, recently their use among several crops such as rice provided improved yield. The production of BGA fertilizers both in small and large scale initiates with the production of starter or mother culture, followed by the large-scale production and mixing with suitable carriers for field application. The production can be carried out in trays, pits, fields, and nursery cum algal production methods. Mass production can also be performed in open and closed systems in the presence of sunlight/artificial light. As BGA biofertilizers attained much interest recently among farmers, the marketing strategies and effective production methods should be improved. Several governmental and private authorities focused on developing marketing networks in rural as well as urban areas to promote their application in field.

References

Abd-Alla MH, Mahmoud ALE, Issa AA (1994) Cyanobacterial biofertilizer improved growth of wheat. Phyton (B Aires) 34(1):11–18

Adesemoye AO, Kloepper JW (2009) Plant–microbes interactions in enhanced fertilizer-use efficiency. Appl Microbiol Biotechnol 85:1–12. https://doi.org/10.1007/s00253-009-2196-0

Apt KE, Behrens PW (1999) Commercial developments in microalgal biotechnology. J Phycol 35: 215–226. https://doi.org/10.1046/j.1529-8817.1999.3520215.x

Atnoorkar AA (2021) Cyanobacterial biofertilizers as an alternative to chemical fertilizers in paddy fields: a review. Appl Biol Chem J 2:49–52. https://doi.org/10.52679/tabcj.2021.0008

Bao J, Zhuo C, Zhang D, Li Y, Hu F, Li H et al (2021) Potential applicability of a cyanobacterium as a biofertilizer and biopesticide in rice fields. Plant Soil 463:97–112. https://doi.org/10.1007/s11104-021-04899-9

Baweja P, Kumar S, Kumar G (2019) Organic fertilizer from algae: a novel approach towards sustainable agriculture, pp 353–370. https://doi.org/10.1007/978-3-030-18933-4_16

Berruti A, Lumini E, Balestrini R, Bianciotto V (2016) Arbuscular mycorrhizal fungi as natural biofertilizers: let's benefit from past successes. Front Microbiol 6:1559. https://doi.org/10.3389/fmicb.2015.01559

Bhardwaj D, Ansari MW, Sahoo RK, Tuteja N (2014) Biofertilizers function as key player in sustainable agriculture by improving soil fertility, plant tolerance and crop productivity. Microb Cell Factories 13:66. https://doi.org/10.1186/1475-2859-13-66

Bhattacharyya PN, Jha DK (2012) Plant growth-promoting rhizobacteria (PGPR): emergence in agriculture. World J Microbiol Biotechnol 28:1327–1350. https://doi.org/10.1007/s11274-011-0979-9

Bhooshan N, Pabbi S, Singh A, Sharma A, Chetan JA et al (2018) Impact of blue green algae (BGA) technology: an empirical evidence from northwestern Indo-Gangetic Plains. 3 Biotech 8(324):324. https://doi.org/10.1007/s13205-018-1345-5

Bhooshan N, Singh A, Sharma A, Verma C, Kumar A, Pabbi S (2020) Cyanobacterial biofertilizer's successful journey from rural technology to commercial enterprise: an Indian perspective. J Appl Phycol 32:3995–4002. https://doi.org/10.1007/s10811-020-02237-7

Britannica (n.d.) Encyclopaedia Britannica blue-green algae, pp 1–5

Cañedo JCG, Lizárraga GLL (2016) Considerations for photobioreactor design and operation for mass cultivation of microalgae. In: Algae—organisms for imminent biotechnology, vol 32. IntechOpen, pp 1854–1858. https://doi.org/10.5772/63069

Chittora D, Meena M, Barupal T, Swapnil P (2020) Cyanobacteria as a source of biofertilizers for sustainable agriculture. Biochem Biophys Reports 22:100737. https://doi.org/10.1016/j.bbrep.2020.100737

Choudhury ATMA, Kennedy IR (2004) Prospects and potentials for systems of biological nitrogen fixation in sustainable rice production. Biol Fertil Soils 39:219–227. https://doi.org/10.1007/s00374-003-0706-2

De Morais MG, Costa JAV (2007) Carbon dioxide fixation by *Chlorella kessleri*, *C. vulgaris*, *Scenedesmus obliquus* and *spirulina* sp. cultivated in flasks and vertical tubular photobioreactors. Biotechnol Lett 29:1349–1352. https://doi.org/10.1007/s10529-007-9394-6

Dhandapani R, Nadu T (2012) Effect of biofertilizers on th growth of Helianthus annuus. Int J Plant Anim Environ Sci 2:143–147

Enagbonma BJ, Babalola OO (2019) Environmental sustainability: a review of termite mound soil material and its bacteria. Sustainability 11(14):3847. https://doi.org/10.3390/su11143847

Essa AMM, Ibrahim WM, Mahmud RM, Elkassim NA (2015) Potential impact of cyanobacterial exudates on seed germination and antioxidant enzymes of crop plant seedlings. Int J Curr Microbiol App Sci 4:1010–1024

Flynn KJ, Greenwell HC, Lovitt RW, Shields RJ (2010) Selection for fitness at the individual or population levels: modelling effects of genetic modifications in microalgae on productivity and environmental safety. J Theor Biol 263(3):269–280. https://doi.org/10.1016/j.jtbi.2009.12.021

Hegde DM, Dwivedi BS, Sudhakara Babu SN (1999) Biofertilizers for cereal production in India—a review. Indian J Agric Sci 69:73–83

Hu L, Robert CAM, Cadot S, Zhang X, Ye M, Li B et al (2018) Root exudate metabolites drive plant-soil feedbacks on growth and defense by shaping the rhizosphere microbiota. Nat Commun 9:2738. https://doi.org/10.1038/s41467-018-05122-7

Ibraheem I (2007) Cyanobacteria as alternative biological conditioners for bioremediation of barren soil. Egypt J Phycol 8:99–117. https://doi.org/10.21608/egyjs.2007.114548

Igiehon NO, Babalola OO (2017) Biofertilizers and sustainable agriculture: exploring arbuscular mycorrhizal fungi. Appl Microbiol Biotechnol 101:4871–4881. https://doi.org/10.1007/s00253-017-8344-z

Intelligence M (2019) Global biofertilizers market—growth, trends and forecast (2020–2025). Ind Reports—Recent Reports 1–7. https://www.mordorintelligence.com/industry-reports/global-biofertilizers-market-industry

Iwaki T, Haranoh K, Inoue N, Kojima K, Satoh R, Nishino T et al (2006) Expression of foreign type I ribulose-1,5-bisphosphate carboxylase/ oxygenase (EC 4.1.1.39) stimulates photosynthesis in cyanobacterium *Synechococcus* PCC7942 cells. Photosynth Res 88(287–97):287. https://doi.org/10.1007/s11120-006-9048-x

Jhala YK, Panpatte DG, Vyas RV (2017) Cyanobacteria: source of organic fertilizers for plant growth. In: Panpatte D, Jhala Y, Vyas R, Shelat H (eds) Microorganisms for green revolution. Microorganisms for sustainability, vol 6. Springer, Singapore. https://doi.org/10.1007/978-981-10-6241-4_13

Joshi H, Shourie A, Singh A (2020) Cyanobacteria as a source of biofertilizers for sustainable agriculture. In: Advances in cyanobacterial biology. Elsevier, pp 385–396. https://doi.org/10.1016/B978-0-12-819311-2.00025-5

Kaushik BD (2014) Developments in cyanobacterial biofertilizer. Proc Indian Natl Sci Acad 80: 379. https://doi.org/10.16943/ptinsa/2014/v80i2/55115

Khatoon N, Pal R (2015) Microalgae in biotechnological application: a commercial approach. In: Bahadur B, Venkat RM, Sahijram L, Krishnamurthy K (eds) Plant biology and biotechnology. Springer, New Delhi. https://doi.org/10.1007/978-81-322-2283-5_2

Kulasooriya S (2012) Cyanobacteria: pioneers of planet earth. Ceylon J Sci (Biol Sci) 40(71):71. https://doi.org/10.4038/cjsbs.v40i2.3925

Kumar K, Mella-Herrera RA, Golden JW (2010) Cyanobacterial heterocysts. Cold Spring Harb Perspect Biol 2(4):a000315. https://doi.org/10.1101/cshperspect.a000315

Kuraganti G, Edla S, Pallaval VB (2020) Cyanobacteria as biofertilizers: current research, commercial aspects, and future challenges, pp 259–278. https://doi.org/10.1007/978-981-15-3204-7_11

Lavens P, Sorgeloos P (1996) Manual on the production and use of live food for aquaculture. In: FAO Fisheries Technical Paper No. 361. FAO, Rome

Llinares-Millán C, Fernández-Plazaola I, Hidalgo-Delgado F, Martínez-Valenzuela MM, Medina-Ramón FJ, Oliver-Faubel I et al (2014) Construction and building research. Springer, Dordrecht. https://doi.org/10.1007/978-94-007-7790-3

Mahanty T, Bhattacharjee S, Goswami M, Bhattacharyya P, Das B, Ghosh A et al (2017) Biofertilizers: a potential approach for sustainable agriculture development. Environ Sci Pollut Res 24(4):3315–3335. https://doi.org/10.1007/s11356-016-8104-0

Malliga P, Uma L, Subramanian G (1996) Lignolytic activity of the cyanobacterium Anabaena azollae ML2 and the value of coir waste as a carrier for BGA biofertilizer. Microbios 86:175–183

Mazard S, Penesyan A, Ostrowski M, Paulsen I, Egan S (2016) Tiny microbes with a big impact: the role of Cyanobacteria and their metabolites in shaping our future. Mar Drugs 14:97. https://doi.org/10.3390/md14050097

Mazid M, Khan TA (2014) Future of bio-fertilizers in Indian agriculture: an overview. Int J Agric Food Res 3:10–23. https://doi.org/10.24102/ijafr.v3i3.132

Mishra U, Pabbi S (2004) Cyanobacteria: a potential biofertilizer for rice. Resonance 9:6–10. https://doi.org/10.1007/bf02839213

Moore AW (1969) Azolla: biology and agronomic significance. Bot Rev 35:17–34. https://doi.org/10.1007/BF02859886

Niedermeyer T (2015) Anti-infective natural products from Cyanobacteria. Planta Med 81:1309–1325. https://doi.org/10.1055/s-0035-1546055

Nosheen S, Ajmal I, Song Y (2021) Microbes as biofertilizers, a potential approach for sustainable crop production. Sustainability 13(4):1868. https://doi.org/10.3390/su13041868

Owen D, Williams AP, Griffith GW, Withers PJA (2015) Use of commercial bio-inoculants to increase agricultural production through improved phosphrous acquisition. Appl Soil Ecol 86: 41–54. https://doi.org/10.1016/j.apsoil.2014.09.012

Parani K, Saha BK (2012) Prospects of using phosphate solubilizing Pseudomonas as bio fertilizer. Eur J Biol Sci 4:40–44. https://doi.org/10.5829/idosi.ejbs.2012.4.2.63117

Pathak J, Rajneesh MPK, Singh SP, Häder D-P, Sinha RP (2018) Cyanobacterial farming for environment friendly sustainable agriculture practices: innovations and perspectives. Front Environ Sci 6:7. https://doi.org/10.3389/fenvs.2018.00007

Percival SL, Williams DW (2014) Cyanobacteria. In: Microbiol of waterborne diseases, 2nd edn. Elsevier, pp 79–88. https://doi.org/10.1016/B978-0-12-415846-7.00005-6

Pereira I, Ortega R, Barrientos L, Moya M, Reyes G, Kramm V (2009) Development of a biofertilizer based on filamentous nitrogen-fixing cyanobacteria for rice crops in Chile. J Appl Phycol 21:135–144. https://doi.org/10.1007/s10811-008-9342-4

Pimratch S, Butsat S, Kesmala T (2015) Application of blue-green algae and mineral fertilizers to direct seeding lowland rice. Sci Asia 41:305. https://doi.org/10.2306/scienceasia1513-1874.2015.41.305

Pirttilä AM, Mohammad Parast Tabas H, Baruah N, Koskimäki JJ (2021) Biofertilizers and biocontrol agents for agriculture: how to identify and develop new potent microbial strains and traits. Microorganisms 9:817. https://doi.org/10.3390/microorganisms9040817

Plavšić M, Terzic S, Ahel M, Van Den Berg CMG (2002) Folic acid in coastal waters of the Adriatic Sea. Mar Freshw Res 53(8):1245–1252. https://doi.org/10.1071/MF02044

Priya H, Prasanna R, Ramakrishnan B, Bidyarani N, Babu S, Thapa S et al (2015) Influence of cyanobacterial inoculation on the culturable microbiome and growth of rice. Microbiol Res 171: 78–89. https://doi.org/10.1016/j.micres.2014.12.011

Qamar H, Hussain K, Soni A, Khan A, Hussain T, Chénais B (2021) Cyanobacteria as natural therapeutics and pharmaceutical potential: role in antitumor activity and as nanovectors. Molecules 26(1):247. https://doi.org/10.3390/molecules26010247

Quesada A, Vincent WF (2012) Cyanobacteria in the cryosphere: snow, ice and extreme cold. In: Whitton BA (ed) Ecology of Cyanobacteria II: their diversity in space and time. Springer Netherlands, Dordrecht, pp 387–399. https://doi.org/10.1007/978-94-007-3855-3_14

Rao SK (2011) Opportunities for biofertilizer industry. Business Planning and Development Unit, National Agricultural Innovation Project, Jawaharlal Nehru Agricultural University, Jabalpur

Research & Markets (2021) Global Biofertilizers Market (2021 to 2026)—increase in the production and yield of crops presents opportunities. Glob Biofertilizers Mark (2021 to 2026). GlobeNewswire, pp 1–6

Rodríguez A, Stella A, Storni M, Zulpa G, Zaccaro M (2006) Effects of cyanobacterial extracellular products and gibberellic acid on salinity tolerance in Oryza sativa L. Saline Systems 2:7. https://doi.org/10.1186/1746-1448-2-7

Roger PA, Reynaud PA (1982) Free—living blue—green algae in tropical soils. In: Dommergues YR, Diem HG (eds) Microbiology of tropical soils and plant productivity. Developments in plant and soil sciences, vol 5. Springer, Dordrecht. https://doi.org/10.1007/978-94-009-7529-3_5

Rosenberg JN, Oyler GA, Wilkinson L, Betenbaugh MJ (2008) A green light for engineered algae: redirecting metabolism to fuel a biotechnology revolution. Curr Opin Biotechnol 19:430–436. https://doi.org/10.1016/j.copbio.2008.07.008

Saadatnia H, Riahi H (2009) Cyanobacteria from paddy fields in Iran as a biofertilizer in rice plants. Plant Soil Environ 55:207–212. https://doi.org/10.17221/384-PSE

Sabier Sae K, Abdulla AS, Ahmaed Has I, Hamed AP (2015) Effect of bio-fertilizer and chemical fertilizer on growth and yield in cucumber (*Cucumis sativus*) in green house condition. Pakistan J Biol Sci 18:129–134. https://doi.org/10.3923/pjbs.2015.129.134

Sadhana B (2014) Arbuscular mycorrhizal fungi (AMF) as a biofertilizer—a review. Int J Curr Microbiol App Sci 3(4):384–400

Singh SP, Montgomery BL (2011) Determining cell shape: adaptive regulation of cyanobacterial cellular differentiation and morphology. Trends Microbiol 19(6):278–285. https://doi.org/10.1016/j.tim.2011.03.001

Singh JS, Pandey VC, Singh DP (2011a) Efficient soil microorganisms: a new dimension for sustainable agriculture and environmental development. Agric Ecosyst Environ 140:339–353. https://doi.org/10.1016/j.agee.2011.01.017

Singh RK, Tiwari SP, Rai AK, Mohapatra TM (2011b) Cyanobacteria: an emerging source for drug discovery. J Antibiot (Tokyo) 64:401–412. https://doi.org/10.1038/ja.2011.21

Singh S, Singh BK, Yadav SM, Gupta AK (2014) Potential of biofertilizers in crop production in Indian agriculture. Am J Plant Nutr Fertil Technol 4:33–40. https://doi.org/10.3923/ajpnft.2014.33.40

Singh JS, Kumar A, Rai AN, Singh DP (2016) Cyanobacteria: a precious bio-resource in agriculture, ecosystem, and environmental sustainability. Front Microbiol 7:529. https://doi.org/10.3389/fmicb.2016.00529

Sinha RK, Valani D, Chauhan K, Agarwal S (2010) Embarking on a second green revolution for sustainable agriculture by vermiculture biotechnology using earthworms: reviving the dreams of Sir Charles Darwin. J Agric Biotechnol Sustain Dev 2:113–128

Thajuddin N, Subramanian G (2005) Cyanobacterial biodiversity and potential applications in biotechnology. Curr Sci 89:47–57

Ugwu CU, Aoyagi H, Uchiyama H (2008) Photobioreactors for mass cultivation of algae. Bioresour Technol 99:4021–4028. https://doi.org/10.1016/j.biortech.2007.01.046

Vejan P, Abdullah R, Khadiran T, Ismail S, Nasrulhaq BA (2016) Role of plant growth promoting rhizobacteria in agricultural sustainability—a review. Molecules 21(5):573. https://doi.org/10.3390/molecules21050573

Venkataraman GS (1972) Algal biofertilizers and rice cultivation. Today & Tomorrow's, New Delhi, p 75

Venkataraman GS (1981) Blue-green algae for rice production: a manual for its promotion. In: FAO/UNDP Regional Project for Improving Soil Fertility through Organic Recycling. Food and Agriculture Organization of the United Nations, Soil Resources, Management, and Conservation Service, Land and Water Development Division, Rome, FAO Soils bulletin, 46.; Field document, no. 2: 102

Verma JP, Yadav J, Tiwari KN, L, Singh V. (2010) Impact of plant growth promoting Rhizobacteria on crop production. Int J Agric Res 5:954–983. https://doi.org/10.3923/ijar.2010.954.983

Verma M, Mishra J, Arora NK (2019) Plant growth-promoting Rhizobacteria: diversity and applications. In: Sobti R, Arora N, Kothari R (eds) Environmental biotechnology: for sustainable future. Springer, Singapore. https://doi.org/10.1007/978-981-10-7284-0_6

Vincent WF (2009) Cyanobacteria. In: Encycl. Inl. Waters. Elsevier, New York, pp 226–232. https://doi.org/10.1016/B978-012370626-3.00127-7

Zahra Z, Choo DH, Lee H, Parveen A (2020) Cyanobacteria: review of current potentials and applications. Environ MDPI 7(2):13. https://doi.org/10.3390/environments7020013

Zhu Y, Anderson DB, Jones SB (2018) Algae farm cost model: considerations for photobioreactors, vol 2018. US Dept Energy, p 16

Chapter 7
Rhizobium Biofertilizers: Mass Production Process and Cost-Benefit Ratio Analysis

Komal A. Chandarana and Natarajan Amaresan

Abstract In the era of demanding food supplies for the increasing populations worldwide, the increasing use of nitrogen fertilizers resulted in adverse effects on human health and the environment. To overcome the rising threat of chemical fertilizers, researchers have sought to find an eco-friendly alternative. Nitrogen-fixing rhizobia are a very important microbial group used to supply nitrogen to leguminous crops, which act as host crops for bacteria. Rhizobial bioformulations have been used as biofertilizers for nearly a century. Currently, various types of rhizobial biofertilizers are commercially available in the market, which may be in solid or liquid form. This chapter deals with mass production and solid carrier-based bioformulations of *Rhizobium* species. Additionally, the calculation of the cost-benefit ratio is also described, aiming to develop the small-scale biofertilizer industry and attract entrepreneurs.

Keywords Nitrogen-fixing · Biofertilizer · Bioformulations · *Rhizobium* · Entrepreneurs

7.1 Introduction

Integrated plant supply systems (IPNS) consist of a judicious combination of organic matter, chemical fertilizer, and biofertilizer, which promises an optimal nutrient supply to crops with simultaneous conservation of soil productivity and ecological health. However, the high cost and environmental and human health issues associated with chemical fertilizers have necessitated the development of eco-friendly and cost-effective alternatives. In this context, biofertilizers have become efficient alternatives to chemical fertilizers. Biofertilizers are living microbial inoculants in the form of bacteria, fungi, or algae that increase plant growth. *Rhizobium* is specifically applied to leguminous crops, and its application leads to a 10–30% enhancement in crop yields through an estimated fixation of 40–250 kg/ha/year nitrogen fixation.

K. A. Chandarana · N. Amaresan (✉)
C.G. Bhakta Institute of Biotechnology, Uka Tarsadia University, Surat, Gujarat, India

N. Amaresan et al. (eds.), *Agricultural Microbiology Based Entrepreneurship*, Microorganisms for Sustainability 39,
https://doi.org/10.1007/978-981-19-5747-5_7

The amount of nitrogen fixed depends on the efficacy of *Rhizobium* strain. *Rhizobium* as a bioinoculant was first commercialized in the USA by private enterprises in the 1930s (Smith 1992).

7.1.1 Rhizobium *Species*

Rhizobium is a Gram-negative root colonizer that generally colonizes leguminous roots. *Rhizobium* species work symbiotically with plants and drive atmospheric nitrogen in plants. The process starts when bacteria attach to the root hair, and the plant protein "lectin" binds the bacteria to the surface. Bacteria then penetrate root hairs, and the infected root cells divide to form nodules that provide an anaerobic environment that is necessary for the nitrogen fixation process (Gomare et al. 2013). It is recommended for crops such as groundnut, soybean, red gram, green gram, black gram, Bengal gram, lentil, fodder legumes, etc. However, colonization by *Rhizobium* strains is very specific to the host plant. Therefore, the industry that opted for *Rhizobium* biofertilizer mass production should produce multiple species of *Rhizobium* for application in various crops as per the farmer's demand and definitely for establishing reliable industry and earnings. The costs and technical details of different species are more or less the same. Some *Rhizobium* species and their host crops are listed in Table 7.1.

7.1.2 Mass Production

Mass production of *Rhizobium* biofertilizer is divided into the following stages:

(a) *Procurement of Standard Strain*

Pure cultures of various strains of *Rhizobium* species maintained in test tubes or vials can be purchased from Agricultural Universities, Indian Agriculture Research Institute, national/regional centers of organic farming, etc. After receiving pure cultures, it should be further subcultured and must be maintained purely for mass production using standard microbiological media and techniques by trained microbiologists.

(b) *Mother Culture Preparation and Maintenance*

Table 7.1 Some of the *Rhizobium* species and its host crop

Rhizobium species	Host crop
R. leguminosarum	Green pea and lentil
R. japonicum	Soybean
R. lupine ornithopus	Lupinus
R. meliloti	Melilotus
R. phaseoli	Phaseoli
R. trifoli	Trifolium

Table 7.2 Selective media for *Rhizobium* culturing

Components	Quantity (g/L)
Mannitol	10.0
K_2HPO_4	0.5
$MgSO_4.7H_2O$	0.2
NaCl	0.1
Yeast extract	0.5
Distilled water	1.0 L
pH	7.0
Agar	15

The selective medium for the mass *Rhizobium* strain is yeast extract mannitol (YEM) agar medium. The compositions are presented in Table 7.2. All ingredients should be mixed well and sterilized by autoclaving at 121 °C for 20 min at 15 lbs pressure. After making the slants, a single colony from the mother culture slant is transferred and incubated at 30 °C for 24 h.

(c) *Starter Culture Preparation*

Starter cultures for mass production can be prepared using the same YEM, excluding agar. The broth is dispersed into conical flasks and autoclave at 15 lbs pressure for 15 min at 121 °C, cool, inoculate with *Rhizobium* colonies, and incubate for 24–48 h at 30 °C in an incubator shaker. Inoculate the flask of 500 mL capacity with this starter culture and incubate it on shaker for 48 h and use this culture as an inoculum for the fermenter.

(d) *Mass Multiplication*

Prepare YEM medium as described and pour into a fermenter, sterilize (15 lbs pressure at 121 °C for 15 min), and cool to normal temperature. Add inoculum from the inoculation point of the fermenter (20 L capacity) at a rate of 5% and allow it to grow under optimum conditions for 4–5 days. Regulate the air flow to 3–10 L of air/hour/lit of the medium. Sterile air provides aeration and agitation for bacterial growth. Periodically, draw samples from the sampling point and analyze for growth and contamination by plating on YEM agar medium. Harvest the broth from the culture outlet for formulation once bacterial growth reaches 10^9 CFUs/mL.

The important considerations that should be kept in mind during the fermentation process are: a) it is not advisable to store the culture after optimum growth is reached, b) there should not be any fungal or other bacterial contamination during and/or after harvesting the culture. The final cost of the product is affected by the medium used for multiplication and the duration of the fermentation process. Therefore, it is essential to choose an appropriate medium and optimize parameters such as pH, temperature, and aeration to reduce the incubation time for different *Rhizobium* species. The ideal pH and temperature for *Rhizobium species* are 7.0 and 30 °C, respectively.

(e) Rhizobium *Bioformulations*

The fermented culture should be harvested in batch culture mode and mixed with the appropriate carrier material at a ratio of 1:5 (v/w). For instance, mix 1 L of enriched culture with 5 kg of carrier material to obtain effective bioformulations of 10^8–10^9CFUs/mL.

(f) *Processing of Carrier Material*

The use of an ideal carrier material is essential for biofertilizer formulations to maintain good quality until application in the field. The selection of an ideal carrier material should be based on: (a) cheaper cost, (b) nontoxicity, (c) high organic matter content, (d) more than 50% water holding capacity, (e) ease of processability, and, most importantly, (f) local availability. Generally, charcoal, press mud, peat soil, lignite, vermiculite, farmyard manure, and soil mixtures can be used as carrier materials for biofertilizer formulations. Neutralized peat soil or lignite is ideal for *Rhizobium* strain formulation. However, more recently, liquid formulations have attracted attention of entrepreneurs because of their low cost, easy storage, low space requirement, and easy maintenance.

(g) *Packaging*

Before packaging, the inoculum-mixed carrier material should be shade dried for 2–3 days at room temperature. Curing should be done by spreading the mixture on polyethylene sheets and keeping it in shallow trays with polyethylene covering. After 2–3 days of drying, the formulations could be packed in the desired quantities (200 g, 500 g, and 1.0 kg) depending on the market demand. Generally, polyethylene bags are used for the packaging of biofertilizers. However, it is important to consider that the bag should be of low-density grade, and the approximate thickness should be 50–75 μm (Sethi et al. 2018). Each packet should contain the product's name, manufacturer's name and address, details about *Rhizobium* strain used, the name of crops for which it can be used, batch number, date of manufacture, expiry date, storage instruction, mode of application, CFUs/g, etc. The population of inoculants in packed products should be checked at monthly intervals for a year to ascertain the quality in terms of CFUs/g count until use (Motsara et al. 1995).

(h) *Quality Control*

The quality of a formulated product is an important factor influencing its success or failure in the market. The correct type of organism in the desired number must be present in the formulations. Microbial processes are sensitive to contamination; therefore, quality should be checked for contamination at every stage of production, including mother culture subculturing, starter culture preparation, fermentation, carrier selection, and mixing with broth during packaging and storage. Quality control must include a serious consultation for microbiologists at each stage of the product. Finally, the product should meet some standards specified by the Indian government, i.e., the formulation must contain at least 10^7cells/g of carrier before the expiry date marked on the packet.

7.1.3 *Calculation of Pilot-Scale Production of* Rhizobium *for Small-Scale Industry*

Various facilities are required for the successful establishment of a biofertilizer unit for the production of *Rhizobium* strains. The infrastructure and laboratory facilities described here are not only applicable for *Rhizobium* production, but the same facilities can also be applied for the production of other nitrogen-fixing biofertilizers such as *Azotobacter* and *Gluconacetobacter*. Most private firms and governmental institutes produce various bacterial biofertilizers using common facilities to ensure the economic viability of project installation. The stepwise calculation for the small-scale industrial production of *Rhizobium* strain is divided into two parts: A) non-recurring cost described in Table 7.3, which includes fixed costs of capital investment in equipment; and B) recurring cost, which includes variable costs for raw materials used for production to packaging, manpower wages, marketing expenses, water and power utilities, and other miscellaneous expenses per year as described in Tables 7.4, 7.5, 7.6, and 7.7.

For small-scale production of *Rhizobium* strains, two 20 L capacity fermenters are sufficient for batch-wise harvesting. 1.0 L of culture is sufficient for mixing with 5.0 kg of carrier material if broth is enriched for 4–5 days under optimal conditions. Thus, 100 kg of *Rhizobium* biofertilizer can be produced from 20 L broth. Therefore, for a one-time harvest in a week from two fermenters, 200 kg of finished product will be formulated. Thus, the monthly production will be 800 kg, and the annual production will be 9600 kg. Thus, the approximate production with minimum

Table 7.3 Non-recurring cost calculation for initial capital investment

Sr. No.	Particular	Total cost in INR (in lakhs)
1.	*Equipment and machinery*	
	Vertical autoclave (600 × 350 mm) × 1	0.8
	Refrigerator (300 L) × 1	0.35
	Laminar air flow (3′ × 2′) × 1	1.0
	BOD incubator × 1	0.6
	Rotary shaker × 1	0.25
	Compound microscope (binocular) × 1	0.60
	Weight balance	0.15
	pH meter × 1	0.3
	Colony counter × 1	0.05
	Stainless steel seed fermenters (20 L cap.) × 2	3.0
	Polyethylene sealer × 1	0.15
	Total for capital investment for equipment	7.25
2.	*Miscellaneous fixed assets (computer, printer, fax, stationary items)*	1.5
3.	*Grand total* (1 + 2)	8.75

Table 7.4 Calculation of cost for production media

Components	Quantity (g/L)	Quantity (g/2000 L)	Approximate Indian rate	Cost for ultimate quantity required to produce 2000 L
Mannitol	10.0	20,000	550/kg	11,000
K_2HPO_4	0.5	1000	1500/kg	1500
$MgSO_4.7H_2O$	0.2	400	340/500 g	340
NaCl	0.1	200	230/500 g	230
Yeast extract	0.5	1000	2200/500 g	4400
Total cost				17,470
Total cost in lakhs (INR)				0.18

Table 7.5 Other miscellaneous raw material cost

Material	Quantity	Approximate rate	Total cost in lakhs (INR)
Carrier material	10,000 kg	4200/ton	0.42
Polyethylene bag and labels	Depending upon market demand for packaging (200 g, 500 g, 1 kg)	Approx.	0.5
Variable cost per annum for consumables (flasks, pipettes, test tubes, measuring cylinders, beakers, loops, gas cylinders, etc.)	Variable	Approx.	0.25
Total cost			1.17

Table 7.6 Man-power wages per annum

Category	Nos	Salary per head (Rs. in lakhs/month) (INR)	Total cost (Rs. in lakhs/month) (INR)	Total cost (Rs. in lakhs/annum) (INR)
Microbiologist	1	0.3	0.3	3.6
Assistant production officer	1	0.15	0.15	1.8
Administrative officer	1	0.15	0.15	1.8
Sales officers	2	0.15	0.3	3.6
Skilled and unskilled labors	3	0.045	0.135	1.62
Total cost			1.035	12.42

instruments and manpower will be approximately 10,000 kg. Therefore, the present model for the cost-benefit ratio is 10,000 kg/year *Rhizobium* biofertilizer production.

(a) Non-recurring cost
(b) Recurring (variable) cost per year

Table 7.7 Total recurring expenses per annum for 10,000 kg production

Particulars/annum	Cost (Rs. in lakhs/annum (INR)
Media cost (cost/10,000 kg/annum)	0.18
Other miscellaneous raw material cost	1.17
Man-power cost	12.42
Building rent/annum	3.6
Utilities—Power (2000 units @ Rs. 5/unit)	1.2
Contingencies	
Marketing and selling expenses	1.0
Repair and maintenance	1.0
Total cost	20.57

Table 7.8 Cost-benefit ratio calculation

Particulars/annum	Cost (Rs. in lakhs/annum (INR)
Expenditure/annum/10,000 kg production	20.57
Loss due to contamination	0.5
Depreciation cost of fixed assets at 5%	0.44
Total expenditure	21.51
Income/annum/10,000 kg selling	35.0
Net benefit	13.49
Profitability (%profit of sale)	38.54%

Raw Material Used for the Production of Rhizobium Species

Production of 2000 L of culture broth is required for 10,000 kg of the finished product of *Rhizobium* biofertilizer per annum. If one considers standard company ingredients then the cost for media (YEM) (2000 L) is given in Table 7.4.

Income per 10,000 kg of Selling

The average cost of solid *Rhizobium* solid formulations in the current market ranges 250–450 INR/kg. Government institutions and organizations may cost less than this, and on the other hand, some private industries may cost double. If we consider an average 350 Rs./kg the income will be 35.0 lakhs/annum (INR). Based on this selling and income, the cost-benefit ratio is calculated, as shown in Table 7.8.

References

Gomare KS, Mese M, Shetkar Y (2013) Isolation of Rhizobium and cost effective production of biofertilizer. Indian J Life Sci 2(2):49

Motsara MR, Bhattacharyya P, Srivastava B (1995) Biofertiliser technology, marketing and usage: a sourcebook-cum-glossary. Fertiliser Development and Consultation Org

Sethi SK, Sahu JK, Adhikary SP (2018) Microbial biofertilizers and their pilot-scale production. In: Microbial biotechnology. CRC, pp 312–331

Smith RS (1992) Legume inoculant formulation and application. Can J Microbiol 38(6):485–492

Chapter 8
Mass Multiplication, Production Cost Analysis, and Marketing of VAM Fungal Biofertilizer

Komal A. Chandarana, Harshida A. Gamit, Sapna Chandwani, Hetvi Y. Naik, and Natarajan Amaresan

Abstract Arbuscular mycorrhizal fungi (AMF)-based biofertilizers are widely accepted because of their role in sustainable agriculture. AMF play a vital ecological role by influencing the microbial environment of the rhizosphere, improving soil aggregation, and strengthening the ability of the plant root system to absorb and translocate minerals across their ambient mycelia. Mass production of AMF can be industrially produced using substrate-free and substrate-based cultivation methods. In this book chapter, we discuss the formulation and commercialization of AMF biofertilizers, including the attributes that endanger them. One critical challenge in the commercialization of AMF formulations is quality assurance during every step of the process.

Keywords AMF (Arbuscular mycorrhizal fungi) · Symbiotic association · AMF biofertilizer · Mass production · Substrate free and substrate-based inoculation · Commercialization

8.1 Introduction

Mycorrhiza are mutual symbiotic associations between a fungus and the roots of vascular plants both of which benefit from each other (Kirk et al. 2001). Plants provide the fungus with photosynthetic products such as sugars, and in return, the fungus provides the plant with minerals and water (Smith and Read 2010). In ectomycorrhiza, the fungal hyphae do not enter or penetrate individual cells in the root, whereas in endomycorrhiza, the fungal hyphae penetrate the cell wall and invaginate the cell membrane. Endomycorrhizae are further classified as arbuscular mycorrhiza, ericoid mycorrhiza, arbutoid mycorrhiza, orchoid mycorrhizae, and monotropoid mycorrhiza (Nicolson and Gerdemann 1968).

K. A. Chandarana · H. A. Gamit · S. Chandwani · H. Y. Naik · N. Amaresan (✉)
C.G. Bhakta Institute of Biotechnology, Uka Tarsadia University, Surat, Gujarat, India

N. Amaresan et al. (eds.), *Agricultural Microbiology Based Entrepreneurship*, Microorganisms for Sustainability 39,
https://doi.org/10.1007/978-981-19-5747-5_8

There are a group of beneficial microorganisms present in the soil or in the rhizosphere regions, the most common type of organism is arbuscular mycorrhizal fungi (AMF). AMF ecologically play an essential role by increasing the ability of the plant root system to absorb and translocate the minerals through their ambient mycelia and improve soil aggregation by releasing hydrophobic glycoproteins from extra radical hyphae (Cavagnaro 2008). AMF provide several ecological advantages such as influencing the microbial and chemical environment of the mycorrhizosphere, more precisely the hyposphere (the zone surrounding individual hyphae), stabilizing soil aggregates, and bioremediation of soil (Johansson et al. 2004). The broadest range of beneficial effects of mycorrhizae is increasing the uptake of immobile nutrients, specifically the acquisition of phosphate (P) from soil (Bolan 1991). AMF hyphae uniquely colonize the root cortex and form a highly branched structure under the cells (Balestrini et al. 2015). Khaliel (1988) reported that pH is an edaphic factor that determines the abundance of AMF; however, according to Sadhana (2014), pH did not influence the mycorrhizal spore density and frequency. The high content of P and N in soil reduces the number of AMF spores (Mosse et al. 1981) infection, and decreases the dependency of the plant on the fungal association (Ojala et al. 1983).

The AMF also reported on the ability to protect plants from biotic and abiotic stress (Cho et al. 2006). Under field conditions and in nurseries, it has been reported that the synergistic interaction of PGPR and AMF alleviates yield and biomass (Tajini et al. 2012). Furthermore, it has been reported that the mycorrhiza helper bacteria enhance mycorrhizal formation in the rhizosphere regions (Frey-Klett et al. 2007). Recently, it was suggested that in natural environments, non-mycorrhizal conditions should be considered abnormal for almost plant species (Smith and Smith 2012).

The use of a microbial formulation termed bio-fertilization in sustainable crop production allows plants to effectively utilize mineral elements such as phosphorus and nitrogen (Alori et al. 2017). Phosphorus is an essential mineral for plants. Alkaline phosphatase (ALP) is the functional and useful enzyme for AM for phosphorus acquisition through plants since it is present in the vacuoles of highly developed arbuscules revealed by ultra-cytochemistry studies (Abdel-Fattah and Wixtrom 2014).

8.2 Cultivation Criteria

Five basic parameters must be satisfied for high-quality production. (1) Light or illumination of the host plant is ideal. (2) The optimal temperature range of the fungi in the substrate should be exceeded for only a short period of time. (3) Avoid waterlogging and long-term drought of the substrate. (4) Plant nutrition should be optimal (5) Plants must be treated against pathogens.

Maize (*Zea mays* L.), Bahia grass (*Paspalum notatum* Flugge), onion and leek (*Allium* pp.) are commonly used for the large-scale production of AMF because of their several advantages such as adequate root system development, short life cycle, low level tolerance of phosphorous, and good characterization level by a large range of AMF. This plant can have tolerance to a wide range of temperature and low sensitivity to pathogen; colonized roots show yellow and uncolonized roots are white in color (leek and maize).

There are three types of AMF inoculants: (1) a substrate-based (sand/soil) system, which is a cheap method for mass cultivation and large-scale production of fungal inoculum, (2) a substrate-free cultivation system (hydroponic and aeroponic), which is used for completely clean AMF production, and (3) In vitro cultivation system, which is based on the excised root, the "root organ culture" (ROC) or whole autotrophic plants.

8.2.1 Popular Substrate-Based Cultivation Methods

Greenhouse Culture Method

This is a widely used conventional method that involves mass multiplication of AM fungi in pots under controlled conditions at greenhouse using trapping plants. Two types of AMF bio-fertilizers can be produced by adopting this technique: a) mixture of AMF species, and b) monosporal AMF species.

1. Mixture of AMF Species Mass Multiplication
   - *Soil Sample Collection*: Soil samples should be collected from the rhizospheric zone of the plant at a depth of 10–15 cm, close to the root zone. In case of cultivated crops, soil should be collected at the flowering stage of the plant, and at the age of 2–3 years of plantation in the case of agro-forestry species. Collect soil in sterile polythene bags, bring it to the laboratory, and keep it at 4 °C until the initial analysis.
   - *Primary Inoculum*: To initialize the production of a mixture of AMF species, fill the collected rhizospheric soil in small cups and sow the surface-sterilized seeds into the cups with the aim of restoring infective structures of the AMF in trapping plants (Sadhana 2014). Keep these cups in a greenhouse for 2 months with regular watering and supplying Hoagland's mineral solution. After incubation, cut the aerial parts of plants and mix the root-soil system thoroughly for preparing new plantation.
   - *Secondary Inoculum*: Fill the larger pots with appropriate soil mixture and inoculate the primary soil-root system along with surface-sterilized seeds in order to obtain more propagules of AMF. A minimum of 12 months period is required to obtain a variety of AMF and different types of propagules.

Two plant species: *Trifolium repens* (Fortin et al. 2015) and *Allium porrum* (Sadhana 2014) are widely used as trapping plants. *Medicago sativa* and *Brachiaria sp.* (Sadhana 2014) can also be used. Evaluation of AMF colonization rate in host plants should be checked at each month of incubation (Moreira et al. 2019), which is described here in the later quality control section. AMF isolation and identification should be performed simultaneously.

2. *Monosporal AMF Mass Multiplication*

   This method is based on the use of one selected AMF species with the appropriate plant species in larger pots under greenhouse conditions (Karima and Samia 2021). For this, one selective AMF species from natural soil is isolated using the wet sieving method (Gerdemann and Nicolson 1963) and cultivated with appropriate trapping plant.

   - *Mother Culture Preparation*: Mother cultures can be produced using single spore inoculation in funnels or in small cups filled with sterilized soil. Incubation time is of 2 months and in-between verification for sporulation and quality checking for the absence of contaminants should be performed. After confirmation through observation and staining, the monospecific spores of AMF are ready to inoculate for subsequent production. Mother cultures should have 100% root colonization and a minimum of 8–10 spores/g of inoculum. It has been reported that *Chloris gayana* (Rhodes grass) is the most appropriate host for *Glomus fasciculatum* (Sreenivasa and Bagyaraj 1988) and *Paspalum notatum* (Bahia grass) is for *Glomus deserticola* (Sadhana 2014).
   - *Mass Production*: Generally, trench (1 m × 1 m × 0.3 m) lined with black polythene sheets are used for commercial mass multiplication instead of single-single pots. Briefly, 5 kg of soil mixed with 50 kg of vermiculite is sterilized before filling in a trench up to a height of 20 cm. Then spread 1 kg of the mother culture 2–4 cm below the vermiculite surface. Sow the surface sterilized (with 4% sodium hypochlorite for 2 min) seeds. Water the plant as and when scheduled and use Hoagland's mineral solution to supply essential micronutrients. Test the quality of the inoculum and the progress of the growth by estimating AM colonization in roots by staining and microscopy after 30 and 60 days. Grow the stock plants for 60 days, then transfer it to the drying and conditioning compartments. After drying, cut the roots of the plants into small pieces and mix it thoroughly with vermiculite in a trench.

     Thus, at the end of 3–4 months, 55 kg inoculum could be produced having mixture of vermiculite, spores, hyphae, AM infected root pieces, and vesicles, which is sufficient to treat 11,000 seedlings grown in the nursery.

In Vitro Mass Multiplication of AMF Using Synthetic Media

In vitro mass multiplication of AMF is performed under aseptic laboratory conditions to trap plant roots cultivated on sterilized synthetic media. First, potentially viable propagules from natural soil should be isolated using the wet sieving method. Spores should be surface sterilized with commercial bleach to optimize the growth conditions for germination under aseptic conditions for particular AMF species. The

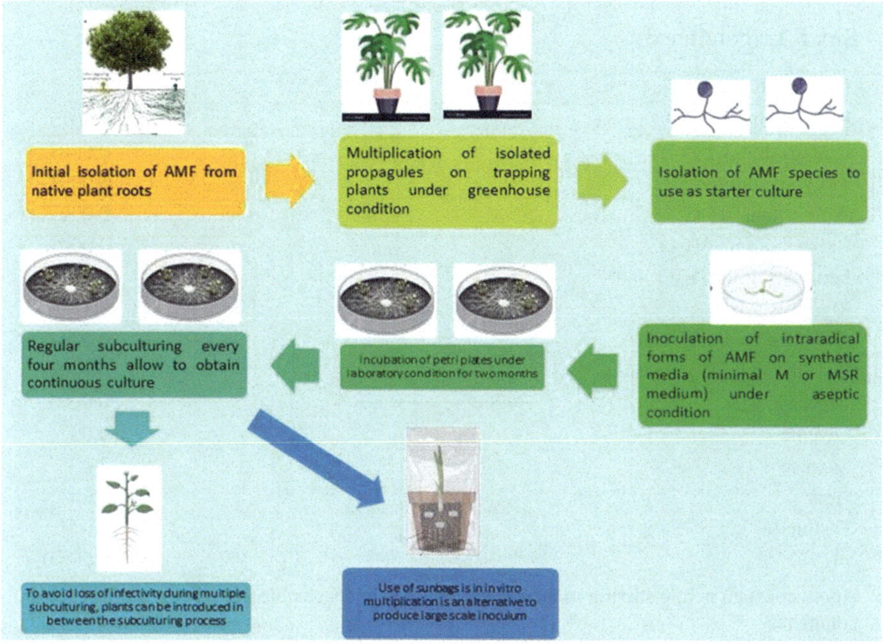

Fig. 8.1 In vitro mass multiplication of AMF on synthetic laboratory agar media

modified white medium (minimal medium M) (Bécard and Fortin 1988) and modified Strullu and Romand medium (MSR medium) (D'Souza et al. 2013) are the two suitable synthetic media utilized for in vitro mass production of AMF. The composition of White's medium, which is widely used is given in the Box 8.1. The pH of the media should be adjusted to 5.5 before autoclaving. The methodology used for the in vitro cultivation is shown in Fig. 8.1.

Box 8.1 Composition of modified White's medium for in vitro mass multiplication of AMF

Ingredients	g/L
Potassium nitrate	0.08
Calcium nitrate	0.222
Magnesium sulfate	0.360
Sodium phosphate monobasic	0.019
Potassium chloride	0.065
Sodium sulfate	0.2
Manganese sulfate H_2O	0.005

(continued)

Box 8.1 (continued)

Ingredients	g/L
Boric acid	0.0015
Potassium iodide	0.00075
Molybdenum trioxide	0.0001
Zinc sulfate $7H_2O$	0.0025
Copper sulfate $5H_2O$	0.0001
Ferrous sulfate $7H_2O$	0.002
Myo-inisitol	0.1
Thiamine hydrochloride	0.0001
Pyridoxine hydrochloride	0.0001
Nicotinic acid	0.0005
Glycine	0.003
Sucrose	20.0
Agar	15.0
Distilled water	1.0 L
pH	5.5–6.0

Apply constant gentle stirring to the solution till all the heat stable ingredients are dissolved completely

Adjust the pH using 1 N HCl/NaOH before autoclaving at 15 Ibs pressure, 121 °C temperature for 15 min

Add the filter sterilize heat labile ingredients after cooling autoclaved medium to 45 °C

8.2.2 Advantages and Disadvantages of the Substrate-Based System

The major advantage of the substrate-based method is that it has the least automated system; hence it is cost effective. The substrate-based production system provides facilities for the preservation of single or consortia of AMF species, and the cultivation criteria can be monitored and regulated correctly. Substrate-based systems provide controlled inocula conditions but there might also be a possibility of unwanted contaminations such as pests and pathogens. This method consumes more space and involves multiple steps such as post-production purification to separate propagules from the substrate and organic debris attached to it (Millner and Kitt 1992).

8.3 Substrate-Free Cultivation System

The production of AM fungi in non-solid matrix is known as a substrate-free cultivation system. However, solid matrices such as soil, sand, or a mixture of both are favorable for AMF cultivation. Substrate-free cultivation systems include hydroponic and aeroponic nutrient film techniques (Ijdo et al. 2011). The substrate-free cultivation method is also termed "solution culture technique." In the static system, the nutrient solution supplement is aerated through an aeration pump to avoid oxygen deprivation. The aeration pump is switched on regularly to reduce the flow of nutrient solution and avoid air bubbles, which may damage the development of delicate extra-radical hyphae (Hawkins and George 1997; Ijdo et al. 2011). The nutrient flow technique is initialized by Moss and Thompson in Mosse and Thompson 1981, which is an alternative technique in which thin nutrient solutions cover the root surface and provide sufficient area for gas exchange and acquire the problem due to insufficient aeration in the channels where the plant roots and AMF grow.

In the aeroponic nutrient film technique, nutrients are provided in the form of fog, which is a hydroponic system in which the host plant roots and AMF propagules dip into the nutrient fog. The liquid film surrounding the roots provides gas exchange and spraying of nutrient solution through the micro-droplets on the host roots, increasing aeration. Different devices, such as atomizing disks, pressurized spray-microirrigated nozzles, and ultrasonic pumps, have been used for spraying nutrients. Jarstfer and Sylavia (Jarstfer and Sylvia 1995) performed an experiment on spraying devices and concluded that the nozzle and spray systems were the most tailored systems for mass production of AMF. Mohammad et al. (2000) compared atomizing devices with the ultrasonic nebulizer method and found that the ultrasonic nebulizer method was the best method for the mass production of AMF.

There are several factors to be considered such as the pH of the nutrient solution, humidity, optimum temperature, and illumination. The optimum pH range for the hydroponic system near neutral and temperature is between 15 and 35 °C for the optimum colonization of the AMF as they provided superficial illumination to the plants for proper growth and hence better colonization by AM fungi. Sylviam and Jarstfer (1997) proposed the ample wavelength (400–700 nm) and high photosynthetic photon flux density.

8.3.1 Aeroponic

In this technique, plants are grown in closed containers and plant roots are continuously exposed to mist of a nutrient solution (Ijdo et al. 2011). The container is made up of opaque plastic material with a thick foam plastic lid. The misting nozzle is centered and firmly attached to the bottom of the container to provide nutrient-solution mist. Pre-colonized seedlings of 6–8 weeks of age are placed in an aeroponic container supported by a compressible polyurethane plug in a foam plastic lid. Plants with the same species colonization are placed in the same container. A diluted Hoagland's nutrient solution (pH 6.5 ± 0.05) is used as nutrient solution for misting the plant roots in the chamber (Hung and Sylvia 1988). Large-scale cultivation of AMF propagules is possible by using this technique. However, building large quantities of aeroponic containers, continuous operation of misting nutrient solutions, and failure to avoid pathogens are some of the drawbacks to this technique.

8.3.2 *Hydroponic*

In this technique, AMF with its symbiotic plant is cultivated in a specialized hydroponic device filled with a nutrient solution (Nurbaity et al. 2019). The plant root lies in a shallow layer of rapidly flowing nutrient solution; as the plant grows, the root mats develop and the upper layer of liquid retains a film of moisture around them. The installation of hydroponic devices under controlled condition is advisable to avoid contamination. Before planting into the hydroponic device, seedling must be pre-inoculated with AMF. Hoagland's mineral solution is used as a nutrient solution. Although it is a better technique for AMF cultivation, refreshing the nutrient solution to avoid microbial contamination is a serious problem. Moreover, submerging in a liquid solution is not a natural condition for AMF, which may result in a low quantity of sporulation of AM fungi.

8.3.3 *Advantages and Disadvantages of the Substrate-Free Production System*

The major advantage of the substrate-free system is the production of substrate-free cultures or inocula. Using this method, root pieces with a high density of infective propagules can be used directly. The liquid method is highly prone to the development of algal contaminants; the spore production rate could be affected by the absence of substrate.

8.4 In Vitro Production System (Root-Organ Culture)

The in vitro production of AMF was first attempted by Mosse in Mosse 1959. Thereafter, in Mosse 1962, he reported the first association of an *Endogone* species with a plant. Bécard and Fortin (1988) described a root organ culture system using T-DNA transformed roots of *Daucus carota* and T-DNA of the Ri plasmid of *Agrobacterium rhizogenes* (Chilton et al. 1982). The split-plate method facilitates access to the AMF and increases the production (St-Arnaud et al. 1996). Tiwari and Adholeya (2003) reported mass scale production of AMF through root organ culture in small containers in an airlift bioreactor and in a mist bioreactor with perlite as a substrate. (Jolicoeur et al. 1999; Fortin et al. 1996).

For the in vitro cultivation,Voets et al. (2005) developed a system in which the Petri plates filled with the gelled medium roots and AMF were associated inside the Petri plate and the shoot was always outside the Petri plate. Declerck et al. (2009) developed in vitro AMF production method in which pre-inoculated host plants produced individually introduced AMF inoculum in sterile growth tubes in a closed system continued with a nutrient solution. De Boulois et al. (2006) developed an in vitro system in which the shoot was developed in a sterile tube vertically connected to the top of a Petri plate, the roots and AMF were closely associated inside the Petri plates.

Briefly, pre-germination of seeds is required to initiate the isolation of plant roots for transformation. The seeds are surface-sterilized with commercially available sodium hypochlorite or hydrogen peroxide and thoroughly washed with sterilized distilled water. Then seeds are transferred to sterilized water agar plates and incubated at room temperature in the dark for 2–4 days. After germination, transfer the seeds with 2 cm long root tips on rich medium such as modified White medium or MSR medium.

8.4.1 Advantages and Disadvantages of the In Vitro Production System

This is a suitable method for the production of high quality and contamination free AMF inoculums. In vitro production system need a regular addition of culture medium and it always requires monitoring and regulating the culture. To make it cost effective, skilled technicians and laboratory equipments are also required.

8.5 Industrial Production of Arbuscular Mycorrhizal Fungi (AMF)

The fundamental technique adopted to accomplish this objective is the inoculation of AMF propagules into the target soil. Because of the obligate symbiotic nature of AMF, they cannot be cultivated in pure culture, away from their host plant, and this obligate nature makes the industrial or large-scale production of AMF inoculum very challenging and complex.

The greenhouse method is commonly used for the industrial production of AMF, because greenhouse experiments are very easy to set up and control. In this method, plants are amended with pure or mixed fungal species in pots. However, the purification of AMF is a major issue in large-scale production of the AMF group. The purity of multiplied AMF can be ensured by next-generation sequencing (NGS), which can be used for AM abundance and composition studies (Thonar et al. 2012; Rodriguez and Sanders 2015). One of the more advanced AMF multiplication techniques is altered root hair culture, which allows AMF multiplication under pure conditions (Vosátka et al. 2012).

8.5.1 Formulation of AM Fungi

AMF biofertilizers are either liquid-based or carrier-based solid formulations. Generally, AMF biofertilizers available widely in the market are solid carrier-based formulations containing different types of propagules, i.e., spores, mycelium, and root fragments. Talk, peat, vermiculite, lignite, clay, charcoal, soil, rice, or wheat straw compost can all be used as carrier materials. Pre-sterilization of carriers is essential to maintain a noncompetitive environment and to exclude pathogens present in the carrier. The addition of nutrients and cell protectants such as glucose, sucrose, maltose, trehalose, glycerol, and molasses is required to maximize cell viability and extend shelf life (Karima and Samia 2021). After mixing with the carrier, the final product of AMF biofertilizer might be in the form of liquid, powder, granules, pralines, or tablets (Karima and Samia 2021). All these formulations have different application methods to achieve efficient infectivity of AMF to host plants for growth promotion.

For example, liquid formulations are generally used to coat seeds or to inject directly into the soil at the base of growing plants. The microgranules (1–4 mm) are mixed with carrier material and applied onto the soil surface near the plant roots. Fine powder formulations can be sprayed on growing media, onto the soil near to roots and can also be used for seed coating. The tablets are placed in the plantation area, which allows slow release. Praline-type formulations are generally suitable for bare-root plants. Regardless of the formulation type, the important factor that takes into consideration the application of AMF biofertilizer is that the application should be in direct contact with plant roots.

8.5.2 Packaging

Packaging is a major constraint driving the market for this business (Keswani et al. 2019). The type of packaging must maintain the quality of the product and ensure the shelf life of AM propagules. Packaging must follow the standards and guidelines and be leak-proof; otherwise, it fails to meet the farmer's expectations. After cutting the trapped plant roots into small pieces, they are mechanically mixed with a sterilized carrier (bulk) under aseptic conditions and packed in polythene bags at the desired quantity. A blending machine can be used to mix the content properly. The bags are sealed with automated sealer. The surface of the bag is disinfected and labeled appropriately before transferring it to the storage room.

8.5.3 Storage

Store the packing bags of AMF at 4 °C, as higher temperature is detrimental to the viability of AMF propagules stored inside the bag. The storage period can reach up to 3 years but it is recommended that preserved culture for up to 3–4 months is the best bioinoculant.

8.5.4 *Quality Control*

In any of the mass multiplication methods, after 1–4 weeks of inoculation with AM spores, the trapped plant root should be aseptically collected and stained to check for AM fungal infection.

- *Staining*: Briefly, collect the host plant roots and wash thoroughly with water. Cut the root in 1 cm size and keep it in 10% KOH solution for 1 h at 90 °C. Wash again with water and acidify with 5 N HCl. Then stain the roots with 0.5% trypan blue in lactophenol. Mount the root bits in lactophenol and observe under a microscope for hyphae/arbuscular/vesicles of AM fungi infected host plant roots.
- Similar root pieces should be examined under a microscope and plated on agar plates to observe pathogenic organisms.

Precaution should be taken to prevent the pathogen propagation along the mycorrhizal roots. Therefore, the field inocula should be drenched several times with pesticides to eliminate contaminating pathogens. For instance, AMF grown using citrus plants should be drenched with nematicides and fungicides to control citrus nematodes and fungi such as *Phytophthora* and *Rhizoctonia*, respectively.

8.5.5 *Product Specification*

The Bureau of Indian Standards is the nodal agency that specifies standards for biofertilizer formulations. They also specified the criteria for the finished product of the AMF formulations, which are described in Box 8.2.

Box 8.2 Product specification for AM fungi biofertilizer

S. No.	Description	Specification
1.	Base	Fine powder/tablets/granules/root biomass mixed with growing substrate
2.	Minimum total viable propagules gram of product	100/g of finished product
3.	pH	6.0–7.5
4.	Moisture content in case of carrier-based formulations	8–12%
5.	Particle size in case of carrier-based formulations	The 90% formulation should pass through 250 micron IS sieve
6.	Infectivity potential	80 infection points in test roots/g of mycorrhizal inoculum used

8.6 Important Considerations to Be Taken During Mass Multiplication

Since isolation and rise of pure cultures of AM fungi is a difficult task, the selection of a suitable host is an important requirement to maintain the pure culture of AMF. Plants with extensive root systems have proven to be good hosts for AM fungal propagation (Sharma et al. 2017). The spinach, sugarbeet, lupine, and mustard family members do not form symbiotic relationships with AM fungi. Therefore, they should not be chosen at any stage of production or even for application. Moreover, the host plant should be from a different family from the inoculated crop to prevent spread of pathogenic microorganisms. For example, inocula from citrus plants should never be produced on citrus plants but it could be on sudan grass. Moreover, the host plant selected for mass production in pot cultures should be suitable for agro-climatic conditions in that particular area and adapted to greenhouse conditions.

Irradiation influences sporulation. Low light irradiation may reduce sporulation. Furthermore, the high temperature may not keep pace with AM colonization; therefore, it can affect overall growth and production. Additionally, proper watering is advisable; non-stressed and unsaturated watering is ideal for increased spore production. The composition of the soil mixture is crucial for ensuring colonization and sporulation. The substrate mixed with soil should be lightweight, possess good water holding capacity, have less leaching of essential nutrients, and be easy to remove from root surfaces.

The container size also matters; it should match the potential volume of the host root system. The trench described for monosporal mass multiplication would be ideal for sufficient growth of the host plant and sporulation of AMF. AM inoculum should not spread harmful pathogenic organisms; therefore, purity must be maintained during production. Quality checking at each point of AMF production is essential to avoid contamination and observe regular increments in AM colonization.

8.7 Economical Details

The installation of biofertilizer unit, especially for AMF, requires a different setup than that of the bacterial biofertilizer production unit. It requires extra care while selecting the area for establishing the AMF production unit in terms of geomorphology, climate, and homogenous soil characteristics that ensure appropriate production from the working capital. The space must be a buffer to reduce the chances of contamination during production and packaging. The technical details provided here are suitable for all types of AMF mass multiplications. However, to extract the cost-benefit ratio, more effort has been made to detail the project on substrate-based mass multiplication methods, as these methods have some advantages over substrate-free methods. These are,

- Economically feasible
- Rapid
- Easy to run
- Used living host, so its infectivity remains viable
- No specialized apparatus installation
- No continuous regulation of nutrients

The cost-benefit ratio is calculated for substrate-based methods only. Aeroponics and hydroponics can also be added to this unit by installing the respective apparatus in an additional separate controlled compartment with additional investment in this basic structural unit.

8.7.1 *Infrastructure*

The infrastructure setting must meet the economic aspects and criteria approved by the WHO (World Health Organization) as well. The unit must be designed in such a way that it follows a sterilization process with forward walking without the possibility of returning, i.e., the ventilation system must ensure interior air movement without recycling. It must be spacious, and electricity must be powerful with an emergency restart system using inverter along with the installation of fire safety systems. The laboratory in the AMF production unit must be compartmentalized into greenhouse for AMF inoculum multiplication, in vitro multiplication compartment, drying and conditioning room, quality control laboratory, and storage room with air conditioner. According to Alamari (2016), the H-shaped architecture is the best design. A minimum of 10,000–12,000 sq.ft. of area is required to build a complete infrastructure of the AMF biofertilizer unit. The brief introduction and importance of each compartment are given below, and the area required to establish the laboratory in its breakup is given in the Box 8.3.

Box 8.3 Break-up of area of production unit

Area allocated to	Compartment	Area (sq.ft)
Administration	Reception	300
	Manager's office	300
Laboratory set up	Greenhouse	2000
	In vitro multiplication	1000
	Aeroponics or hydroponics establishment	1000
	Drying and conditioning room	2000
	Quality control laboratory	1000
	Cold storage room	1000
Dispatch	Sales office	500
Common facilities	Pantry, changing room, etc.	500

- *Greenhouse for AMF Multiplication*
  This is the most important compartment used to take care of young plants to maximize productivity. Therefore, the following criteria must be followed to setup a greenhouse for AMF multiplication:
  - It must be behind the unit having a clear space.
  - Easily available solar energy, consistent moisture content, and low air pollution.
  - Wind direction is an important climatic problem, that's why the orientation of greenhouses must be compromised by wind orientation.
  - Must be double-sloped with natural exposure to sunlight.
  - Must have big windows to ensure ventilation; an automatic ventilation system might be installed, if financial conditions allow.
  - The water supply must be favorable and inexpensive.
- *Drying and Conditioning Room*
  This compartment is utilized for drying the contents of pots and containers for conditioning; therefore, it should be located just before the greenhouse. As the trapping plants are ready to harvest, the pots should be transferred to this shed area for 2–3 weeks for drying. After drying, the trapping plants are harvested, and the roots are separated and mixed with a suitable substrate. The AMF inoculum mixed with substrate is filled in sealed bags and kept aside in stacking conditions for conditioning.
- *In Vitro Multiplication Room*
  Separate and aseptic rooms to avoid contamination and to maintain sterility.
- *Quality Control Laboratory*
  This compartment should be designed to facilitate the checking of the growth and quality of AMF at each stage of production. It must have a suitable architecture that supports microscopy, staining, glassware and media storage, refrigeration and, of course, data recording.
- *Storage Room*
  The room must contain metal racks with shelves equipped with wheels to facilitate their movement. The room must be maintained at 4 °C because high temperatures affect the long-term storage of AMF spores. AMF spores can be stored for up to 3 years.

8.7.2 Non-recurring Cost (Tables 8.1, 8.2, 8.3, 8.4, and 8.5)

Table 8.1 Cost of land and infrastructure

Particulate	Rate	Cost in INR (in lakhs)
Land (12,000)	Approx.	30.0
Building construction	hvApprox. (1000 Rs./sq.ft)	80.0
Greenhouse construction cost	Approx. (800 Rs./sq.m)	1.5
Furniture	Approx.	4.0
Total cost		115.5

Table 8.2 List and cost of equipment required for the production of AMF

Item	Number	Rate/unit cost in INR (in lakhs)	Total cost in lakhs
Weighing balance	1	0.2	0.2
Vertical autoclave	1	0.8	0.8
Refrigerator	1	0.35	0.35
BOD incubator	2	1.5	3.0
Laminar air flow	1	1.0	1.0
Microscope (binocular)	1	0.7	0.7
RO plant (200–500 L/h)	1	0.35	0.35
Carrier blending tanks	1	0.5	0.5
Polybag sealer	1	0.15	0.15
Labeling machines and other miscellaneous items for packaging	–	–	3.0
Air conditioners	4	0.35	1.4
Generator (65 kVA, diesel generator)	1	1.5	1.5
Glassware and plasticwares for microbiological works	–	Approx.	1.5
Other miscellaneous requirements for lab use	–	Approx.	1.5
Total cost			15.95

Table 8.3 Miscellaneous fixed assets

Item	Numbers	Cost (Rs. in lakhs) (INR)
Computer	1	0.4
Printer	1	0.1
Fax	1	0.1
Others	–	0.2
Total cost		0.8

Table 8.4 Non-recurring cost for establishment of AMF production unit

Components	Cost in INR (in lakhs)
Land and infrastructure development	115.5
Equipment purchase	15.95
Miscellaneous fixed assets	0.8
Total cost	132.25

Table 8.5 Means of finance on capital non-recurring investment

Particulars	Contribution in percentage	Total contribution in cost (INR)
Owner's contribution	25%	33.0625
Bank loan	50%	66.125
Subsidy by government	25% of total project cost or 40 lakhs whichever is less	33.0625
Total cost		132.25

8.7.3 Recurring Cost (Tables 8.6, 8.7, 8.8, 8.9, and 8.10; Box 8.4)

Table 8.6 Manpower specification and salary particulars

Category	Duty	Nos	Salary per head (INR in lakhs/ month)	Total cost (INR in lakhs/ month)
Manager	Management of entire plant	1	0.4	0.4
Senior microbiologist	Isolation, separation, in vitro mass production, quality checking, and maintenance of mother culture	1	0.25	0.25
Assistant microbiologist		2	0.15	0.3
Administrative officer	Reception and administration	1	0.15	0.15
Sales officers	Marketing	5	0.15	0.75
Skilled and unskilled labors	Preparation of soil mixture, potting, inoculating, seedling, moving pots from greenhouse to drying room, pruning, watering, harvesting, grinding, and packaging	10	0.045 (0.03–0.06 depending upon the work allotment)	0.45
Total cost				2.3

Table 8.7 Cost for raw materials

Material	Cost in INR (in lakhs)
Media and chemicals (modified White's media, Hoagland's mineral solution, other chemicals for quality control, seeds of host plant)	1.0
Carrier material (vermiculite, lignite, perlite, peat, or other). Approximate rate of vermiculite is Rs. 18/kg	0.9
Containers (funnels, cups, pots, trench)	0.5
Packaging material (polythene bags, labels, stamps)	0.25
Total	2.65

Table 8.8 Overhead expenses

Reasons for extra expenses/annum	Cost in INR (in lakhs)/annum
Maintenance allowance	0.5
Office supplies	0.1
Loss due to undeveloped plants	0.1
Loss due to contamination	0.1
Taxes and insurance	2.31
Total expenditure	3.11

Table 8.9 Total recurring expenses

Particulars	Cost (INR in lakhs/annum)
Raw material and packaging	2.65
Man-power cost	27.6
Utilities—Power (4000 units @ Rs. 5/unit)	2.4
Marketing and selling expenses	6.0
Total recurring investment for first year	38.65

Table 8.10 Capital investment for first year of establishment

Particular	Total cost (INR in lakhs)
Fixed assets	132.25
Technology know-how	10
Recurring expenses per annum	38.65
Overhead expenses per annum	3.11
Grand total	184.01

Box 8.4 Production of AMF/year through monosporal mass multiplication greenhouse technique

Particular	Product in kg
Soil	500
Vermiculite	5000
Trenches can be filled	100
Mother culture (1 kg/trench)	100
Finished product at the end of 3 months	5500
Finished product per year	22,000

8.7.4 Income Calculation

The approximate rate of AM fungal biofertilizer application in India is 390 Rs./kg. The rate may vary depending on the state, region, market utility, packaging, and owner's investment. The price ratio decreased with an increase in packaging size.

If nearly 22,000 kg of finished product will be sold throughout the year in different packings (1, 5, 25, and 50 kg), then the average cost per kg might be approximately 340 Rs.

Therefore, the total annual income may be 22,000 × 340 = 74,80,000 Rs. Based on this, the cost-benefit ratio is calculated in Table 8.11.

The profit from mycorrhizal biofertilizer business can reach up to 60% or more after 4–5 years of industry establishment; as the raw material cost for production is very low, only the initial investment is high.

Table 8.11 Calculation of cost-benefit ratio

Total production/year (approx.)	22,000
Total expenditure/annum (recurring expenses, overhead expenses)	41.76
Total income	74.80
Total benefit	33.04
Profit in %	44.17%

8.8 Conclusion

In agriculture, the application of mycorrhizal-based biofertilizers and biopesticides is mainly dependent on scientific knowledge derived from applied and fundamental research, or a combination of both mycorrhizal industry and mycorrhizal science. For the production of mycorrhizal fungi-based inocula, a number of enterprises were involved not only in the developed world but also in the emerging markets. Various techniques such as hydroponics, aeroponics, and nutrient flow techniques have been widely used for the mass production of pure AM inocula. Currently, an increasing number of private and scientific communities are collaborating, thereby improving the development of the private sector to conduct research activities. At last but not the least, the growing pull of the market, increasing employment opportunities, and reduction in usage of agrochemical inputs are alternative strategies in planting and plant production. Due to such circumstances, mycorrhizal inocula production, its application, and maturation of its industry is possible; however, the major drawback or bottleneck effect for its commercialization is inoculum purity testing and quality control due to the obligate nature of AM fungi. Another issue is the lack of knowledge of the origin of mycorrhizal fungal strains used for the commercial production of inocula. As a number of low-quality inocula are available in the market, accessible and appropriate methods for quality products and certified products are required. To exploit the positive effects of mycorrhizal symbiosis, the protection and proper management of native AMF populations in soil is required. Achievements in the mycorrhizal industry are very promising, but there are still a number of regulatory issues that need to be resolved. Therefore, to facilitate the commercial use of mycorrhizal fungi, further research is required to overcome these drawbacks and ensure that the joint efforts of farmers and industry are required for better plant production.

References

Abdel-Fattah TM, Wixtrom A (2014) Catalytic reduction of 4-nitrophenol using gold nanoparticles supported on carbon nanotubes. ECS J Solid State Sci Technol 3(4):M18

Alamari R (2016) Installation d'une unité de production de biofertilisants. Mémoire d'ingéniorat en agronomie, Institut National Agronomique de Tunisie, p 74

Alori ET, Dare MO, Babalola OO (2017) Microbial inoculants for soil quality and plant health. In: Sustainable agriculture reviews. Springer, Cham, pp 281–307

Balestrini R, Lumini E, Borriello R, Bianciotto V (2015) Plant-soil biota interactions. In: Paul EA (ed) Soil microbiology, ecology and biochemistry. Academic, London, pp 311–338

Bécard G, Fortin JA (1988) Early events of vesicular–arbuscular mycorrhiza formation on Ri T-DNA transformed roots. New Phytol 108(2):211–218

Bolan NS (1991) A critical review on the role of mycorrhizal fungi in the uptake of phosphorus by plants. Plant Soil 134(2):189–207

Cavagnaro TR (2008) The role of arbuscular mycorrhizas in improving plant zinc nutrition under low soil zinc concentrations: a review. Plant Soil 304(1–2):315–325

Chilton M-D, Tepfer DA, Petit A, David C, Casse-Delbart F, Tempé J (1982) Agrobacterium rhizogenes inserts T-DNA into the genomes of the host plant root cells. Nature 295(5848): 432–434

Cho I, Tada T, Shinozaki Y (2006) Centerless circular array method: inferring phase velocities of Rayleigh waves in broad wavelength ranges using microtremor records. J Geophys Res Solid Earth 111(B9)

D'Souza J, Rodrigues KM, Rodrigues BF (2013) Modified Strullu and Romand (MSR) medium devoid of sucrose promotes higher in vitro germination in Rhizophagus irregularis. J Mycol Plant Pathol 43(2):240–242

De Boulois HD, Voets L, Delvaux B, Jakobsen I, Declerck S (2006) Transport of radiocaesium by arbuscular mycorrhizal fungi to Medicago truncatula under in vitro conditions. Environ Microbiol 8(11):1926–1934

Declerck S, IJdo M, Fernandez K, Voets L, de la Providencia I (2009) Method and system for *in vitro* mass production of arbuscular mycorrhizal fungi. In: Patent No. WO/2009/090220

Fortin JA, St-Arnaud M, Hamel C, Chaverie C, Jolicoeur M (1996) Aseptic *in vitro* endomycorrhizal spore mass production. In: US Patent No. 5554530

Fortin JA, Plenchette C, Piché Y (2015) Les mycorhizes: la nouvelle révolution verte. Éditions MultiMondes

Frey-Klett P, Garbaye JA, Tarkka M (2007) The mycorrhiza helper bacteria revisited. New Phytol 176(1):22–36

Gerdemann JW, Nicolson TH (1963) Spores of mycorrhizal endogone species extracted from soil by wet sieving and decanting. Trans Br Mycol Soc 46(2):235–244

Hawkins HJ, George E (1997) Hydroponic culture of the mycorrhizal fungus Glomus mosseae with Linum usitatissimum L., Sorghum bicolor L. and Triticum aestivum L. Plant Soil 196(1): 143–149

Hung LLL, Sylvia DM (1988) Production of vesicular-arbuscular mycorrhizal fungus inoculum in aeroponic culture. Appl Environ Microbiol 54(2):353–357

Ijdo M, Cranenbrouck S, Declerck S (2011) Methods for large-scale production of AM fungi: past, present, and future. Mycorrhiza 21(1):1–16

Jarstfer AG, Sylvia DM (1995) Aeroponic culture of VAM fungi. In: Mycorrhiza. Springer, Berlin, Heidelberg, pp 427–441

Johansson JF, Paul LR, Finlay RD (2004) Microbial interactions in the mycorrhizosphere and their significance for sustainable agriculture. FEMS Microbiol Ecol 48(1):1–13

Jolicoeur M, Williams RD, Chavarie C, Fortin JA, Archambault J (1999) Production of Glomus intraradices propagules, an arbuscular mycorrhizal fungus, in an airlift bioreactor. Biotechnol Bioeng 63(2):224–232

Karima B, Samia T (2021) Native arbuscular mycorrhizal fungi and agro-industries in arid lands: productions, applications strategies and challenges. In: Mycorrhizal fungi: utilization in agriculture and forestry. IntechOpen, p 77

Keswani C, Dilnashin H, Birla H, Singh SP (2019) Regulatory barriers to agricultural research commercialization: a case study of biopesticides in India. Rhizosphere 11:100155

Khaliel AS (1988) Incidence of VAM in some desert plants and correlation with edaphic factors. In: Mahadevan A, Raman N, Natarajan K (eds) Mycorrhizae for Green Asia. First Asian conference on Mycorrhizae. University of Madras, pp 156–159

Kirk PM, Cannon PF, David JC, Stalpers JA (eds) (2001) Ainsworth and Bisby's dictionary of the fungi, 9th edn. CABI

Millner PD, Kitt DG (1992) The Beltsville method for soilless production of vesicular-arbuscular mycorrhizal fungi. Mycorrhiza 2(1):9–15

Mohammad A, Khan AG, Kuek C (2000) Improved aeroponic culture of inocula of arbuscular mycorrhizal fungi. Mycorrhiza 9(6):337–339

Moreira BC, Prates P, Jordão TC, Silva MDCSD, Ribeiro APF, Stürmer SL, Salomão LCC, Otoni WC, Kasuya MCM (2019) Effect of inoculation of pineapple plantlets with arbuscular mycorrhizal fungi obtained from different inoculum sources multiplied by the on-farm method. Revista Brasileira de Ciência do Solo 43

Mosse DT (1959) Slenderizing machine. In: US Patent 2,869,539

Mosse B (1962) The establishment of vesicular-arbuscular mycorrhiza under aseptic conditions. Microbiology 27(3):509–520

Mosse B, Thompson JP (1981) National Research Development Corp UK. Production of mycorrhizal fungi. In: U.S. Patent 4,294,037

Mosse B, Stribley DP, LeTacon F (1981) Ecology of mycorrhizae and mycorrhizal fungi. In: Advances in microbial ecology. Springer, Boston, pp 137–210

Nicolson TH, Gerdemann JW (1968) Mycorrhizal endogone species. Mycologia 60(2):313–325

Nurbaity A, Istifadah N, Haryantini BA, Ilhami MF, Habibullah MI, Arifin M (2019) Optimization of hydroponic technology for production of mycorrhiza biofertilizer, IOP conference series: earth and environmental science, vol 347, No. 1. IOP, p 012017

Ojala JC, Jarrell WM, Menge JA, Johnson ELV (1983) Influence of mycorrhizal fungi on the mineral nutrition and yield of onion in saline soil. Agron J 75(2):255–259

Rodriguez A, Sanders IR (2015) The role of community and population ecology in applying mycorrhizal fungi for improved food security. ISME J 9(5):1053–1061

Sadhana B (2014) Arbuscular mycorrhizal fungi (AMF) as a biofertilizer-a review. Int J Curr Microbiol App Sci 3(4):384–400

Sharma S, Sharma S, Aggarwal A, Sharma V, Singh MJ, Kaushik S (2017) Mass multiplication of arbuscular mycorrhizal fungi. In: Aggarwal A, Yadav K (eds) Mycorrhizal fungi. Astral International, pp 157–174

Smith SE, Read DJ (2010) Mycorrhizal symbiosis. Academic press

Smith SE, Smith FA (2012) Fresh perspectives on the roles of arbuscular mycorrhizal fungi in plant nutrition and growth. Mycologia 104(1):1–13

Sreenivasa MN, Bagyaraj DJ (1988) Chloris gayana (Rhodes grass), a better host for the mass production of *Glomus fasciculatum* inoculum. Plant Soil 106(2):289–290

St-Arnaud M, Hamel C, Vimard B, Caron M, Fortin JA (1996) Enhanced hyphal growth and spore production of the arbuscular mycorrhizal fungus *Glomus intraradices* in an *in vitro* system in the absence of host roots. Mycol Res 100:328–332

Sylviam DM, Jarstfer AG (1997) Distribution of mycorrhiza on competing pines and weeds in a southern pine plantation. Soil Sci Soc Am J 61(1):139–144

Tajini F, Trabelsi M, Drevon JJ (2012) Combined inoculation with Glomus intraradices and Rhizobium tropici CIAT899 increases phosphorus use efficiency for symbiotic nitrogen fixation in common bean (*Phaseolus vulgaris* L.). Saudi J Biol Sci 19(2):157–163

Thonar C, Erb A, Jansa J (2012) Real-time PCR to quantify composition of arbuscular mycorrhizal fungal communities—marker design, verification, calibration and field validation. Mol Ecol Resour 12(2):219–232

Tiwari P, Adholeya A (2003) Host dependent differential spread of Glomus intraradices on various Ri T-DNA transformed roots in vitro. Mycol Prog 2(3):171–177

Voets L, Dupré de Boulois H, Renard L, Strullu DG, Declerck S (2005) Development of an autotrophic culture system for the in vitro mycorrhization of potato plantlets. FEMS Microbiol Lett 248(1):111–118

Vosátka M, Látr A, Gianinazzi S, Albrechtová J (2012) Development of arbuscular mycorrhizal biotechnology and industry: current achievements and bottlenecks. Symbiosis 58(1–3):29–37

Chapter 9
Mass Multiplication and Cost Analysis of *Frankia* Biofertilizer

Thirumagal Thirugnanam and Dhanasekaran Dharumadurai

Abstract Soils that have lost primary production owing to abiotic or biotic stressors are classified as degraded lands. Drought, salt, and heavy metals are the most dangerous abiotic stressors in tropical environments. These pressures have an impact on plant development and output. Nitrogen-fixing microbial biofertilizers that can thrive in poor and disturbed soils are commonly used to reclaim damaged land. Plant growth, biomass, shoot and root N content, and survival rate after transplantation in fields are all improved when actinorhizal plants are inoculated with *Frankia*. However, the selection of functional *Frankia* strains is critical to the success of actinorhizal plantation formation in degraded environments. This chapter summarizes mass production and cost-benefit ratio of *Frankia* biofertilizer production.

Keywords *Frankia* · Biofertilizer · Cost–benefit ratio · Carrier · Mass culture

9.1 Introduction

The human population has doubled in the previous four decades, while food production has also doubled. Plant nutrition has made a significant contribution to the enormous increase in food demand and supply. Increased agricultural productivity has been made possible by the advent of commercially accessible man-made fertilizers. Agrochemicals build belowground due to irresponsible application and their incapacity to biodegrade, creating unwanted changes in soil qualities such as structure, fertility, and water holding capacity. Excessive use of synthetic fertilizers has also been linked to water eutrophication, the greenhouse effect, and harmful heavy metal accumulations such as arsenic, cadmium, and plumbum. Mineral fertilizers used in excess can deplete soil nutrients and make crops more susceptible to disease (Mącik et al. 2020). Biofertilizers, which are natural, useful, and environmentally and user-friendly, can be used to address these issues. Commercial

T. Thirugnanam · D. Dharumadurai (✉)
Department of Microbiology, School of Life Sciences, Bharathidasan University, Tiruchirappalli, Tamil Nadu, India

N. Amaresan et al. (eds.), *Agricultural Microbiology Based Entrepreneurship*, Microorganisms for Sustainability 39,
https://doi.org/10.1007/978-981-19-5747-5_9

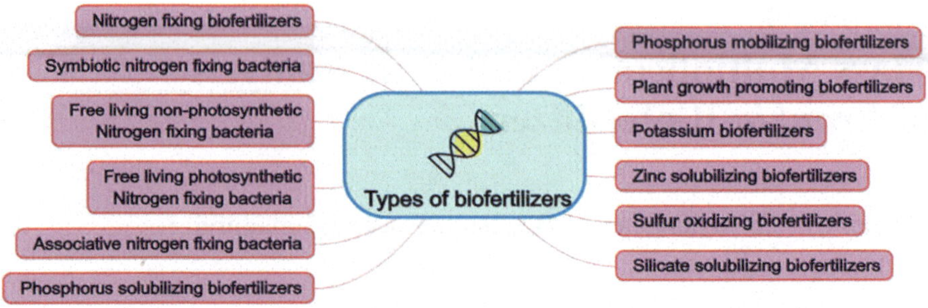

Fig. 9.1 Types of biofertilizers

biofertilizer production and application date back to 1895, with the first biofertilizer patent (no. GB 189511460) filed for registration in the United Kingdom in 1896 (Raimia et al. 2021). Plants receive nutrients from biofertilizers, which also help to combat soil borne illnesses and preserve soil structure. Microbial biofertilizers are critical to the long-term viability of agriculture. Microbial biofertilizers are biological preparations containing suitable densities of microorganism strains that aid plant development in rhizospheres. Living or dormant (metabolically inactive) cells are used in biofertilizer formulations. Some microbes used as biofertilizers are *Azotobacter, Anabaena, Clostridium, Aulosira Beijerinckia, Nostoc, Klebsiella, Stigonema, Desulfovibrio, Rhodospirillum,* and *Rhodopseudomonas, Rhizobium, Frankia, Anabaena azollae, Trichodesmium, Azospirillum* spp, *Herbaspirillum* spp, *Alcaligenes, Enterobacter, Azoarcus spp Acetobacter diazotrophicus, Bacillus circulans, B subtilis, Pseudomonas striata, Penicillium* spp, *B. polymyxa Microccocus Agrobacterium, Aereobacter, Flavobacterium, Penicillium* spp, *Aspergillus awamori, Trichoderma, Arbuscular mycorrhiza, Glomus* spp, *Gigaspora* spp, *Acaulospora* spp, *Scutellospora* spp and *Sclerocystis* spp, *Bacillus. mucilaginosus, B. circulanscan, B. edaphicus,* and *Arthrobacter* spp, *Aspergillus niger, Bacillus* spp, *Thiobacillus* spp, *Mycorhiza Pseudomonas* spp, *Pseudomonas* spp, *Agrobacterium, Pseudomonas fluorescens, Arthrobacter, Erwinia, Bacillus, Rhizobium, Enterobacter, Streptomyces,* and *Xanthomonas*. They are applied to the soil or for seed treatment (Saeid and Chojnacka 2019). Biofertilizers used throughout organic agriculture are divided into several varieties based on the microorganisms they contain (Mącik et al. 2020) (Fig. 9.1).

For plant development, nitrogen is the most limiting nutrient. The biological nitrogen fixers are a specialized category of bacteria that fix nitrogen and make it accessible to the plant (BNFs). Free-living bacteria (*Azotobacter* and *Azospirillium*), blue-green algae, and symbionts (*Rhizobium*, *Frankia*, and *Azolla*) are the three types (Nosheen et al. 2021). *Frankia* is a nitrogen-fixing actinobacterium having a gram-positive symbiotic relationship with actinorhizal plants. Actinorhizal plants and *Frankia* have a symbiotic relationship that results in the production of a perennial root organ termed a nodule, which houses bacteria and fixes nitrogen (Diagne et al. 2013). Because these microbes are important for nitrogen fixation, nutrient solubilization and mobilization, phytohormone production, microbial

Table 9.1 Microorganisms used as biofertilizers

Biofertilizers	Mechanism	Examples	Reference
Phosphorus mobilizing	Transfer phosphorus from the soil to the root cortex	*Arbuscular mycorrhiza, Glomus* spp., *Gigaspora* spp., *Acaulospora* spp., *Scutellospora* spp. and *Sclerocystis* spp	Chang and Yang (2009)
Potassium solubilizing	Solubilize potassium (silicates) by producing organic acids that decompose silicates and help in the removal of metal ions and make it available to plants	*Bacillus. Mucilaginosus, B. circulanscan, B. edaphicus,* and *Arthrobacter* spp. *Aspergillus niger*	Etesami et al. (2017)
Potassium mobilizing	They mobilize the inaccessible forms of potassium in the soil	*Bacillus* spp. and *Aspergillus niger*	
Plant growth promoting	Produce hormones that promote root growth, improve nutrient availability, and improve crop yield	*Pseudomonas* spp. *Agrobacterium, Pseudomonas fluorescens, Arthrobacter, Erwinia, Bacillus, Rhizobium, Enterobacter, Streptomyces,* and *Xanthomonas*	Backer et al. (2009)
Nitrogen fixing	They transform the inert N_2 into plant-usable organic form	*Rhizobium, Frankia, Azolla, Mesorhizobium, Azorhizobium, Bradyrhizobium, Sinorhizobium, Achromobacter, Alcaligenes, Arthrobacter, Acetobacter, Azomonas, Beijerinckia, Clostridium, Bacillus, Enterobacter, Erwinia, Desulfovibrio, Derxia, Corynebacterium, Campylobacter, Herbaspirillum, Klebsiella, Lignobacter, Mycobacterium, Rhodospirillum, Rhodopseudomonas, Xanthobacter, Mycobacterium,* and *Methylosinus*	Choudhury and Kennedy (2004)

community diversification, and soil physicochemical property improvement, it is gaining interest among agronomists and soil scientists for biofertilizer production. Countries such as Argentina, Canada, China, Europe, India, and the United States are driving the global biofertilizer industry forward. When compared to chemical fertilizers, the manufacture of biofertilizer is both cost-effective and simple. During biofertilizer manufacture, however, important parameters like as microbial strains, formulation type, carrier materials, and field uses must be taken into account (Table 9.1, Fig. 9.2).

Frankia is a nitrogen-fixing actinobacterium having a gram-positive symbiotic relationship with actinorhizal plants. It's a free-living filamentous bacterium that may be found in root nodules or soil (Diagne et al. 2013). There are presently

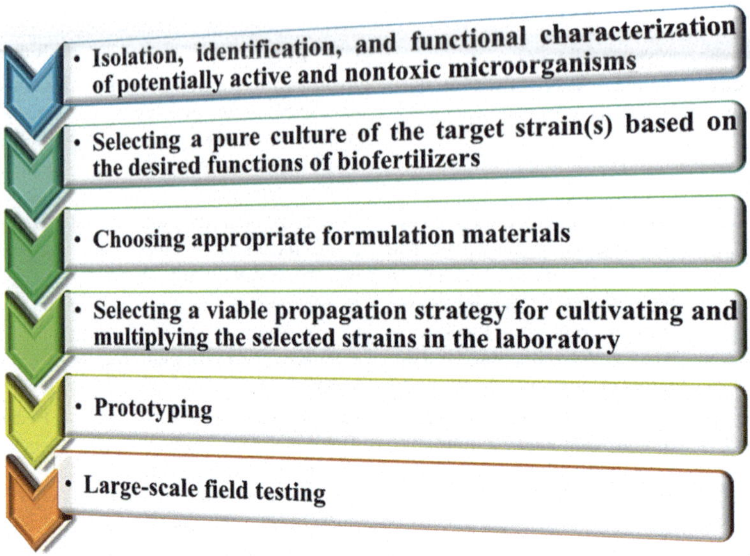

Fig. 9.2 Mass production method of *Frankia* biofertilizer

11 officially recognized species in the genus. *Frankia alni*, the type species, *Frankia asymbiotica*, *Frankia canadensis*, *Frankia casuarinae, Frankia coriariae, Frankia discariae, Frankia elaeagni, Frankia inefficax, Frankia irregularis, Frankia saprophytica,* and *Frankia torreyi* (Nouioui et al. 2019). *Frankia* is the only Frankeniaceae species with the ability to fix nitrogen. A high GC percent content and a sluggish growth rate define these microaerophilic bacteria. *Frankia* develops hyphae and multilocular sporangia on either terminal or intercalary hyphae in liquid culture, depending on the culture setting. The hyphae, or free-living structures, are septate, and the sporangia, or multilocular structures, contain the spores, or bacteria's effective propagules. Vesicles are the sites of nitrogenous activity, and they occur when nitrogen in the media is scarce. Actinorhizal plants and *Frankia* have a symbiotic relationship that results in the production of a perennial root organ termed a nodule, which houses bacteria and fixes nitrogen. The *alder-Frankia* symbiosis has been shown to increase remediation capacities and soil quality by increasing soil nutrients, pH, and cation exchange capacity, as well as boosting plant performance under these adverse conditions. In 1971, R. D. Haller began a planting to restore the landscape of the Bamburi Cement Factory in Mombasa, Kenya, by testing 26 species. Only three species, *C. equisetifolia* sp., *Conocarpus lancifolius* sp., and coconut palm, survived after 6 months. *Casuarina* plants injected with *Frankia* had a better survival percentage following transplantation in this reclamation effort (Diagne et al. 2013)

9.2 Mass Multiplication and Production of *Frankia* Biofertilizer

Mass production of *Frankia* bioferilizer involves five stages.

1. Culturing of *Frankia*
2. Selection of suitable strain of *Frankia*
3. Processing of carrier material
4. Mixing the carrier and the *Frankia*
5. Packaging and storage

9.2.1 *Culturing of* Frankia

Field-collected fresh root nodules of Casuarina are cleansed with tap water to eliminate any dirt and organic particles. The nodules are held in place using forceps, and the roots with connected nodules are sliced 2–3 mm on each side (Narayanasamy et al. 2020).

Intact and undamaged nodules are immersed in 95% ethanol or isopropanol for 10 s (to break the surface tension and remove air bubbles from the tissue); then transferred to a 2.5–3% (v/v) sodium hypochlorite or chlorox (commercial bleach) 1: 1 (v/v) solution and soaked for 4–5 min. The segments are then transferred using sterile forceps after being rinsed in five changes of sterile water. Sterilization of nodules can be done using mercuric chloride solution (0.1% weight/volume) or hydrogen peroxide solution (3% W/V). The nodule is crushed in a sterile tube with sterile glass rod and sterile water. The slurry is than diluted and then streaked on the surface of P medium plates. The *Frankia* culture was plated in DPM medium and incubated at 25 °C for 3–4 weeks (Narayanasamy et al. 2022).

One liter of DPM medium was prepared as follows: 10 g $CaCl_2$ $2H_2O$ 1.4 mM, $MgSO_47H_2O$ 0.2 mM, FeNa EDTA 0.195 mM, K_2HPO_4 3.2 mM, NH_4Cl 0.5 mM, Pyruvate 1 g, Propionte 1 g, Salinity 2% and micro-elements 1 mL (H_3BO_4, $ZnSO_47H_2O$, $MnCl_24H_2O$, $CuSO_45H_2O$, $Na_2MoO_42H_2O$, $CaCl_26H_2O$, $NaH_2PO.2H_2O$) and vitamins 0.1 g (pyridoxine and biotin) and 8 g agar. The medium's pH was adjusted to 8.5. Frankia growth was detected as fluffy white hazy colonies on DPM medium plates after 12–15 days of incubation. For mass multiplication, these colonies were put to DPM medium broth (Fig. 9.3).

Fig. 9.3 Isolation and cultivation of *Frankia* from root nodules of *Casuarina*

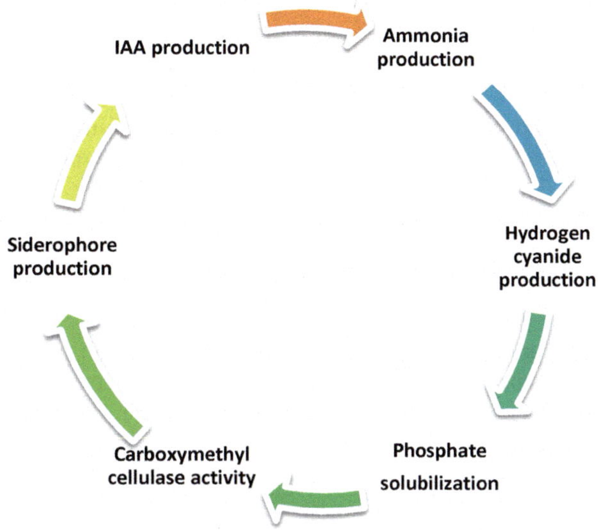

Fig. 9.4 Plant growth promoting property of *Frankia*

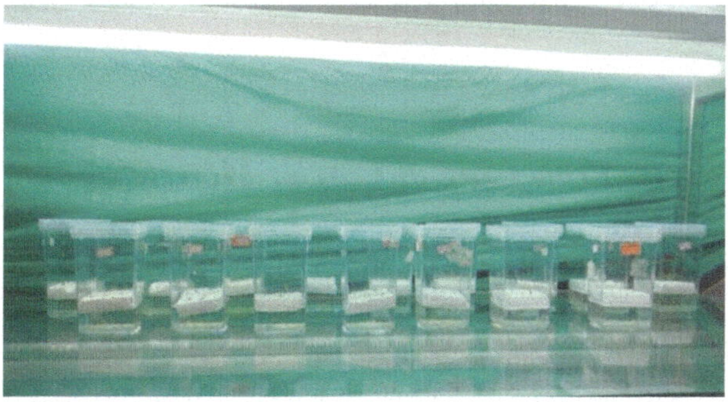

Fig. 9.5 In vitro nodulation study of *Frankia* under hydroponics

9.2.2 Selection of Suitable Strain of Frankia

Highly efficient strains of *Frankia* isolates were screened by plant growth promoting tests and nodulation experiment under hydroponic system. Figure 9.4 illustrates the various Plant growth promoting traits tests for screening potential culture (Fig. 9.5).

Screened Frankia Isolates Were Tested for Nodulation Kinetics

In particular, soil inoculation with PGPB is a promising tool of integrated management systems to increase the efficiency of plants' use of nutrients (from either soil or fertilizers) through microbial technology and the sustainability of the cropping systems.

Casuarinaceae irregularly bear nitrogen $(N)^{-}$ fixing root nodules induced by a symbiotic actinomycete of the genus *Frankia*. *Frankia* cultures grown in vitro are used to artificially inoculate Casuarinaceae. *Frankia* strain was grown in axenic culture. For inoculation, the culture was dispersed in four times its volume of distilled water, and the solution applied in 5-mL aliquots directly to the roots of Casuarinaceae seedlings growing in low-N potting media.

9.2.3 *Mass Culture of* Frankia

The mass production of *Frankia*, efficient strain was grown on slants for preservation as per need culture from slant were transferred to liquid broth(DPM medium) of selective as well as optimized medium in the rotary shaker for 1214 days to prepare starter culture. Later on the starter culture is transferred to the fermenter in batch culture mode with proper maintenance of 30 °C and continuous agitation for 14 days. When cell count reached to 108–109 cells/mL, the broth is used as inoculant. The *Frankia* liquid culture may be applied as such on to the plant root system or by mixing it with the habitat soil and the mixture further spread in the plants.

9.2.4 *Processing of Carrier Material*

The following are the properties of a good carrier material for seed inoculation:

1. Nontoxic to inoculant bacterial strain
2. Good moisture absorption capacity
3. Easy to process and free of lump-forming materials
4. Easy to sterilize by autoclaving or gamma-irradiation
5. Available in adequate amounts
6. Inexpensive
7. Good adhesion to seeds
8. Good pH buffering capacity
9. Nontoxic to plant

The carrier material, which helps to maintain the microbial contents throughout manufacture and ensures the transport of a proper number of viable microorganisms to the field, is an important component of biofertilizer. Solids/tablets, granules, and powders are all examples of carrier materials. It may be organic, inorganic, or synthetic (Rai 2006).

Carrier Sterilization (Amutha et al. 2014)

1. There are two common methods of sterilization.
2. Autoclave (High temperature + high pressure) is used popular due to low cost.
3. Irradiation is a promising alternative method for carrier sterilization.

Carrier Preparation (Amutha et al. 2014)

1. Biofertilizer from associative N_2_fixing bacteria come in three forms: liquid, solid, and lyophilized.
2. For liquid and lyophilized ones, only solution medium is used.
3. But for solid form, carriers such as peat and chicken dung are needed.
4. Peat and chicken dung are dried to just dryness and ground into small particles and sieved at 0.18 mm.

9.2.5 *Mixing of Carrier and* Frankia

Sterilized carrier-based inoculants production: A totally sterile carrier in a sterile container is required for the manufacture. Mixing the sterile carrier with *Frankia* culture is the easiest method. A syringe connected with a sterile needle is used to inject culture into each pre-sterilized carrier bag aseptically. An automatic dispensing machine (auto syringe) might be utilized for large manufacturing. Ethanol must be used to sterilize the puncture site. The amount of broth used should be enough to moisten 40% of the carrier's weight in soup. If the carrier material has nutrients that the integrating bacteria may use to thrive (e.g., mineral soil), injecting a starting culture of bacterial cells into the carrier package along with sterile water for moisture adjustment would suffice.

After that, the puncture hole is sealed with a preprinted self-sticking label. The bags are then kneaded or shaken until the liquid inoculums are evenly distributed throughout the carrier. The inoculants' ultimate moisture content should be 45–50 %. After the injection, the carrier package should be kept in a temperature-controlled environment for a length of time to allow the bacterial cells to reach their maximum population. After 2 weeks, the inoculants are ready for use in non-sterile carrier-based inoculant production. In this mode of manufacturing, broth culture is blended or sprayed over the carrier in a non-sterile mixer. The type and moisture retention ability of the carrier determine the quantities of broth and carrier.

During this time, the moisture balances out and any heat from the soaking dissipates. Following the curing stage, when some *Frankia* multiplication occurs, the inoculants is run through a coarse filter or hammer mill to eliminate lumps and produce a consistent product. The final inoculants can be sealed in moisture-resistant polyethylene bags and stored for testing and use.

Liquid inoculant production (for scale 1500 mL)

Step 1: Inoculate a *Frankia* loop into a 500 mL Erlenmeyer flask containing 150mL sterilized DPM broth.

Step 2: Culture on a rotary shaker at 28 °C and 200 revolutions per minute until late log phase (cell concentration of around 10^8cells/mL). This culture will be used as a beginning culture for the development of liquid inoculants.

Step 3: Inoculate 150 mL of starting culture into a 2000 mL Erlenmeyer flask as a simple fermenter containing sterilized 1500 mL of modified DPM medium as a baseline media for liquid inoculant made with selected suitable ingredient for each *Frankia* genus.

Step 4: Culture at 28 °C with air continually pumped into the media through a 0.45 m filter until maximum cell concentrations of 10^9 cells/mL are reached.

Step 5: Place a 20 mL aliquot of cell culture in a sterile polypropylene bag and seal it. Before using a liquid inoculant, make sure it's kept at the right temperature.

Step 6: Before planting, inoculate 1 kg of seed (for medium seed size, such as *Casuarina*) with 20 mL of liquid inoculants without using a sticker.

DPM media composes of (g/L) 10 g $CaCl_2$ $2H_2O$ 1.4 mM, $MgSO_47H_2O$ 0.2 mM, FeNa EDTA 0.195 mM, K_2HPO_4 3.2 mM, NH_4Cl 0.5 mM, Pyruvate 1 g, Propionte 1 g, Salinity 2% and micro-elements 1 mL (H_3BO_4, $ZnSO_47H_2O$, $MnCl_24H_2O$, $CuSO_45H_2O$, $Na_2MoO_42H_2O$, $CaCl_26H_2O$, $NaH_2PO.2H_2O$) and Vitamins 0.1g (Pyridoxine and biotin). pH is adjusted at 8.5 before autoclaving

9.2.6 *Packaging of* Frankia *Biofertilizer*

Depending on the form, *Frankia* biofertilizer made from associative nitrogen-fixing bacteria is produced and packaged differently. Bacteria in solution media are directly transported into 1 kilogramme, 5 kg, or 10 kg plastic bottles or glass bottles after fermentation, occasionally in large plastic barrels, for liquid biofertilizer. The fermentation liquid is promptly freeze dried and then vacuum packed into finger-shaped glass tubes for lyophilization. After autoclaving (or irradiation) for solid form, the fermentation solution is thoroughly mixed with carriers and packed into tiny polyethylene bags in an axenic atmosphere. All small packages should be put into big paper boxes, tied tightly, and labeled with product name, brand, standard number, producer, address, production date, log number, and net weight after information sheets with product name, brand, standard number, available bacterium number, production date, period of validity, technical specification, manual and producer address sealed inside paper boxes.

9.2.7 *Transportation*

As long as the goods are protected from the sun and rain, common vehicles can be utilized to carry associative N_2-fixing bacteria biofertilizer. However, rain and sunlight are required. The temperature of the vehicle must not exceed 35 °C, and if the temperature is below 0 °C, protective measures must be taken. (c) Storage Frankia biofertilizer containing symbiotic N_2-fixing bacteria should be kept in a shaded, dry, and well-ventilated storage area. It should not be left out in the open. Temperatures between 10 and 25 °C are ideal, with temperatures below 0 °C and above 35 °C being avoided.

9.3 Field Application of *Frankia* Biofertilizer

In fields, inoculation with *Frankia* is commonly carried out with crushed nodule, *Frankia* suspension, *Frankia* enrobed in alginate bead, soil containing *Frankia*, or leaf litter from around nodulated plants. The response to *Frankia* inoculation is strongly linked to factors such as provenance source, *Frankia* strain, and nutrients status of the site such as nitrogen. *Frankia* inoculation in nursery and field conditions is beneficial to *Casuarina* given that it reduces transplantation shock. Another study showed that *Casuarinas* generally do not form nodule outside their zone of origin; however, nodulation occurs when they are inoculated with *Frankia* from the plants' zone of origin. When *Casuarina* trees are planted in sites where they have not previously been planted, inoculation with *Frankia* is recommended for the successful establishment of the plantation given that *Casuarina* strains are generally absent in zone without the host plant. However, for some actinorhizal species infective *Frankia* strains can be found in zone without the host plant.

9.4 Analysis of the Cost-Benefit Relationship

On a daily basis, businesses are forced to make critical financial decisions. Choosing where to put your money is a crucial financial choice that may make or ruin a company. The greatest method to prevent making a poor investment is to use a financial analysis tool that is appropriate for the scenario or investment. A cost-benefit analysis is one of the most extensively utilized financial analysis methods.

Five steps make up a cost-benefit analysis:

1. Establishing the program or investment to be analyzed.
2. Describing the alternatives to the program or investment to be studied.
3. Calculating the program or investment's cash flows.
4. Quantification and monetization of benefits and costs.
5. Analyzing the benefits in relation to the costs (Rogers 1997).

S.No	Requirement	No.	Amount
1	Autoclave	1	45,000
2	Chamber	1	65,000

(continued)

S.No	Requirement	No.	Amount
3	Gas connection	1	8000
4	Bunsen burner	1	500
5	Fermenter	1	100,000
6	Rotatory shaker	1	45,000
7	Incubator	1	30,000
8	Conical flask 1000 mL	10	1200
9	Petri plates	10	1500
10	DPM medium	1	4567
	Total		3,00,767

The production cost for *Frankia* biofertilizer unit permanent asset is Rs. 3, 00,767. Therefore, to produce 50 kg of *Frankia* biofertilizer, we have to invest around Rs. 10,000–15,000.

The IRR is the final measurement that must be determined. To assess if an investment is worth the money spent on it, the IRR is frequently compared to the cost of capital, or the cost of not investing the money and allowing it to accrue interest.

The IRR is calculated using the formula:

$$-\text{Initial Investment} + \text{CF}_i\ (1 + \text{IRR})_i.$$

$i = 1n \sum = 0.13$ and solving for IRR, where CF_i is the cash flow for each year and n is the number of years past the base year.

The average return per unit of time utilized in the computation will be displayed here. If the IRR is higher than the calculated discount rate, it is deemed cost-beneficial. The discount rate is expected to stay constant during the investment's lifetime, which is a disadvantage of computing the IRR.

9.5 Conclusions

As a supplemental, sustainable, and environmentally acceptable supply of plant nutrients, biofertilizers are a potential tool in agricultural environments. Actinorhizal symbioses are a biological tool generally used for the remediation and re-vegetation of soils affected by salt, heavy metal, oil, and so forth. Inoculation with the nitrogen-fixing bacteria *Frankia* improves the nutrient status and enhances actinorhizal plant development. Some *Frankia* strains are very tolerant to salinity and can be used as biofertilizers in land affected by salt. It has been demonstrated that nodulation occurred under saline conditions until 300 mM, approximately 28 dSm^{-1}. Given that *Frankia* improves plant performance in stressed conditions, inoculation with *Frankia* has a beneficial effect on restoration and reforestation of bauxite mine spoils. Their studies showed that in bauxite mine spoils the growth and nutrients uptake (N, P, K) of plants inoculated with *Frankia* was higher than those of non-inoculated plants. Besides, the important role of *Casuarina* on lands reclamation, the symbiotic relationship between *Frankia* actinorhizal plant can also be used as a biocontrol tool against diseases such as bacterial wilt (*Ralstonia*), cataplexy (*Rhizoctonia* sp.), powdery mildew (*Oidium* sp.), Hexenbesen (mycoplasma-like organism; bacteria-like organism), and canker (*Phomopsis* sp.) (Diagne et al. 2013).

Acknowledgments The authors are grateful for the instrumentation facilities provided by the DST-FIST (Department of Science and Technology-Fund for Improvement of Science and Technology Infrastructure), New Delhi (DST Sanction Order No.: SR/FIST/LSI-013/2012/ Dt.13.08.2012). We gratefully acknowledge the Department of Science and Technology's (DST) financial support for the research facility, which was provided through the DST Promotion of University Research and Scientific Excellence (PURSE) scheme-Phase II, Rashtriya Uchchatar Shiksha Abhiyan (RUSA)-2.0-Biological Sciences (TN RUSA: 311/RUSA (2.0)/2018 dt. 02/12/ 2020).

References

Amutha R, Karunakaran S, Dhanasekaran S, Hemalatha K, Monika R, Shanmugapriya P, Sornalatha T (2014) Isolation and mass production of biofertilizer (Azotobacter and Phosphobacter). Int J Latest Res Sci Technol 3(1):79–81

Backer R, Rokem JS, Ilangumaran G, Lamont J, Praslickova D, Ricci E, Subramanian S, Smith DL (2009) Plant growthpromoting rhizobacteria: context, mechanisms of action, and roadmap to commercialization of biostimulants for sustainable agriculture. Front Plant Sci 9:1473

Chang C-H, Yang S (2009) Thermo-tolerant phosphate-solubilizing microbes for multi-functional biofertilizer preparation. Bioresour Technol 100:1648–1658

Choudhury A, Kennedy I (2004) Prospects and potentials for systems of biological nitrogen fixation in sustainable rice production. Biol Fertil Soils 39:219–227

Diagne N, Arumugam K, Ngom M, Nambiar-Veetil M, Franche C, Narayanan KK, Laplaze L (2013) Use of *Frankia* and actinorhizal plants for degraded lands reclamation. BioMed Res Int 2013:948258. https://doi.org/10.1155/2013/948258

Etesami H, Emami S, Alikhani HA (2017) Potassium solubilizing bacteria (KSB) mechanisms, promotion of plant growth, and future prospects A review. J Soil Sci Plant Nutr 17:897–911

Mącik M, Gryta A, Frąc M (2020) Biofertilizers in agriculture: An overview on concepts, strategies and effects on soil microorganisms. Adv Agron 162:458–500. https://doi.org/10.1016/bs.agron.2020.02.001

Narayanasamy M, Dhanasekaran D, Thajuddin N, Akbarsha MA (2020) Morphological, molecular characterization and biofilm inhibition effect of endophytic *Frankia* sp. from root nodules of Actinorhizal plant *Casuarina* sp. S Afr J Bot 134:72–83. https://doi.org/10.1016/j.sajb.2020.02.039

Narayanasamy M, Dhanasekaran D, Thajuddin N (2022) Isolation of *Frankia* from *Casuarina* root nodule. In: Practical handbook on agricultural microbiology, Springer Protocols Handbooks. Springer, pp 320–350. https://doi.org/10.1007/978-1-0716-1724-3_2

Nosheen S, Ajmal I, Song Y (2021) Microbes as biofertilizers, a potential approach for sustainable crop production. Sustainability 13:1868. https://doi.org/10.3390/su13041868

Nouioui I, Cortés CL, Castro JF, Gtari M, Ghodhbane-Gtari F, Klenk H-P, Tisa LS, Sangal V, Goodfellow M (2019) Genomic insights into plant-growth-promoting potentialities of the genus *Frankia*. Front Microbiol 10:1457. https://doi.org/10.3389/fmicb.2019.01457

Rai MK (2006) Handbook of microbial biofertilizers. The Haworth Press

Raimia A, Roopnarainb A, Adelekea R (2021) Biofertilizer production in Africa: current status, factors impeding adoption and strategies for success. Sci Afr 11:e00694. https://doi.org/10.1016/j.sciaf.2021.e00694

Rogers ES (1997) Cost-benefit studies in vocational services. Psychiatr Rehabil J 25–34

Saeid A, Chojnacka KW (2019) Organic farming fertlizers: need for new strategies. Elsevier, pp 345–360. https://doi.org/10.1016/B978-0-12-813272-2.00004-5

Chapter 10
Mass Multiplication, Production Cost Analysis and Marketing of *Azospirillum*

S. Gomathi, V. Ambikapathy, and A. Panneerselvam

Abstract Biofertilizers are natural fertilizers that are microbial inoculants of bacteria, algae, and fungi. Biofertilizers like *Rhizobium, Azotobacter, Azospirillum*, and Blue green algae are known to be responsible for nitrogen fixation in grasses, cereal crops, vegetables, and plantation crops. *Azospirillum* and *Trichoderma harzianum* have also been reported to increase rice yield up to 20% compared to that of standard inorganic fertilizer. *Azospirillum brasilense* Az39 strain was isolated in Argentina in the 1980s within the context of a program carried out to identify microorganisms with potential to be used as agricultural bioinputs. In this chapter we have discussed about *Azospirillum* morphology characters, its species list, isolation of *Azospirillum*, and mass production. Finally, there is a discussion on the marketing, uses, and profits of *Azospirillum* biofertilizer production.

Keywords *Azospirillum* · Mass production · Biofertilizer · Marketing

10.1 Introduction

The escalating cost of synthetic fertilizers with continuous and indiscriminate use makes agriculture expensive, which disturbs the soil environment. A little amount of these fertilizers are utilized by the plants and the remaining is left unused in the soil. Hence to reduce this burden these nutrients needed by plants are made available by the use of biofertilizers, especially those bacteria involved in nitrogen-fixation, plant growth promotion (PGP), phosphate solubilization, potassium solubilization, and fungi like arbuscular mycorrhiza (AM fungi). Biofertilizers are natural fertilizers, which are microbial inoculants of bacteria, algae, and fungi. The biofertilizers like *Rhizobium, Azotobacter, Azospirillum*, and Blue green algae are known to be responsible for nitrogen fixation in grasses, cereal crops, vegetables, and plantation crops. The work on *Azospirillum* during the past 20 years showed increasing

S. Gomathi (✉) · V. Ambikapathy · A. Panneerselvam
Department of Botany, A.V.V.M Sri Pushpam College (Autonomous) Poondi, Thanjavur, Tamil Nadu, India

N. Amaresan et al. (eds.), *Agricultural Microbiology Based Entrepreneurship*, Microorganisms for Sustainability 39,
https://doi.org/10.1007/978-981-19-5747-5_10

scientific interest because of the nitrogen-fixing capacity and their association close to the roots of the forage and cereal crops (Raffia and Charyulu 2021; Sagadevan et al. 2014).

Presently, the global biofertilizer market is expanding due to the rising acceptance of efficient soil nutrient management practices such as the application of biofertilizers among farmers. Biofertilizers are preferred to chemical fertilizers because they are cost effective, ecologically friendly, and guarantee sustainable agricultural production. However, the biofertilizer industry is underdeveloped in many African countries due to several challenges; thus, the full adoption and benefits of biofertilizer are yet to be fully realized compared to the developed nations (Raimi et al. 2020). Plant growth-promoting rhizobacteria have first been used for agricultural purposes in the former Soviet Union and India in the early twentieth century and are now being tested worldwide (Lucy et al. 2004). *Azospirillum* and *Trichoderma harzianum* have also been reported to increase rice yield up to 20% compared to that of standard inorganic fertilizer. *Azospirillum* species were considered as nitrogen fixers that made them to be used as biofertilizers. Biofertilizers can replace chemical fertilizers to promote the plant growth without causing any pollution. The study of hydrogen-oxidizing bacteria (HOB) enrichment as biofertilizers from mixed culture is scarce (Khan 2018; Wei Zhang et al. 2020).

One of the main mechanisms that explain plant growth promotion by *Azosprillum* is its ability to produce or metabolize compounds such as phytoregulators (Di Salvo et al. 2021). *Azospirillum* species influence plant growth through versatile mechanisms; they include N_2 fixation, phytohormone production (e.g., auxins, cytokinins, and gibberellins), increased nutrient uptake, enhanced stress resistance, vitamin production, siderophores and biocontrol, and some of them do P solubilization (Dobbelaere et al. 2003; Rodriguez et al. 2004; Massena Reis et al. 2011).

Some *Azospirillum* strains have species mechanisms to interact with roots and colonize even the root interior, while others colonize the mucigel layer or injured root cortical cells. Using root surface sterilization methods, it was demonstrated that certain strains of *Azospirillum* spp. in fact colonize the root interior of wheat. An *Azospirillum* plant root association can only be successful if the bacterium is able to survive in the soil and attain significant populations on the host root system (Baldani et al. 1986).

10.2 *Azospirillum*

Azospirillum was first described as *Spirillum lipoferum* by Beijerinck in 1925 as a nitrogen-fixing bacterium. *Azospirillum* is recognized as a ubiquitous soil organism capable of colonizing effectively not only the roots of a wide variety of plants but also their above ground portions forming apparently an associative symbiosis (Tilak et al. 1981). Okon and Labandera Gonzalez (1994) summarized the results of several field experiments performed in different countries for 20 years and reported that *Azospirillum* inoculation positively affected plant growth in 60–70% of the cases analyzed, with grain yield increases in the range of 5–30%. *Azospirillum brasilense* Az39 strain was isolated in Argentina in the 1980s within the context of a program carried out to identify microorganisms with potential to be used as agricultural bioinputs. *Azospirillum* utilizes glucose, lactate, succinate, fructose, malate, pyruvate, fumarate, as carbon source, reduced nitrate, and does not require biotin. The Nitrogen source is used by *Azospirillum* for their growth: ammonium, nitrate, amino acids, nitrogen elements.

10.3 Morphology

The genus *Azospirillum* genus belongs to the alphaproteobacteria (Baldani et al. 2005), the species of which belong to associative, facultative, endophytic microaerophilic diazotrophs isolated from the roots and aboveground parts of a variety of crop plants like forage grasses, cereals, legumes, millets, and soils, which colonize the surface and interior of roots. *Azospirillum* represents a lineage that might have transitioned to terrestrial environments much later than the precambrian split of "hydrobacteria" and "terrabacteria." *Azospirillum* species are described as Gram-negative, rod-shaped, 1 mm in diameter, and very motile. Cells are about 1.0 μm × 3.5 mm in size single flagellum when grown in MPSS broth while lateral flagella when grown on MPSS agar at 30 °C. They also form wrinkled, dark pink colonies when grown on MPSS agar. A formation of a white veil or bacteria band, is visible when inoculated into an Nfb and Dobereiner's liquid medium (Wu et al. 2020).

Azospirillum spp. are highly adaptable, being able to grow under:

- Anaerobic conditions (nitrate used as electron acceptor).
- Microaerobic (elemental or ammonia used as N source).
- Fully aerobic conditions (ammonia, nitrate, amino acid, or combined N only) (Figs. 10.1 and 10.2).

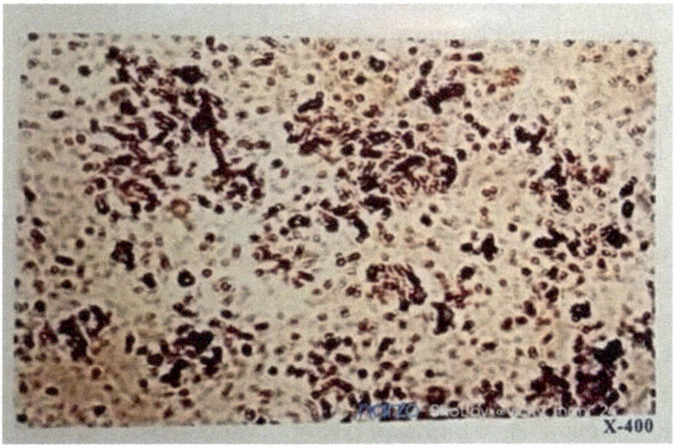

Fig. 10.1 *Azospirillum* oval-shaped cells

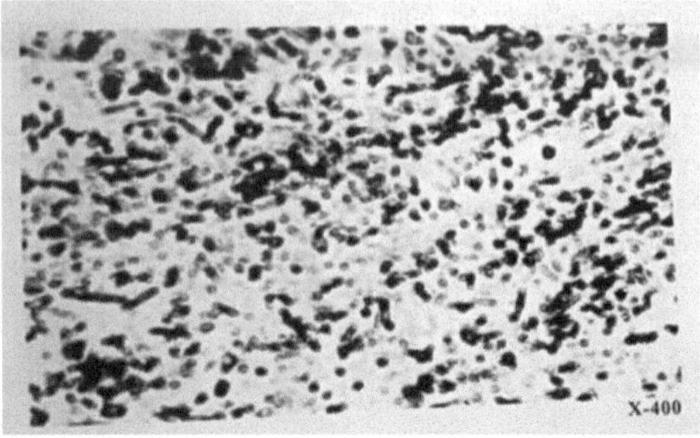

Fig. 10.2 *Azospirillum* in rod-shaped cells

Table 10.1 Some selected *Azospirillum* species list

Bacterial species	Origin of isolation	Reference
Azospirillum brasilense sp245	Surface-sterilized wheat roots, Brazil	Baldani et al. (1986)
Azospirillum lipoferum	Roots of wheat and maize, Brazil	Tarrand et al. (1978)
Azospirillum amazonense	Roots and rhizosphere soil of Gramineae, Amazon region, Brazil	Magalhaes et al. (1983)
Azospirillum halopraeferens	Roots of Kallar grass, grown in saline soils in Pakistan	Reinhold et al. (1987)
Azospirillum irakense	Roots and rhizosphere of rice, Iraq	Khammas et al. (1989)

10.4 Species List of *Azospirillum*

The occurrence of *Azospirillum* species is widespread in the environment, including tropical, subtropical, and temperate regions; most of them were described from plant roots and soil samples. However, besides its association with plants, *Azospirillum* spp. have also been associated with other environments under extreme conditions of temperature or contamination. Some species have significant agricultural importance, specifically as aerobic nitrogen-fixing species, with considerable plant growth-promoting abilities (Oda Steenhoudt and Jos Vanderleyden 2000) (Table 10.1).

Until now, a total of 22 species have been described: *A. lipoferum*, *A. brasilense*, *A. amazonense* (now reclassified as *Nitrospirillum amazonense*; Lin et al. 2014), *A. halopraeferens*, *A. irakense* (now reclassified as *Niveispirillum irakense*; Lin et al. 2014), *A. largimobile* (a reclassification of *Conglomeromonas largomobilis* sub sp. *Largomobilis*; Sly and Stackebrandt, 1999), *A. doebereinerae*, *A. oryzae*, *A. melinis*, *A. canadense*, *A. zeae*, *A. rugosum*, Candidatus *A. massiliensis* (a provisional name for a well-characterized but as-yet uncultured organism), *A. picis*, *A. palatum* (not validated), *A. thiophilum*, *A. formosense*,

A. fermentarium, A. humicireducens, A. himalayense (not validated), *A. soli* and *A. agricola*. However, in terms of physiology, genetics, and agricultural utilization, the most studied ones are *A. brasilense* and *A. lipoferum* described by Tarrand et al. (1978), associated with forage grasses, maize, wheat, rice, sorghum, sugarcane, and several other plants (Hartmann and Baldani 2006; Zambrano et al. 2007; Pedraza et al. 2020).

10.5 Isolation of *Azospirillum*

10.5.1 Principle

When root sample containing *Azospirillum* is introduced into the N-free malic acid medium, *Azospirillum* starts utilizing malic acid present in the medium as carbon source, which resulted in the color change from yellowish green to blue color due to the change in pH of the medium from acidic to alkaline.

10.5.2 Materials Required

- Freshly collected root samples
- 70% ethanol or 0.1% mercuric chloride
- Sterile water blanks
- N-free malic acid semisolid medium in test tubes
- Sterile forceps and petridishes

10.5.3 N-Free Semisolid Malic Acid Medium

- Malic acid: 5.0 g
- Potassium hydroxide: 4.0 g
- Dipotassium hydrogen orthophosphate: 0.5 g
- Magnesium sulfate: 0.2 g
- Sodium chloride: 0.1 g
- Calcium chloride: 0.2 g
- Fe-EDTA (1.64% w/v aqueous): 4.0 mL
- Trace element solution: 2.0 mL
- BTB (0.5% alcoholic solution): 2.0 mL
- Agar: 1.75 g
- Distilled water: 1000 mL pH: 6.8

10.5.4 Trace Element Solution

- Sodium molybdate: 200 mg
- Manganous sulfate: 235 mg
- Boric acid: 280 mg
- Copper sulfate: 8 mg
- Zinc sulfate: 24 mg
- Distilled water: 200 mL

10.5.5 Procedure

- Prepare semisolid malic acid medium taken in test tubes with 5 mL quantity and sterilize 121 °C (15 psi) for 15 min.
- Collect fresh root samples from any graminaceous plant.
- Wash the roots in tap water to remove the adhering soil particles.
- Using sterilized knife/blade, cut the roots into small bits of 0.5–2 cm size.
- Surface sterilize the root bits by immersing them in either 70% ethanol or 0.1% mercuric chloride for 1 min.
- Wash the root bits with sterile distilled water 3–4 times to remove the excess ethanol or mercuric chloride.
- Using sterile forceps transfer aseptically 2–3 root bits to the test tubes containing N-free semisolid malic acid medium.
- Incubate the tubes under room temperature 28 ± 2 °C for 2–3 days.
- Maintain one tube as control without root bits (Fig. 10.3).

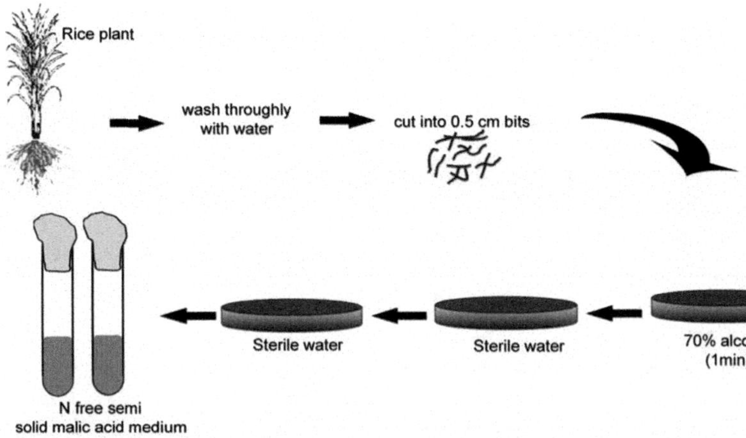

Fig. 10.3 Isolation of *Azospirillum* from cereal roots

10.5.6 Observation

- At the end of incubation time observe the tubes for the change of the color of the medium from yellowish green to blue and for the formation of subsurface white pellicle.
- Collect the positive tubes and take a loopful of the growth from the pellicle and streak on the Malate medium in Petri dish; purify and maintain in slants.

10.6 Carrier Making

(a) *Carrier Preparation*
Biofertilizer from associative N_2-fixing bacteria come in three forms: liquid, solid, and lyophilized. For liquid and lyophilized ones, only solution medium is used, but for solid form, carriers such as peat and chicken dung are needed. Peat and chicken dung are dried to just dryness and ground into small particles and sieved at 0.18 mm.

(b) *Carrier Sterilization*
There are two common methods of sterilization. Autoclave (High temperature + high pressure) is used popularly due to low cost. Irradiation is a promising alternative method for carrier sterilization. All procedures are described previously.

10.7 Mass Production of *Azospirillum*

- Select and dry carrier materials.
- Grind carrier materials.
- Sift carrier materials and select suitable sizes for granular and powdered inoculants.
- Neutralize carrier materials.
- Sterilize the carriers.
- Examine the carriers for sterility after sterilization.
- Inoculate carriers with broth cultures.
- Plate inoculant for quality control.

The water holding capacity of a carrier determines the amount of liquid inoculum that can be added to it. Carriers vary greatly in their water holding capacity. The first step is to determine the inherent moisture level of the carrier. Weigh 10 g accurately on glass weighing dish and place it into the oven at 70 °C for 24 h. Weigh and return to the oven. Another weighing at 48 h will confirm the endpoint of moisture loss.

$$\text{Moisture content} = \frac{(W1 - W2) \times 100\%}{W2}$$

W_1 = Weight of carrier before drying.
W_2 = Weight of carrier after drying 70 °C.

Carrier materials are chosen based on criteria mentioned earlier. The pH of an inoculant carrier should be around 6.5–7.0. Test the sterility of carrier materials by aseptically removing a 10 g sample from each bag and transferring into 90 mL of sterile water in dilution bottles. Prepare serial dilutions from 10–1 to 10–4. Perform Miles and Misra drop plate method on specific media. Check the plates daily for 7 days for signs of growth and appearance of microorganisms, which survived the sterilization.

The sterilized carrier materials in sealed bags are injected aseptically with a suitable amount of broth culture. Swap a small area in a corner of the carrier bag with 70% ethanol. Cut open the bag and inoculate the desired amount of inoculum. Seal the hole with labeling tape. Work the broth into the peat by kneading the bags until the liquid inoculum has been uniformly absorbed by the carrier. Incubate at 30–32 °C for 1–2 weeks.

10.8 Packaging and Preservation of Biofertilizer with Associative N_2 Fixer

(a) *Packing*

Production and packaging of biofertilizer from associative nitrogen-fixing bacteria is different, depending on its forms. For liquid biofertilizer, bacteria in solution medium is directly transferred into 1, 5, or 10 kg plastic bottles or glass bottles after fermentation, sometimes in big plastic barrels. For lyophilized form, fermentation liquid is immediately freeze dried and then packaged into finger-shaped glass tubes under vacuum. For solid form, fermentation solution is mixed thoroughly with carriers after autoclaving (or irradiation), which is then packed into small polyethylene bags under axenic environment. All small packages should be put into big paper boxes, tied tightly, and labeled with product name, brand, standard number, producer, address, production date, log number, and net weight after information sheets with product name, brand, standard number, available bacterium number, production date, period of validity, technical specification, manual and producer address sealed inside paper boxes.

(b) *Transportation*

Common vehicles can be used for transportation of associative N_2-fixing bacteria biofertilizer, as long as the products are sheltered from sunshine and rain. But rain and sunshine shelf are needed. Temperature of transportation must not be over 35 °C, protection measures should be used if temperature is below 0 °C.

(c) *Storage*

Biofertilizer with associative N_2-fixing bacteria should be stored under shade, dry, and air circulated storeroom. It should not be stored in open areas. The best temperature is 10–25 °C, avoiding temperature below 0 and above 35 °C (FNCA 2006).

10.9 Marketing

10.9.1 Field Demonstration

The farmers do what they see because "Seeing is believing" and therefore result as well as method demonstration are very effective tools in promoting biofertilizer usage. The producers may synergize their efforts on this front as bio-fertilizers are new and it is very crucial to show the impact of bio- fertilizer usage to farmers and educate them the economics/returns. Therefore a demonstration farm may be developed jointly, at different locations, defining a catchment area, which could be shown to farmers at different crop stages.

10.9.2 Market Segmentation and Product Positioning

The segmentation is primarily dividing market into various groups of buyers. First of all the organic producers will be the most important buyers as organic production without bio fertilizers will not be possible. Among non-organic producers, the market can be segmented by "specific crop grower (Fruits/Vegetable/Oilseed/Pulses/Sugarcane/Cereals), institutional buyers (Cane/Tea/Coffee/cotton/oilseeds/pulses federations & research-farms, SFCI, Agro-industries etc.).

Biofertilizers can be easily positioned as environment friendly growth enhancer manure with long term benefits such as enrichment of soils, similarly other benefits for example: (a) "Save cost through reduced dosage of chemical fertilizers"(b) "Improves resistance power against disease" (c) "Enhance sugar recovery percent in sugarcane" etc. need to be highlighted.

10.9.3 Pricing

Being a price-sensitive input, the pricing needs to be kept at penetrative level, slightly lower than the competitors. However, real advantage to the units will come from reduction in logistics costs being near to the consuming areas.

10.9.4 Publicity and Training

The POS (Point of Sales) material giving details of proper method of application must be made available to all dealer/distributors and also needs to be ensured that product is displayed visibly. To deploy Extension Executives for promoting bio fertilizers with constant visits and developing a close connect with farmers and undertaking demonstrations with its replication in nearby villages.

10.9.5 Marketing Linkages

With the promotion of alternate sources of nutrition management, there is already awareness among the farmers related to bio fertilizer and it is becoming popular gradually. Now Bio fertilizers of many brands are readily available in the market through the regular dealer/distributor network. So it is not very difficult to promote the appropriate crop specific products manufactured inside any state. Moreover these products will have added advantage of lower transportation and marketing cost. The marketing of the products can therefore be done through the existing marketing network. The farmer co-operatives and farmer groups can also be contacted for bulk selling.

10.9.6 Marketing Challenges and Options in Biofertilizer Business

In spite of being cost effective input, the biofertilizers have not been accepted by the farmers completely till now. Some of the reasons/constraints for low acceptance of biofertilizer are narrated below. However, the "Liquid form" has overcome few limitations and has provided opportunities for Marketers.

- Biofertilizers are live microorganisms, which dies in case of high temperature.
- The shelf life of biofertilizer is limited to 6–12 months in powder form.
- The biofertilizers are used before sowing and delay in dispatches leads to inventory carry over and expiry of product.
- Some of the bio-fertilizers are crop-specific as well as location-specific and therefore its efficacy does not remain the same at different locations due to difference in agro-climatic conditions and soil ediphic factors.
- Soil characteristics like high nitrate, low organic matter, less available phosphate, high soil acidity or alkalinity, high temperature, as well as presence of high agro-chemicals or low micro-nutrients contribute to failure of inoculants or adversely affect its efficacy.

Table 10.2 Outline for expentidure of *Azospirillum* production

Scale level	Particulars	Price
Lab scale	Glasswares	2000
	Chemicals	2000
	Miscellaneous	1000
	Total	5000
Small scale	Building cost	7000
	Glasswares	3000
	Machinery	7000
	Carrier materials	1000
	Miscellaneous	2000
	Total	20,000
Large scale	Building cost	10,000
	Machineries	15,000
	Glasswares	5000
	Chemicals	5000
	Carrier materials	5000
	Travel expense	7000
	Miscellaneous	3000
	Total	50,000
Net profit	Cost for 1 kg	₹170
	Cost for 1 tonne	₹17,000

- Supply of substandard or spurious material by some of the manufacturers also adversely affect the credibility of the biofertilizers, being a new product.
- Some firms are selling organic manures as biofertilizers. Some organizations mention shelf life as 2 years/1 year despite norm of maximum 3–6 months.
- Lack of awareness of the farmers regarding benefits of bio-fertilizer.
- There is no magic effect of bio-fertilizer & its impact is not visible in standing crop and therefore farmer is not convinced with the benefits of bio-fertilizer use (Table 10.2).

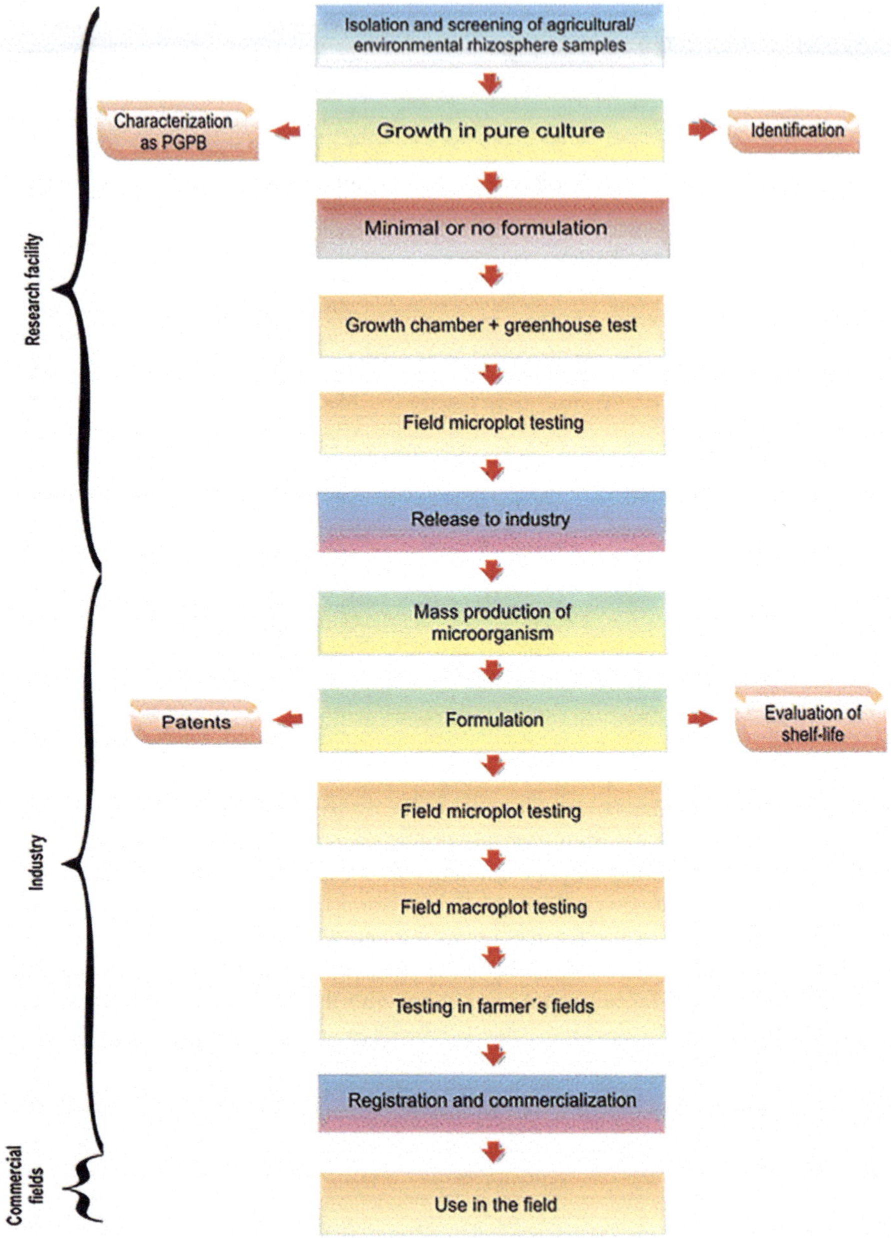

10.10 Uses of *Azospirillum*

The benefits of biofertilizer with associative N_2 fixing *Azospirillum* bacteria can be seem as follows:

- Enhances shoot growth and root development.
- Improves yield of host plants.
- Increased uptake of N, P, K, and microelements.
- Improved water status of plants.
- Replaces 30–50% of the total amount of N requirement.
- Makes plants resistant to drought and pests.
- Increased dry matter accumulation and grain yield.
- Reduces incidence of rice tungro and corn ear-worm attack.
- Increases yield and milling recovery of rice.

10.11 Profits

In general, biofertilizer from associative N_2-fixing bacteria could be used especially for cereal crops such as rice and wheat, but also used for cash crops such as vegetables, fruits, flowers, tobacco, cotton, oilseed, tea, and medicinal or herbal crops. BIO-N in Philippines is a microbial-based fertilizer for rice, corn, and other agricultural crops like tomatoes, pepper, eggplant, okra, lettuce, pechay, and ampalaya. It is a breakthrough technology that promises very significant impact on the country's farmers in terms of increasing farm productivity and income as well saving the country's dollar reserve due to decreased importation of inorganic nitrogenous fertilizers. In China and other FNCA-Countries, associative nitrogen fixing bacteria biofertilizer increased yield by 10–30% and saved chemical N fertilizer by 15–25%.

10.11.1 Cereal Crops

Liquid form is good for rice. At transplanting, immerse rice roots into liquid biofertilizer for 10–15 min before transplanting and spread on paddy soil at regreening stage at rate of 1.5–3.0 L per ha. For wheat, immerse seeds into liquid biofertilizer overnight before sowing, and spread onto wheat leaf at rate of 1.5–3.0 L per ha with water.

10.11.2 Rice

(a) *Solid Inoculant for Direct-Seeded Rice or for Sowing on Dapog Bed*
Soak seeds overnight in clean water. Pre-germinate the seeds in gunny sacks or suitable containers. When radicles (embryonic root) come out, place the germinants in suitable container. Pour required amount of BIO-N and mix

thoroughly until germinants are evenly coated. Sow directly over field or on prepared beds.

(b) *Liquid Inoculant for Dapog Bed*

Suspend the required amount of bio-N in sufficient volume of clean water (e.g., 1 packet bio-N to 1 gallon water) and evenly drench the seed/seedling-lined dapog bed.

(c) *Slurry for Transplant Seedling*

In a suitable container, mix BIO-N with clean water to form a slurry or thick preparation. Prune the roots of seedlings into uniform length and dip for at least 30 min or 1 h before transplanting.

10.11.3 Corn

Place seeds in a suitable container and moisten with water. Pour sufficient amount of inoculants, one packet of BIO-N, which is sufficiently treated with 3 kg of seeds. Mix thoroughly until the seeds are evenly coated. Sow coated seeds immediately. Be sure not to expose the inoculated seeds to direct sunlight. Depending on the soil analysis, very marginal soils may require a basal application of at least a bag or two of 14-14-14 to a hectare as side dress.

10.11.4 Vegetables

Solid biofertilizer is spread, band-spread, and hole applied as basal or top dressing. For leafy vegetables such as celery, spinach, and cabbage, apply at rate of 3.75–15.0 kg per ha. For fruits, vegetables such as cucumber, eggplant, tomato, and melon, apply at rate of 7.5 kg per ha. For root vegetable such as sweet potato, potato, ginger, and garlic, apply at rate of 3.75–15.0 kg per ha.

10.11.5 Fruits

The amount of fertilizers such as mg/g10–20, 20–30, or 30–50 per plant will be applied to those respectively with plant yield less than 50, 50–100, and over 100 kg.

For those where biofertilizer with associative N_2-fixing bacteria applied, N fertilizer should be reduced by 20–25%. Mixed application with organic manure should be encouraged, because organic manure will benefit microbes.

10.12 Conclusion

In using biofertilizers in the field of agriculture some cautions and limitations should be followed, such as never mixing it with chemical nitrogen fertilizers; never applying with fungicides, plant ash, etc., at the same time; never directly exposing to sunlight; not keeping the used solution overnight; and storing at room temperature, not below 0 °C and over 35 °C. Biofertilizer with associative N_2-fixing bacteria only serves as supplement for nitrogen requirement of corn, rice, and sugarcane. It is still necessary to apply 30–50% of the recommended inorganic forms to meet the requirements for other nutrients such as phosphorous and potassium.

References

Baldani VLD, de Alvarez B MA, Baldani JI, Döbereiner J (1986) Establishment of inoculated *Azospirillum* spp. in the rhizosphere and in roots of self grown wheat and sorghum. Plant Soil 90:35–46

Baldani JI, Krieg NR, Baldani VLD, Hartmann A, Deobereiner J (2005) Genus II. *Azospirillum*. In: Brenner DJ, Krieg NR, Staley JT (eds) Bergey's manual of systematic bacteriology, vol 2. Springer, New York, pp 7–26

Di Salvo LP, García JE, Puente ML (2021) The drop plate method as an alternative for *Azospirillum* spp viable cell enumeration within the consensus protocol of the REDCAI network. Rev Argent Microbiol 54(2):152–157. https://doi.org/10.1016/j.ram.2021.05.002

Dobbelaere S, Vanderleyden J, Okon Y (2003) Plant growth-promoting effects of diazotrophs in the rhizosphere. Crit Rev Plant Sci 22:107–149

FNCA Biofertilizer Project Group (2006) Biofertilizer manual. Asia Cooperation Center, Japan Atomic Industrial Forum (JAIF). http://www.fnca.jp/english/index.html

Hartmann A, Baldani JI (2006) Chapter 3.1.5. The genus Azospirillum. PRO 5:115–140. https://doi.org/10.1007/0-387-30745-1_6

Khammas KM, Ageron E, Grimont PAD, Kaiser P (1989) Azospirillum irakense sp. Nov., a nitrogen-fixing bacterium associated with rice roots and rhizosphere soil. Res Microbiol 140: 679–693

Khan HI (2018) Appraisal of biofertilizers in rice: to supplement inorganic chemical fertilizer. Rice Sci 25(6):357–362. https://doi.org/10.1016/j.rsci.2018.10.006

Lin SY, Hameed A, Shen FT, Liu YC, Hsu YH, Shahina M, Lai WA, Young CC (2014) Description of *Niveispirillum fermenti* gen. Nov., sp. Nov., isolated from a fermentor in Taiwan, transfer of *Azospirillum irakense* (1989) as *Niveispirillum irakense* comb. Nov., and reclassification of *Azospirillum amazonense* (1983) as *Nitrospirillum amazonense* gen. Nov. Antonie Van Leeuwenhoek 105:1149–1162

Lucy M, Reed E, Glick BR (2004) Applications of free living plant growth-promoting rhizobacteria. Antonie Van Leeuwenhoek 86(1):1–25

Magalhaes FM, Baldani JI, Souto SM, Kuykendall JR, Dobereiner J (1983) A new acid-tolerant *Azospirillum* species. An Acad Bras Cienc 55:417–429

Okon Y, Labandera Gonzalez CA (1994) Agronomic applications of *Azospirillum* an evaluation of 20 years worldwide field inoculation. Soil Biol Biochem 26(1):591–601

Pedraza RO, Filippone MP, Fontana C, Salazar SM, Ramirez-Mata A, Sierra-Cacho D, Baca BE (2020) Azospirillum. In: Beneficial microbes in agro-ecology. Academic Press. https://doi.org/10.1016/B978-0-12-823414-3.00006-X

Raffia MM, Charyulu PBBN (2021) *Azospirillum*—biofertilizer for sustainable cereal crop production: current status. In: Viswanath B (ed) Recent developments in applied microbiology and biochemistry. Academic Press, pp 193–209. https://doi.org/10.1016/B978-0-12-821406-0.00018-7

Raimi A, Roopnarain A, Adeleke R (2020) Biofertilizer production in Africa: current status, factors impeding adoption and strategies for success. Sci Afr 11:e00694. https://doi.org/10.1016/j.sciaf.2021.e00694

Reinhold B, Hurek T, Fendrik I, Pot B, Gillis M, Kersters K, Thielemans S, De Ley J (1987) *Azospirillum halopraeferens* sp. Nov., a nitrogen-fixing organism associated with roots of Kallar grass (*Leptochloa fusca* (L.) Kunth). Int J Syst Bacteriol 37:43–51

Reis VM, dos Santos Teixeira KR, Pedraza RO (2011) What is expected from the genus *Azospirillum* as a plant growth-promoting bacteria? In: Maheshwari DK (ed) Bacteria in agrobiology; plant growth responses. Springer, pp 123–138. https://doi.org/10.1007/978-3-642-20332-9_6

Rodriguez H, Gonzalez T, Goire I, Bashan Y (2004) Gluconic acid production and phosphate solubilization by the plant growth-promoting bacterium *Azospirillum* spp. Naturwissenschaften 91:552–555

Sagadevan PR, Rajakumar SN, Suresh R, Ranjithkumar P, Karthikeyan P, Rathish Kumar S (2014) Isolation and mass inoculum production of *Azospirillum* from paddy field soil. Int J Biosci Nanosci 1(6):141–145

Sly LI, Stackebrandt E (1999) Description of Skermanella parooensis gen. nov., sp. nov. to accommodate Conglomeromonas largomobilis subsp. parooensis following the transfer of Conglomeromonas largomobilis subsp. largomobilis to the genus Azospirillum. Int J Syst Evol Microbiol 49:541–544

Steenhoudt O, Vanderleyden J (2000) *Azospirillum*, a free living nitrogen-fixing bacterium closely associated with grasses: genetic, biochemistry and ecological aspects. FEMS Microbiol Rev 24(4):487–506. https://doi.org/10.1111/j.1574-6976.2000.tb00552.x

Tarrand JJ, Krieg NR, Dobereiner J (1978) A taxonomic study of the *Spirillum lipoferum* group, with descriptions of a new genus, *Azospirillum* gen. Nov. and two species, *Azospirillum lipoferum* (Beijerinck) comb. Nov. and *Azospirillum brasilense* sp. Nov. Can J Microbiol 24: 967–980

Tilak KVBR, Singh CS, Rana JPH (1981) Effect of combined inoculation of *Azospirillum brasilense* with *Rhizobium trifoli, R. melioti* and *Rhizobium* sp. (cowpea miscellany) on nodulation and yield of clover (*Trifolium repens*), Lucorne (*Mdeicago sativa)* and chick pea (*Cicer arientinum)* Zentralb Bakteroil Parasitenkd infekxtionskr Hyg, Zweite Naturwiss. Zentralbl Mikrobiol 136:22–30

Wu D, Zhang XJ, Liu HC, Zhou YG, Wu XL, Nie Y, Kang YQ, Cai M (2020) *Azospirillum oleiclasticum* sp. Nov, a nitrogen-fixing and heavy oil degrading bacterium isolated from an oil production mixture of Yumen Oilfield. Syst Appl Microbiol 44(1):126171. https://doi.org/10.1016/j.syapm.2020.126171

Zambrano ER, Jiménez-Salgado T, Tapia-Hernández A (2007) Estudio de bacterias asociadas a orquídeas (Orchidaceae). Lankesteriana 7(1–2):322–325

Zhang W, Li Y-X, Niu Y, Zhang F, Li Y-B, Zeng RJ (2020) Two-stage enrichment of hydrogen-oxidizing bacteria as biofertilizers. Chemosphere 266:128932. https://doi.org/10.1016/j.chemosphere.2020.128932

Chapter 11
Mass Multiplication, Production Cost Analysis, and Marketing of *Azotobacter*

Prittesh Patel, Sajjan Grover, and Pushpendra Singh Chauhan

Abstract *Azotobacter* is a nitrogen-fixing, free-living bacteria used as a biofertilizer for eco-friendly and sustainable crop production. These organisms are polymorphic, have peritrichous flagella, and synthesize polysaccharides; they are sensitive to acidic pH, high salts, and temperatures above 35 °C, and they can survive on a nitrogen-free media, allowing them to use atmospheric nitrogen (N_2) for cell protein synthesis. The manufacturing of large-scale *Azotobacter* sp. inoculants begins with the isolation and screening of viable strains, then progresses to mass production using a suitable culture medium, and ultimately, the formulation selection based on the application mode required. Depending on their particle size, solid formulations can be categorized into powders and granules. *Azotobacter* is inoculated into crops using seed, seedling root dip, and direct soil application. While working to improve the nutritional characteristics of *Azotobacter* as a biofertilizer, it's also important to think about cost-effective techniques that can give a cheaper source of biofertilizer to the agricultural industry. Total approximate production cost including chemical media and bottle for 1 L of biofertilizer would be 38 INR and the expected profit would be 210 INR. While India is recognized for its agriculture, it shows that beginning a bio-fertilizers firm in India might be lucrative. The most effective way to reach fertilizer distributors and farmers is through social media applications and websites. One can now register a product and sell it directly through online buying sites. Through field demonstrations, research, and financial help to investors, the national program aims to popularize the innovative biofertilizer-based technology.

Keywords *Azotobacter* · Biofertilizer · Diazotrophs · N fixation · Plant · Soil

---

P. Patel (✉)
C. G. Bhakta Institute of Biotechnology, Uka Tarsadia University, Surat, Gujarat, India

S. Grover
Department of Entomology, University of Nebraska—Lincoln, Lincoln, NE, USA

P. S. Chauhan
Crop Life Science Ltd, Ankleshwar GIDC, Ankleshwar, Gujarat, India

N. Amaresan et al. (eds.), *Agricultural Microbiology Based Entrepreneurship*,
Microorganisms for Sustainability 39,
https://doi.org/10.1007/978-981-19-5747-5_11

11.1 Introduction

Biofertilizers are biologically active products that contain single or composite cultures of bacteria, algae, or fungi. The majority of biofertilizers have the ability to transform atmospheric nitrogen into forms suitable for plant uptake. In general, there are two types: symbiotic and free living. Rhizobium represents the former, which require a symbiotic relationship with plants. *Azotobacter, Azospirillium*, blue green algae (BGA), and Azolla are among the latter, which can fix nitrogen on their own (Fig. 11.1).

One of the limiting variables in agriculture is provisioning plant nutrients, notably nitrogen (N), to crops sustainably. As a result, inoculation of diazotrophs such as *Azotobacter*, Rhizobium, and Azospirillum has been proposed as an environmentally acceptable approach for elevating crop output (Choudhury and Kennedy 2004). *Azotobacter* is a heterotrophic, negative, rod-shaped, aerobic, free-living, nitrogen-fixing bacterium that lives in slightly acidic soil. Chemical fertilizers have a negative influence on soil, water, and air. However, *Azotobacter* inputs can make abundant nutrients present in the atmosphere and soil available for plant use. Because its inoculation benefits a wide range of crops, it is one of the most intensively studied plant growth-promoting bacteria. *Azotobacter* colonizes the rhizosphere of plants, where it remains in close proximity to roots, influencing their growth.

Biological nitrogen fixation (BNF) is a microbial-mediated process that involves an enzymatic "Nitrogenase" conversion of atmospheric nitrogen (N_2) into ammonium that roots may readily absorb. N_2-fixing bacteria, often known as "diazotrophs," can fix N_2 biologically in conjunction with plant roots (Aasfar et al. 2021). *Azotobacter* species possess the important enzymes such as ferredoxin,

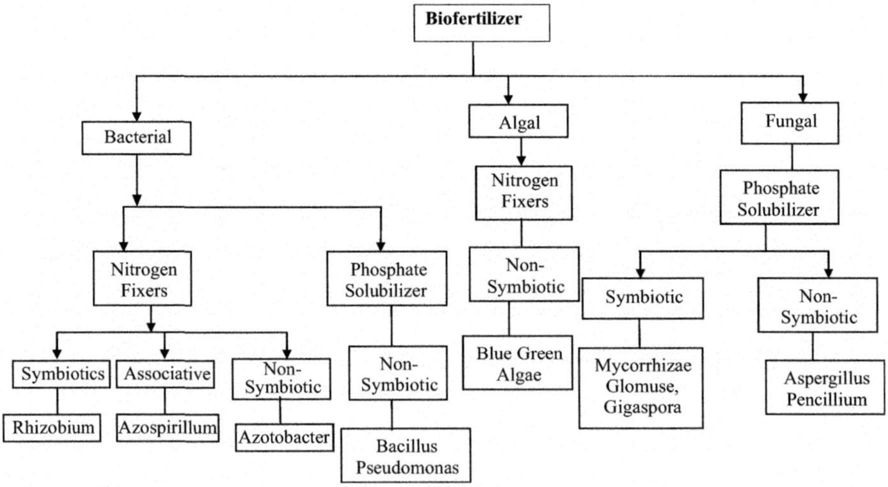

Fig. 11.1 Classification of biofertilizers with respect to different microbial classes, their ability to fix nitrogen, solubilize the phosphorus and their nature of symbiosis

hydrogenase, and the key enzyme nitrogenase, required for nitrogen fixation. An infusion of energy in the form of adenosine triphosphate (ATP) is required for nitrogen fixing. Because nitrogen fixation is particularly sensitive to the presence of oxygen, *Azotobacter* evolved a unique defense mechanism against it: a considerable increase in metabolism, which lowers the concentration of oxygen in the cells (Shank et al. 2005).

The plant rhizosphere is a complex niche in which soil, plant, and microbes interact, and plant roots house a variety of useful and harmful bacteria. Exudation comprises a wide range of organic molecules, including acids, carbohydrates, nucleic acid derivatives, vitamins, enzymes, and a variety of other growth-promoting compounds. Aerobic bacteria of the genus *Azotobacter* play a critical role in sustaining soil fertility in this niche by exerting a variety of positive effects in the rhizosphere of cereals and grasses. Several types of *Azotobacter*ia have been found in the soil and in the rhizosphere: *A. chroococcum* (Beijerinck 1901), *A. paspali* (Hala and Ali 2019), *A. vinelandi* (Upadhyay et al. 2015), *A. armenicus* (Chennappa et al. 2014), *A. salinestris* (Page and Shivprasad 1991). A specific strain of *Azotobacter chroococcum* has been identified in the rhizosphere of the mulberry plant and is recommended for use at a rate of 20 kg per hectare per year, resulting in significant nitrogen fertilizer savings (about 50% reduction in N requirement). This indirectly lowers the costs of silkworm rearing, and enhanced microbial activity improves soil health (www.nrdcindia.com).

11.2 Utilization of *Azotobacter* as Potent Biofertilizer

Azotobacter is one of the best options as a biofertilizer for eco-friendly and sustainable crop production because of its ability to improve plant health through nitrogen fixation, growth hormone biosynthesis, phosphate solubilization, plant disease management, and reclamation of better soil health. Understanding and managing all of *Azotobacter*'s favorable traits may prove to be a crucial priority in future crop development efforts (Kyaw et al. 2019). While working to improve the nutritional characteristics of *Azotobacter* as a biofertilizer, it's also important to think about cost-effective techniques that can give a cheaper source of biofertilizer to the agriculture industry. When considering large-scale *Azotobacter* production, it is vital to optimize cultural and nutritional parameters in order to improve its fermentation development as well as its biofertilizer ability (Gauri et al. 2012).

11.3 Mode of Action

Azotobacter develops flat, slimy, paste-like colonies with a diameter of 5–10 mm that, when isolated, can form films on growth media. Temperatures of 20–30 °C are ideal for growth. *Azotobacter* germination begins with the cysts and takes around

4–6 h. During germination, the central body expands and collects the volutin granules that were previously in the intima (the innermost layer). The exine then explodes, releasing the vegetative cell from the exine, which has a distinct horseshoe form. Metabolic alterations occur as a result of this process. The cysts begin to absorb oxygen and exhale carbon dioxide as soon as they are given a carbon source; the pace of this activity steadily increases and reaches saturation after 4 h. Protein and RNA synthesis happens in parallel, although it becomes more intense only 5 h after the carbon source is added. The synthesis of DNA and nitrogen fixation are initiated 5 h after the addition of glucose to a nitrogen-free nutrient medium. For every gram of glucose ingested, *Azotobacter* can fix at least 10 mg of nitrogen. Molybdenum ions are required for nitrogen fixation, while vanadium ions can partially or entirely substitute them. If atmospheric nitrogen is not fixed, nitrates, ammonium ions, or amino acids can be used as a nitrogen source. *Azotobacter* can also grow mixotrophically, albeit this growth mode is hydrogen-dependent in a molecular-nitrogen-free media. Because hydrogen is present in the soil, this growth mode could occur naturally.

11.4 *Azotobacter* Biofertilizer Production

11.4.1 Inoculum Preparation

A variety of sources, including agricultural universities and government-affiliated groups, provide production technology. A starter culture is a little amount of bacterial suspension that is given to the medium to kick-start the bacterium's growth. Good biofertilizer inoculant, a large number of inoculant cells, and good carriers, as well as the method of carrier sterilization, are among the most critical elements that influence biofertilizer quality. The growth medium used to develop the microbial cells exerts a significant effect in the amount of inoculant cells in carriers. For mass production of inoculant microbial cells, the medium must be able to support rapid growth with a large number of inoculant microbial cells while being reasonably inexpensive.

11.4.2 Mass MultiplicationProcess

Production cost analysis and packaging details are presented in Tables 11.1, 11.2, 11.3, and –11.4. For mass microbe cultivation, a fermenter is utilized. Any standard liquid biofertilizer manufacturing process has three stages (Fages 1992; Gódia and López 1998; Prabavathy et al. 2007):

1. *Bacteria Propagation*: Mass multiplication of the selected strain until the desired inoculum concentration and volume.

Table 11.1 Information of product *Azotobacter* biofertilizer production

Formulation details	
Type of formulation	Liquid/powder/granules
Active organism(s)	*Azotobacter* (common name/s)
CFU	CFU minimum 5×10^7 cell/gm of powder, granules or carrier material, or 1 × cell/mL of liquid
Production	The manufacturing process in short involves selection of suitable strain of the organism for which market demand is identified Mass multiplication: 1. Culture augmentation—Culture has to be mass multiplied in two levels at (a) Primary level, using shakers in flasks (b) Secondary stage, multiplication in fermenter 2. Preparation of growing medium 3.Sterilization of media 4.Inoculation in fermenter It is inoculated using the starter cultures multiplied in the flasks at definite ratios usually 5% The bacterium-growing medium is called broth and it is continuously aerated by passing sterile air from compressors 5. Incubation period: After about 3–4 days of fermentation period, the broth will be ready for packing in a carrier material 6. Addition of carrier material: While the broth is getting ready in the fermenter, the carrier material, which is usually the carbon source for the cultures to survive, is sterilized in autoclaves and kept ready for mixing with the broth 7. Quality check: At various stages the quality is tested by drawing samples, including the quality of raw materials 8. Packaging and storage: Description is made on the label of the product including all the contents, cell counts, and batch number in proper packing material
Shelf life	1–2 year
Mode of application	Seed treatment: Mix 10 mL of *Azotobacter* with 10 g. of organic joggery/crude sugar in sufficient water to make a slurry and coat 1 kg of seeds, dry in shade and sow/broadcast/dibble in the field Seedling/nursery treatment: Mix 100 ml of *Azotobacter* with sufficient quantity of water and organic manure to form a slurry. The seedlings are dipped in this slurry for 30 min prior to planting so that the bacteria get attached to the roots Soil application: Mix 0.5–1 L/acre of *Azotobacter* with compost and apply to 1 acre of soil Drip irrigation: Mix 1 L/acre of *Azotobacter* in drip stream for one acre of land
Packaging	Type of packaging material/container: HDPE bottles Pack size(s): 500 mL, 1 L, 5 L, 10 L

2. *Bacteria Cultivation*: Fermentation of bacterial strains in a large, industrial-sized fermenter until the appropriate cell concentration.
3. *Bacteria Recovery/Formulation/Packing*: The bacteria in the fermentation broth can be extracted using centrifugation or filtration, and then formulated with formulation chemicals. Finally, the latent cells containing liquid is put into plastic

Table 11.2 Media components for production media and costing

Media components for production media	g/L	INR/kg	INR/g	Per liter cost
Di-potassium hydrogen phosphate (K_2HPO_4)	1	340.2	0.3402	0.3402
Magnesium sulfate($MgSO_4$)	0.5	121.8	0.1218	0.0609
Sodium chloride (NaCl)	5	16.8	0.0168	0.084
Ferrous sulfate ($FeSO_4$)	0.1	178.5	0.1785	0.01785
Calcium carbonate ($CaCo_{3)}$	2	159.6	0.1596	0.3192
Sodium molybdate (Na_2MoO_4)	0.005	13,230	13.23	0.06615
Sucrose	1	149.73	0.14973	0.14973
Soyabean meal	1	351.75	0.35175	0.35175
Glycerol (mL)	10	224	0.224	2.24
Antifoam (mL)	0.4	268	0.268	0.1072
pH	6.8–7.5			
Total cost/L = INR 3.73				

Table 11.3 Approximate overall cost of machinery required and sources

Sr. no.	Machine	Number	Approx. cost (INR, Lakhs)
1	Fermentor	1	30.00
2	Mass mixer	2	3.00
3	Wooden vessel for fermentation with lid, cap. 50 and 100 L	2	1.50
4	Reactor vessels—M.S. vat cap. 750 and 1000 kg	2	1.00
6	Tray driers cap 96	1	1.00
7	Bottle filling machine	1	0.50
8	Bottle sealing machine	1	0.50
9	S.S. mixing vessel with stirrer cap. various capacities	3	1.00
10	Hot air oven with 24 trays	1	1.00
11	Bottle washing and drying machine	2	1.00
14	Water treatment plant, 100 L capacity	1	2.50
15	QC and testing equipments	—	2.00
		Total	45

bottles. The manufacturing of large-scale *Azotobacter* sp. inoculants begins with the isolation and screening of viable strains, then progresses to mass production using a suitable culture medium, and ultimately, the formulation selection based on the application mode needed. Depending on their particle sizes, solid formulations can be categorized into powders and granules. They are commonly used as seed coverings or soil supplements (Bashan et al. 2014). Liquid formulations, on the other hand, are ideal for a wide range of application technologies; they can be adhered directly to the seed, right before sowing, or used as a coating for chemical fertilizers (Hindersah et al. 2020). They can also be applied to the soil in the furrow during planting or via fertigation systems at a later time (Malusá et al. 2012).

Inoculants of *Azotobacter* can be transferred to a flask containing sterile Jensen's medium and cultured for a few days at 30 °C on a rotary shaker. The broth culture serves as a beginning culture for inoculant production in large quantities. When the density of the inoculums reaches 10^8–10^9 cell/mL broth, it should be harvested and carrier-based inoculants created. The gathered inoculants broth is mixed with carrier and curing is carried out for a week with a moisture content of 40%. The carrier-based inoculants are then packaged in polythene bags for storage or shipment to the market for sale. In addition, liquid formulations enable for the treatment of above-ground plant sections, such as foliar sprays (Jambhulkar et al. 2016).

11.5 Costs and Profit in *Azotobacter* Biofertilizer Production

The details of chemical required in *Azotobacter* biofertilizer production media is given in Table 11.2. The cost of the chemicals such as di-potassium hydrogen phosphate (K_2HPO_4), magnesium sulfate ($MgSO_4$), sodium chloride (NaCl), ferrous sulfate ($FeSO_4$), calcium carbonate ($CaCo_3$), sodium molybdate (Na_2MoO_4), sucrose, soyabean meal, glycerol, and antifoam would be INR 3.73 for 1 L *Azotobacter* biofertilizer production. Apart from this the approximate overall cost of machinery required and sources would be 4.5 million INR (Table 11.3). Overall cost and profit of *Azotobacter* biofertilizer at different volume is presented in Table 11.4. The total production cost including chemical media and bottle for 1 L of biofertilizer would be 38 INR and the expected profit would be 210 INR. At large scale with 1000 L production the expected profit would be 210,000 INR.

11.6 Marketing

While India is recognized for its agriculture, it shows that beginning a biofertilizers firm in India might be lucrative. Since more than two decades, bio-fertilizers have been used in the country. Bio-fertilizers, on the other hand, make up a minor percentage of the total chemical fertilizer production and sales. The entire production of our fertilizer in the country is anticipated to be between 12,000 and 15,000 tons

Table 11.4 Overall cost and profit of *Azotobacter* biofertilizer at different volume

Production cost	Liter (INR)	500 L (INR)	1000 L (INR)
Total production cost chemical media + bottle	38	19,000	38,000
Finished goods expenses (including prod cost)	120	60,000	120,000
Billing rate	330	165,000	330,000
Difference/expected profit	210	105,000	210,000

per annum. Biofertilizers are produced in tiny quantities by a number of chemical fertilizer facilities. Farmers' unawareness of biofertilizers, insufficient instructions, a lack of technical support, working marketing techniques, poor product quality, market segmentation, and seasonal demand are some of the marketing obstacles. However, with the promotion of alternative sources of nutrition management, farmers are becoming more aware of biofertilizers, which are gradually gaining popularity. Biofertilizers from a variety of manufacturers are now widely available through the traditional dealer/distributor network. As a result, the products can be marketed using the existing marketing network. Bulk sale can also be done through agricultural cooperatives and farmer groups. The most effective way to reach fertilizer distributors and farmers is through social media applications and websites. One can now register a product and sell it directly through online buying sites. Different state governments also provide subsidies, which can amount to up to 50% of the sales revenue; however the method of subsidization is quite unsystematic. Discrimination and manipulation in subsidies have resulted in a lot of intra-industry price fluctuation in several circumstances. The government also has a significant role in biofertilizer marketing, with three possible channels: (a) state government via district level officers and village level workers to farmers, (b) state marketing federation via cooperative bodies to farmers, and (c) state agro-industries corporations via Agro Service Center to farmers. Producers, on the other hand, have the option of selling through their own sales network or through the market (i.e., wholesalers and private dealers) (Ghosh 2004).

As liquid biofertilizers offer significant potential as an application using this technology, marketing partnerships with technology providers such as "Drip Irrigation" companies may be established. Similarly, collaborations with export-oriented crops such as turmeric, ginger, spices, fruits, and vegetable growers could be pursued, as organic products are desired by this segment due to the importing nation's requirement of permitted chemical residual limits in the produce.

11.7 Statutory/Government Approvals

It is recommended that you follow the Department of Biotechnology's basic guidelines. Additional MSME and GST registration, IEC Code for end-product export, and local government clearance may be necessary for shops and establishments, as well as registration for ESI, PF, and labor regulations, if applicable, and approval from the Pollution Control Board.

References

Aasfar A, Bargaz A, Yaakoubi K et al (2021) Nitrogen fixing *Azotobacter* species as potential soil biological enhancers for crop nutrition and yield stability. Front Microbiol 12:354

Bashan Y, de-Bashan LE, Prabhu S et al (2014) Advances in plant growth-promoting bacterial inoculant technology: formulations and practical perspectives (1998–2013). Plant Soil 378:1–33

Beijerinck M (1901) Uber oligonitrophile mikroben. Zentralbl Bakterol Parasitenkd Infektionskr Hyg Abt II(7):561–582

Chennappa G, Adkar-Purushothama C, Naik M et al (2014) Impact of pesticides on PGPR activity of *Azotobacter* sp. isolated from pesticide flooded paddy soils greener. J Biol Sci 4:117–129

Choudhury A, Kennedy I (2004) Prospects and potentials for systems of biological nitrogen fixation in sustainable rice production. Biol Fertil Soils 39:219–227

Fages J (1992) An industrial view of Azospirillum inoculants: formulation and application technology. Symbiosis 13:15–26

Gauri SS, Mandal SM, Pati BR (2012) Impact of *Azotobacter* exopolysaccharides on sustainable agriculture. Appl Microbiol Biotechnol 95:331–338

Ghosh N (2004) Promoting biofertilizers in Indian agriculture. Econ Polit Wkly 39:5617–5625

Gódia F, López J (1998) Ingeniería bioquímica. Síntesis, Madrid

Hala Y, Ali A (2019) Isolation and characterization of *Azotobacter* from neems rhizosphere. J Phys Conf Ser 1:012019

Hindersah R, Setiawati MR, Asmiran P et al (2020) Formulation of Bacillus and *Azotobacter* consortia in liquid cultures: preliminary research on microbes-coated urea. Int J Agric Syst 8:1–10

Jambhulkar PP, Sharma P, Yadav R (2016) Delivery systems for introduction of microbial inoculants in the field. In: Microbial inoculants in sustainable agricultural productivity. Springer, pp 199–218

Kyaw EP, Soe MM, San San Yu ZKL et al (2019) Study on plant growth promoting activities of *Azotobacter* isolates for sustainable agriculture in Myanmar. J Biotech Biores 1:1–6

Malusá E, Sas-Paszt L, Ciesielska J (2012) Technologies for beneficial microorganisms inocula used as biofertilizers. Sci World J 2012:491206

Page WJ, Shivprasad S (1991) *Azotobacter salinestris* sp. Nov., a sodium-dependent, microaerophilic, and aeroadaptive nitrogen-fixing bacterium. Int J Syst Evol Microbiol 41: 369–376

Prabavathy V, Rengalakshmi R, Nair S (2007) Decentralised production of biofertilizers–Azospirillum and Phosphobacteria. JRD Tata Ecotechnology Centre, Chennai, p 36

Shank Y, Demin O, Bogachev A (2005) Respiratory protection of nitrogenase complex in *Azotobacter* vinelandii. Success Biol Chem Sat 45:205–234

Upadhyay S, Kumar N, Singh V et al (2015) Isolation, characterization and morphological study of *Azotobacter* isolates. J Appl Nat Science 7:984–990

Chapter 12
Mass Multiplication, Production Cost Analysis, and Marketing of *Pseudomonas*

Komal A. Chandarana, Harshida A. Gamit, Hetvi Y. Naik, Sapna Chandwani, and Natarajan Amaresan

Abstract *Pseudomonas fluorescens* is the most widely used biopesticide accounting for 71% of the bacterial biopesticide market share in India. It is effective against a range of phytopathogens that cause root rot and wilt diseases in different crops such as rice, cotton, chick pea, peas, tomato, banana, soybean, and cucumber. The Indian biopesticide market has increased from US \$127.5 to US \$362.1, almost triple from 2014 to 2020, indicating a rising demand for chemical-free crop products. The Indian Government has encouraged the use of biopesticides by conducting several farmer awareness programs and by releasing agricultural schemes to encourage the development of small- and large-scale industries to meet increasing consumption. A small effort has been made here to attract entrepreneurs to enter agri-business and contribute to maintaining sustainability along with money making. The present chapter deals with the technological details of the mass production and cost-benefit ratio to establish a small-scale biopesticide (*P. fluorescens*) production unit.

Keywords *Pseudomonas fluorescens* · Biopesticide · Agri-business · Entrepreneurs · Cost-benefit ratio

12.1 Introduction

India loses nearly 30% of its crops every year due to diseases caused by the microbial community, which ultimately results in a loss of farmers' income and food crises. To counter loss, huge efforts have been made by the government since the 1970s (Chakravarti 1973) to enhance crop yield through the use of chemical pesticides and a high yielding variety of crops, the so-called Green Revolution (GR). Despite enhancing crop yield, GR has proven unsuccessful in sustainable agriculture. Later in the 1980s, the darker side of GR was realized by the scientists that the methods which were used in GR led to serious health issues and environmental problems

K. A. Chandarana · H. A. Gamit · H. Y. Naik · S. Chandwani · N. Amaresan (✉)
C.G. Bhakta Institute of Biotechnology, Uka Tarsadia University, Surat, Gujarat, India

N. Amaresan et al. (eds.), *Agricultural Microbiology Based Entrepreneurship*, Microorganisms for Sustainability 39,
https://doi.org/10.1007/978-981-19-5747-5_12

(Pingali 2012). The serious causes associated with the GR, such as loss of biodiversity, land degradation, environmental pollution, and severe health risks to humans (Benbi 2017) have led to the development of an alternative approach for sustainable agriculture, i.e., the use of biopesticides. Biopesticides are living organisms or their extracts that control agricultural pests by producing secondary metabolites, for instance, toxins, antibiotics, or by detrimentally affecting them (Samada and Tambunan 2020). Organic farming using biopesticides is gaining popularity globally because of the target specificity and environmentally benign nature. Biopesticides have several advantages over chemical pesticides because they are eco-friendly, cheap, effective in small quantities, biodegradable, and compatible with microbial biofertilizers (Fenibo et al. 2021). Despite these advantages, developing countries like India still have less usage of biopesticides compared to chemical pesticides due to lower adaptability and awareness among farmers. The Indian Government encourages farmers to use biopesticides by placing them in various agricultural schemes. However, the technological challenges in mass multiplication, application in the field, and marketing are some of the questions raised regarding its long-term usage for sustainability. The main bottleneck in biopesticide production is usually to maintain the strain and obtain contamination-free inoculum throughout production. The methodology for mass production of various biopesticides has been addressed frequently by researchers, but the detailed procedure for mass production in a cost-effective manner has not been discussed. Therefore, the present chapter deals with the cost-effective calculation of bacterial biopesticide production and marketing at small scale to attract entrepreneurs to adopt biopesticide technology as an agri-business.

12.2 *Pseudomonas fluorescens*

The most prominent bacterial biopesticide that is commercially available in India is *Pseudomonas fluorescens*, which is effective against root rot and wilt diseases in a range of major crops (rice, cotton, and vegetables). *P. fluorescens* is a free-living bacterium that establishes robust and long-lasting colonization on root surfaces and penetrates the epidermis. It is effective against several soil and seed borne pathogens including *Rhizoctonia* (Anupriya et al. 2019), *Sclerotium* (Hamarawati et al. 2018), *Sclerotinia* (Aeron et al. 2011), *Fusarium* (Alzandi and Naguib 2019; Suresh et al. 2021), *Pythium* (Prabhukarthikeyan et al. 2018), *Ralstonia* (Elsayed et al. 2020), *Macrophomina* (Suresh et al. 2021), and *Colletotrichum* (Chlebek et al. 2020). It can be used in cereals, millets, pulses, vegetables, and commercial plantation crops. *Pseudomonas* suppresses pathogens by producing antibiotics and siderophores (low molecular weight chelators that transport iron into bacterial cells), which limits the availability of iron to pathogens and supresses the growth of pathogens. Apart from protecting plants from pests, it enhances root growth and development, translocation of essential nutrients from the soil, crop productivity, and resistance to biotic and abiotic stresses (David et al. 2018). *Pseudomonas fluorescens* formulations can be stored for up to 1 year. This is compatible with other biofertilizer applications. There are no reports on its toxicity or adverse effects on nontarget organisms. Depending on the method of application, 200 g–4 kg of culture formulation is required for 1 acre, which costs approximately Rs. 10–200/acre.

12.3 Technological Details

Technological development in biopesticide production includes four core areas of its production and selling; (1) mass production (2) formulation type (3) packaging process, and (4) application method. Over the past few years, the technology and research in this field have been upgraded, and this has led to an increase in the reliability of biocontrol products. *P. fluorescens* takes the lead over other biopesticidal bacterial products. However, mass production still faces some technical challenges and other barriers, such as contamination, poor shelf life, application methods used by farmers, and low faith of farmers. For better introduction of *P. fluorescens* into the market and to encourage more entrepreneurs to adopt this technology for sustainable agriculture, the methodological aspects along with detailed cost-benefit ratios have been discussed. The first section focuses on the production process will be focused and in the next section, the project cost and technical details are discussed, which are essential for the establishment of small- or large-scale industries to widen the use of *P. fluorescens* in local markets for sustainability.

12.4 Production Process

The production process of *P. fluorescens* involves; (1) preparation of starter culture (2) mass multiplication (3) product formulation, and (4) packaging. An outline of this process is shown in Fig. 12.1.

12.4.1 Preparation of Starter Culture

Pure cultures of *P. fluorescens* can be purchased from the State Agriculture University, authorized research institutes, or Research Laboratories of Government of India in the form of slants or vials. To maintain the mother culture, it is necessary to subculture the strain before starting mass multiplication using standard microbiological techniques. For this, pick up the well-isolated colony from the slant with the use of wire-loop and transfer it onto sterile King's B agar plate (Box 12.1), or transfer the pinch of the material from the vial into the sterile King's B medium

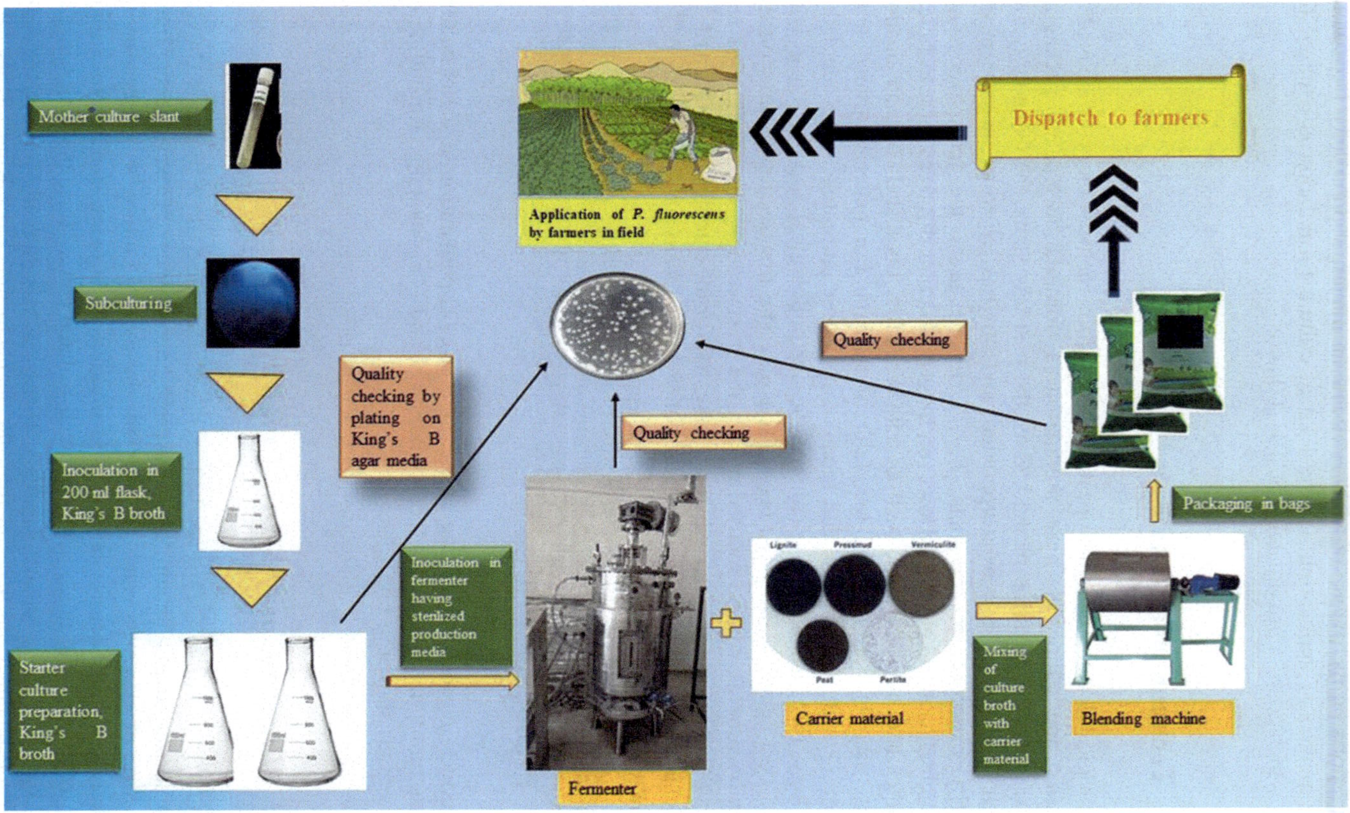

Fig. 12.1 Schematic representation of mass production process of *P. fluorescens*

followed by streaking on King's B agar plate after 24 h of incubation. Plates with well-isolated colonies should be stored in refrigerator at 4 °C for further use.

Box 12.1 The composition of King's B agar medium

Ingredients	g/L
Proteose peptone	20.0
Dipotassium hydrogen phosphate	1.5
Magnesium sulfate heptahydrate	1.5
Agar	20.0
pH	7.2 ± 0.2

Suspend total 43.0 g of above ingredients in 1 L distilled water containing 15 mL of glycerol
Mix well and sterilize it by autoclaving at 15 lbs pressure at 121 °C for 15 min

Starter cultures for mass production can be prepared using same King's B medium, excluding agar. The composition of the medium is shown in Box 12.1. The broth should be dispersed into conical flasks and autoclaved at 15 lbs pressure for 15 min at 121 °C, cooled, inoculated with *P. fluorescens* colonies and incubated for 48 h at 37 °C in an incubator shaker. Inoculate the flask of 5-L capasity from this mother culture and incubate for 48 h at 37 °C to prepare a starter culture. This culture can be used as an inoculum for the fermenter.

12.4.2 Mass Multiplication

Prepare King's B medium as described and pour it into a fermenter, sterilize (15 lbs pressure at 121 °C for 15 min) and cool it to normal temperature by circulating cool water. Instead of King's B medium, a production medium that includes beef extract, sugar, ammonium sulfate, and micronutrients (quantity and cost described in project cost details) can be used at the fermentation level. Add inocula from inoculation point of the fermenter at a rate of 1–2% and allow to grow under optimum conditions for 4–5 days. Regulate the air flow to 3–10 L of air/h/L of the medium. Sterile air provides aeration and agitation for bacterial growth. Periodically, collect samples from the sampling point and analyze for growth and contamination by plating on King's B agar medium. Harvest the broth from the culture outlet for formulation once bacterial growth reaches 10^9 CFUs/mL.

There are some important factors should be considered during the fermentation process. Fungal or other bacterial contamination should not be detected during or after harvesting of the culture. It is not advisable to store the broth after fermentation for more than 24 h. The final cost of product is affected by the medium used for multiplication and the duration of the fermentation process. Therefore, it is essential to choose appropriate medium and optimize parameters such as pH, temperature, and aeration to reduce incubation time. The ideal pH and temperature for *P. fluorescens* are 7.5 °C and 25 °C–32 °C, respectively.

12.4.3 Product Formulation

Product formulations are prepared by blending living organisms with inert ingredients, referred to as carriers. The formulations should be safe enough, effective, have longer shelf life, and be easy to use according to the CIBRC (Central Insecticides Board and the Registration Committee) guidelines that mandate the biopesticide firm to maintain the quality parameters before entering the product into the market.

Currently, biopesticides are formulated as solid carriers that include talc, peat, lignite, clay, wheat husk, rice bran, fly ash, and sawdust (Mishra et al. 2020). In India, wettable granules (WG), wettable powder (WP), aqueous solutions (AS), and suspension concentrations (SC) are commonly used formulations. WP and WG are the first choices for *P. fluorescens* (Mishra et al. 2020). However, owing to the difficulties associated with solid formulations, i.e., the requirement of more space to store the carrier material, shorter shelf life, and the need for milling and drying, AS has become a more popular and frequently utilized technology. In the case of *P. fluorescens,* liquid formulations with 0.5–5% active ingredients have been prepared by different firms.

12.5 Carrier Sterilization

When the broth is ready in the fermenter after 4–5 days, the carrier material should also be ready for mixing with the broth. Generally, the carrier material is contaminated; therefore, it is necessary to sterilize it in the autoclave following standard procedures. The carrier is either sterilized in bulk or first filled in packets and then sterilized, whichever is easy and standardized by a technologist.

12.5.1 Packaging

Packaging is a vital component of biopesticide manufacturing and one of the major constraints driving the market of this business. The packaging type must maintain product quality. It must ensure the shelf life of living ingredients (*P. fluorescens*) and should be leak-proof. Ultimately, packaging must follow the standards and guidelines, but unfortunately, most private firms do so as per their convenience and overlooked quality of the packaging material, which in turn affects the quality of the product and fails to meet the farmer's expectations.

There are two methods for packaging solid carrier-based formulations;

1. The broth is harvested from fermenter and mixed mechanically with a sterilized carrier (bulk) under aseptic conditions, packed in polythene bags at the desired quantity and sealed.
2. For following a completely sterile system, high-quality polypropylene bags are filled with carrier and sterilized by autoclaving. Culture from the broth is directly injected into sterile packets using an auto-syringe and dispenser of the desired volume. The hole is immediately sealed, and the packets are kept for 1 week incubation after manual mixing.

For liquid formulations, high-density polyethylene (HDPE) bottles are used. Usually, if the culture is fully grown, a 15 mL culture of *P. fluorescens* per 1 L of RO is sufficient.

12.6 Quality Control and Product Specification

The quality of a formulated product is an important factor that influences its success or failure in the market. The right type of organism in the desired number must be present in the formulations, resulting in acceptance or rejection by farmers. Microbial processes are sensitive to contamination; therefore, quality should be checked for contamination at every stage of production, including mother culture preparation, starter culture preparation, fermentation, carrier selection, and mixing with broth during packaging and storage. Quality control must include a microbiologist at each stage of the product. Finally, the product should meet the standards specified by the government.

Quality and maintenance of *P. fluorescens* are covered under reconstituted schedule lists of the Insecticides Act 1968 of Government of India. The Central Insecticides Laboratory (CIL) in India is responsible for checking the quality of biopesticides supplied in the market. The table (Box 12.2) below comprises some valuable considerations for the end-product of *P. fluorescens*.

Box 12.2 Specifications for *P. fluorescens*

Sr. No.	Description	Specification
1.	Base	Carrier material or liquid
2.	Viable cell count	5×10^7 CFUs/mL for carrier-based formulation and 1×10^8 CFUs/mL for liquid formulations
3.	Contamination level	No contamination at 10^5 dilution
4.	Moisture content in case of carrier-based formulations	30–40%
5.	Particle size in case of carrier-based formulations	The formulation should pass through 0.15–0.212 mm IS sieve

12.7 Project Details

Various facilities are required for the successful establishment of the biopesticide unit and implementation of the production project. The infrastructure and laboratory facilities described here are not only applicable for *P. fluorescens* production, but the same facilities can also be applied for the production of other biofertilizers as well as biopesticides. Most private firms and governmental institutes produce both biofertilizers and biopesticides using common facilities that ensure the economic viability of the project installation. The projects set up in India vary from 10 to 2000 tunes per annum (TPA) for the production of particular biofertilizers or biopesticides. The present model for setting up the unit for *P. fluorescens* is of 200 TPA units. However, the capacity utilization for the newly established industry is nearly 50%. Therefore, the cost for each particular is counted accordingly for 100 tons/annum production.

Land and Location: Land is prerequisite for establishing manufacturing business. To set up a laboratory, other facilities, and offices for setting up a 200 TPA unit, 6000–9000 sq.ft of land is required.

The location of the biopesticide unit should be far away from the residential areas. It should not be set up in areas with high temperature and humidity as this increases the production cost for maintaining the same.

Building Infrastructure: The building should be divided into different compartments such as the microbiological (quality control) laboratory, culture preparation room, fermenter arrangement facility, carrier preparation room, packaging and dispatching area, storage facilities, administrative office, staff room, pantry, etc. A floor plan design is ideal to facilitating maximum land use. The infrastructure design should facilitate the optimum temperature and humidity required for the culture maintenance. Around 6000–7000 sq.ft of area is sufficient for civil works; the rest of the area will be utilized for plantation, parking, and fencing or for future expansion of units. The entire site should be either fenced with a barbed wire or with a compound wall and gates (Box 12.3 and 12.4).

Box 12.3 The breakup of the area is given below

Compartment type	Area (sq.ft)
Administration	500
Culture preparation and inoculation laboratory	500
Quality control laboratory	500
Incubation chamber with shakers and incubator	1000
Fermenter platform arrangements	1000
Miscellaneous space for RO plant, chiller, steam generator, etc.	1000
Carrier blending and packaging (with internalize aseptic compartment)	1000
Storage and dispatch	1000
Total area covered	6500

Box 12.4 Infrastructure cost

Description	Rate	Cost (Rs. in lakhs, INR)
Land (9000 sq.ft)	Approx.	20.0
Site development (land leveling, fencing)	Approx.	5.0
Civil works (6500 sq.ft)	1000/sq.ft	65.0
Furniture	—	3.0
Total cost		93.0

12.8 Utilities

1. *Power:* Usually, three phase supply is required. Depending on the machinery operated for 200 TPA plants, the requirement is nearly about 40–75 KVA. A standby generator may be required as an option for the power supply.
2. *Water:* At each production step, water supply is mandatory. Broth preparation, steam generation, chelation, demineralization equipment operation, cleaning of glassware used during the production process, and fermenter cleaning are some of the basic requirements that need water supply. The average requirement of water for 200 TPA is approximately 3000–5000 L/day.

Machinery: Although a minimum number of instruments can facilitate biopesticide (*P. fluorescens*) production, a large number of machineries are required to fulfill the standards and achieve the maximum production capacity. The instrumentation for each production step is presented in the following table (Tables 12.1 and 12.2).

Manpower: The manpower requirement with average salary cost is given in the table below (Table 12.3).

12.9 Raw Material

Cost of raw materials used for starter culture preparation and mass production of 100 tons of *P. fluorescens* is given in the table below (Table 12.4):

12.10 Packaging

Packaging cost per 100 ton production of *P. fluorescens* is given in detail in the table below (Tables 12.5, 12.6, 12.7, 12.8, and 12.9).

Table 12.1 Instruments required and their cost

Equipment name	Nos	Cost in lakhs/ Nos (INR)	Total cost in lakhs (INR)
Vertical autoclave (600 × 350 mm)	2	0.8	1.6
Horizontal autoclave (2 × 2 × 4 ft chamber size)	1	4.5	4.5
Hot air oven (24″ × 24″ × 24″)	1	0.25	0.25
Refrigerator (300 L)	1	0.3	0.3
Deep freeze (300 L)	1	0.35	0.35
Laminar air flow (3′ × 2′)	1	1.0	1.0
Orbital incubator shaker	1	1.5	1.5
Rotary shaker	1	0.25	0.25
Compound microscope (binocular)	1	0.65	0.65
Electronic weight balance	1	0.2	0.2
pH meter	1	0.3	0.3
Colony counter	1	0.05	0.05
Stainless steel seed fermenters (50 L cap.)	2	2.5	5.0
Stainless steel fermenters (1000 L, 750 L working cap.)	2	10	20
Air compressor oil free type (2000 L air/min cap.)	2	0.5	1.0
Automated steam generator (100 kg/h cap.)	1	3.0	3.0
Chiller (1 ton cap.)	2	1.0	2.0
RO plant (200–500 L/h)	1	0.35	0.35
Carrier blending tanks	2	0.5	1.0
Polybag sealer	3	0.15	0.45
Capping and Labeling machines and other miscellaneous items for packaging	3	1.0	3.0
Air conditioners	4	0.35	1.4
Generator (65 kVA, diesel generator)	1	1.5	1.5
Glassware and plasticwares for microbiological works	—	Approx.	2.5
Other miscellaneous requirements for lab use	—	Approx.	2.5
Grand total			54.65

Table 12.2 Miscellaneous fixed assets

Item	Numbers	Cost (Rs. in lakhs, INR)
Computer	2	0.8
Printer	1	0.1
Fax	1	0.1
Others	—	0.2
Total cost		1.2

Table 12.3 Manpower specification and salary particulars

Category	Nos	Salary per head (Rs. in lakhs/month, INR)	Total cost (Rs. in lakhs/month, INR)
Chief microbiologist	1	0.3	0.3
Assistant Microbiologist	2	0.15	0.30
Production officer	2	0.25	0.50
Technical staff	2	0.1	0.2
Administrative officer	1	0.15	0.15
Sales officers	5	0.15	0.75
Skilled and unskilled labors	10	0.045 (0.03–0.06 depending upon the work allotment)	
Total cost			2.65

Table 12.4 Raw material specification and cost

Medium and specification	Raw material	Quantity	Amt. (Rs., INR)	Total cost (Rs. in lakhs)/100 ton (INR)
King's B medium (for subculturing, mother culture and starter culture preparation, quality checking)	Ready to use media from standard company (500 g)	2 kg	10,500	0.11
	Glycerol	500 mL	500	
Production media (fermentation process)	Beef extract	5 kg	9000	0.14
	Dextrose	50 kg	2000	
	Ammonium sulfate	50 kg	1500	
	Micronutrient		1500	
Carrier	Talc for solid formulations (50%)	50,000 kg	750,000	7.5
	RO for liquid formulations (50%)	50,000 L	100,000	1.0
Total cost				8.75

12.11 Income per 100 ton of Selling

The cost per unit given below in the table is the average market price for private companies. Government institutions and organizations may cost less than this, and on the other hand, some private factories may cost double (Tables 12.10 and 12.11).

Table 12.5 Packaging particulates and cost

Particular	Packaging (bag or bottle)	Approximate cost (Rs./piece)	Market utility for particular pack (%)	Total estimated cost for 100 ton (Rs. in lakhs)
Solid base (50%)	1 kg	11	30	3.3
	5 kg	18	20	0.72
Liquid base (50%)	1 L	21	30	6.3
	5 L	45	15	1.35
	20 L	155	5	0.39
Other miscellaneous items for packaging				2.0
Grand total for 100 ton packaging				14.06

Table 12.6 Model project cost outlay for 200 TPA

Particulars	Cost
Infrastructure cost including land	93.0
Machinery and equipment	54.65
Other miscellaneous fixed asset cost	1.2
Preliminary and preoperational expenses	2.5
Total project outlay cost	148.85

Table 12.7 Means of finance

Particulars	Contribution in percentage	Total contribution in cost (Rs., INR)
Owner's contribution	25%	37.2125
Bank loan	50%	74.425
Subsidy by Government	25% of total project cost or 40 lakhs whichever is less	37.2125

Table 12.8 Recurring expenses per month

Particulars	Cost (Rs. in lakhs/month, INR)	Cost (Rs. in lakhs/annum (INR)
Raw material and packaging (cost/100 ton/annum)	1.90	22.81
Man-power cost	2.65	31.8
Utilities—Power (4000 units @ Rs. 5/unit)	0.2	2.4
Contingencies		
Marketing and selling expenses	0.5	6.0
Repair and maintenance	0.4	4.8
Total	4.95	67.81

Table 12.9 Capital investment

Particular	Total cost (Rs. in lakhs, INR)
Fixed assets	148.85
Technology know-how	10
Recurring expenses per annum	67.81
Grand total	226.66

Table 12.10 Approximate income per 100 ton of selling *P. fluorescens*

Sr. No.	Packing	Cost in Rs./unit (INR)	Market utility (%)	Approximate total income/100 ton/annum (Rs. in lakhs, INR)
1.	1 kg	160	30	48.0
2.	5 kg	800	20	32.0
3.	1 L	200	30	60.0
4.	5 L	1000	15	30.0
5.	20 L	3000	5	7.5
Grand total				177.5

Table 12.11 Total income and expenditure for model project of 200 TPA biopesticide unit (Rs. in lakhs)

Sr. No.	Particulars	Values
1.	Installed capacity in ton	200
2.	Capacity utilization in %	50%
3.	Capacity utilized in ton	100
4.	Income (Rs. in lakhs/annum)	177.5
Expenditure		
1.	Recurring expenses	67.81
2.	Administrative expenses	1.8
3.	Taxes and insurance	3.24
4.	Interest on bank loan (16%/annum)	11.908
5.	Loss due to contamination (5%)	8..875
6.	Depreciation for fixed assets @ 5%	7.382
Total expenditure		101.1
Net benefit		76.4
Profitability		
Net profit		76.4 INR in lakhs
% Profit of sale		43.0%

References

Aeron A, Dubey RC, Maheshwari DK, Pandey P, Bajpai VK, Kang SC (2011) Multifarious activity of bioformulated Pseudomonas fluorescens PS1 and biocontrol of Sclerotinia sclerotiorum in Indian rapeseed (Brassica campestris L.). Eur J Plant Pathol 131(1):81–93

Alzandi AA, Naguib DM (2019) Pseudomonas fluorescens metabolites as biopriming agent for systemic resistance induction in tomato against Fusarium wilt. Rhizosphere 11:100168
Anupriya N, Devi AA, Anushya R, Apoorva M, Anusiya R, Raj TS (2019) Antifungal activity of Pseudomonas fluorescens against Rhizoctonia solani under in vitro condition. Int J Adv Res Ideas Innov Technol 3(2)
Benbi DK (2017) Nitrogen balances of intensively cultivated rice–wheat cropping systems in original green revolution states of India. In: The Indian nitrogen assessment. Elsevier, pp 77–93
Chakravarti AK (1973) Green revolution in India. Ann Assoc Am Geogr 63(3):319–330
Chlebek D, Pinski A, Żur J, Michalska J, Hupert-Kocurek K (2020) Genome mining and evaluation of the biocontrol potential of Pseudomonas fluorescens BRZ63, a new endophyte of oilseed rape (Brassica napus L.) against fungal pathogens. Int J Mol Sci 21(22):8740
David BV, Chandrasehar G, Selvam PN (2018) Pseudomonas fluorescens: a plant-growth-promoting rhizobacterium (PGPR) with potential role in biocontrol of pests of crops. In: Crop improvement through microbial biotechnology. Elsevier, pp 221–243
Elsayed TR, Jacquiod S, Nour EH, Sørensen SJ, Smalla K (2020) Biocontrol of bacterial wilt disease through complex interaction between tomato plant, antagonists, the indigenous rhizosphere microbiota, and Ralstonia solanacearum. Front Microbiol 10:2835
Fenibo EO, Ijoma GN, Matambo T (2021) Biopesticides in sustainable agriculture: a critical sustainable development driver governed by green chemistry principles. Front Sustain Syst 5: 141
Hamarawati E, Mugiastuti E, Manan A, Loekito S, Soesanto L (2018) Research article applications of Pseudomnas fluorescens P60 in controlling basal stem rot (Sclerotium rolfsii Sacc.) on dragon fruit seedlings. Asian J Plant Pathol 12(1):1–6
Mishra J, Dutta V, Arora NK (2020) Biopesticides in India: technology and sustainability linkages. 3 Biotech 10(5):1–12
Pingali PL (2012) Green revolution: impacts, limits, and the path ahead. Proc Natl Acad Sci 109(31):12302–12308
Prabhukarthikeyan SR, Keerthana U, Raguchander T (2018) Antibiotic-producing Pseudomonas fluorescens mediates rhizome rot disease resistance and promotes plant growth in turmeric plants. Microbiol Res 210:65–73
Samada LH, Tambunan USF (2020) Biopesticides as promising alternatives to chemical pesticides: A review of their current and future status. OnLine J Biol Sci 20(2):66–76
Suresh P, Vellasamy S, Almaary KS, Dawoud TM, Elbadawi YB (2021) Fluorescent pseudomonads (FPs) as a potential biocontrol and plant growth promoting agent associated with tomato rhizosphere. J King Saud Univ Sci 33(4):101423

Chapter 13
Mass Production, Formulation, and Cost-Benefit Ratio Analysis of *Bacillus thuringiensis* Bioinsecticide

Komal A. Chandarana, Harshida A. Gamit, Hetvi Y. Naik, Sapna Chandwani, and Natarajan Amaresan

Abstract *Bacillus thuringiensis* (Bt) is the most widely used bioinsecticide, accounting for 97% of the global market share. It is effective against a range of insect pests such as Lepidoptera, Coleoptera, and Diptera. The Indian Government has encouraged the use of microbial-based bioinsecticidal products by conducting several farmer awareness programs and releasing agricultural schemes to encourage the development of small- and large-scale industries to meet the increasing consumption. A small effort has been made here to attract entrepreneurs to enter agribusiness and contribute to maintaining sustainability along with money making. This chapter deals with the technological details of mass production and the cost-benefit ratio to establish a small-scale Bt production unit.

Keywords Agribusiness · *Bacillus thuringiensis* · Bioinsecticide · Entrepreneurs · Money-making

13.1 Introduction

A significant increase in population demands a higher amount of crop productivity. However, simultaneous industrialization with an increase in population increases the pollution of soil, air, and water and affects agricultural ecosystems by decreasing soil fertility, deforestation, and increase in phytopathogens. To balance this two-sided situation, the indiscriminate use of chemical fertilizers, pesticides, insecticides, fungicides, and herbicides increases crop production and protects plants from phytopathogenic loss to higher productivity. However, the use of such chemical products poses a serious threat to human health and the environment (Rodrigo et al. 2014). Despite their tremendous harmful effects, such chemicals are widely used because of their price, ease of use, and rapid visual positive representation of plant

K. A. Chandarana · H. A. Gamit · H. Y. Naik · S. Chandwani · N. Amaresan (✉)
C.G. Bhakta Institute of Biotechnology, Uka Tarsadia University, Surat, Gujarat, India

N. Amaresan et al. (eds.), *Agricultural Microbiology Based Entrepreneurship*, Microorganisms for Sustainability 39,
https://doi.org/10.1007/978-981-19-5747-5_13

growth and control of phytopathogens. However, the use of different microorganisms such as bacteria, fungi, viruses, and algae as biofertilizer, biopesticides, biofungicides, and bioinsecticides provides an alternative to agrochemicals.

Biopesticides and bioinsecticides are naturally occurring microorganisms that control agricultural pests in a nontoxic manner (Stankus 2013; Kumar 2015). They are eco-friendly, biodegradable, pose less threat to human health and the environment, and are often effective in small quantities. The growing demand for the organic food market and easier registration than chemical pesticides as well as governmental support in terms of subsidies, are the important driving factors for growing the biopesticide market (Glare et al. 2012).

Among bioinsecticides, *Bacillus thuringiensis* (Bt)-based bioinsecticides account for 97% of the global bioinsecticide market (Osman et al. 2015). Commercially, for Bt, two different strategies are used to overcome the disease challenges in plants: first, the development of Bt transgenic crops and second, development of Bt bioinsecticide products (Kashyap et al. 2017). The present chapter deals with the product formulations of Bt in an economical way to develop entrepreneurship and catch the market for growing demands of this particular species.

13.1.1 Bacillus Thuringiensis

B. thuringiensis is rod-shaped, Gram-positive, aerobic in nature, and a sporulating bacterium found in soils, insects, plants, and aquatic environments (Bizzarri and Bishop 2008; Al-Momani and Obeidat 2013). This is a unique species of *Bacillus* due to the presence of crystal proteins formed during sporulation (Höfte and Whiteley 1989). These crystal proteins are composed of delta-endotoxins, which are lethal to pests. They bind to the epithelial cells of larvae, create a pore in the cell membrane, and thereby destabilize the epithelial cells, leading to paralysis of the larval digestive system and cessation of feeding (Elleuch et al. 2016). Moreover, viable spores of Bt germinate in the midgut of budworms and produce Bt cells and crystal proteins, resulting in septicemia of larvae and ultimately death (Vu et al. 2012). Bt is the most successful bioinsecticide used to control pests such as Lepidoptera, Coleoptera, and Diptera (Elleuch et al. 2016). Bt-type strains are effective in controlling one specific target pest, or in some cases, they affect few target organisms. The different types of Bt strains and its target organisms are listed in Table 13.1 (Montesinos 2003). According to the Indian Government insecticide Act, 1968, under Section 9(3), four *Bacillus* species are recommended for use as bioinsecticides, as listed in Table 13.2.

Table 13.1 *Bacillus thuringiensis* strains and its target phytopathogenic pests

Bt strain	Target
B. thuringiensis	Dipteran larvae
B. thuringiensis var. kurstaki	Lepidopteran larvae
B. thuringiensis var. aizawai	Galleria melonella
B. thuringiensis var. xentari	Lepidopteran larvae
B. thuringiensis var. tenebrionis	Coleopteran larvae
B. thuringiensis var. San Diego	Coleopteran larvae
B. thuringiensis EG2348	Lymantria dispar
B. thuringiensis EG2371	Lepidopteran larvae
B. thuringiensis EG2424	Coleopteran larvae

Table 13.2 Bacterial bioinsecticide registered under Section 9(3) of the insecticides Act, 1968 for use in India

Sr. no.	Name of bioinsecticide
1.	*Bacillus thuringiensis var. israellensis*
2.	*Bacillus thuringiensis var. kurstaki*
3.	*Bacillus thuringiensis var. galleriae*
4.	*Bacillus sphaericus*

13.2 Technological Details

13.2.1 Mass Production Process

Mass production technology for Bt is divided into five major steps: (1) derivation and mother culture preparation, (2) seed culture preparation, (3) mass fermentation, (4) product recovery, and (5) formulation.

Derivation, Maintenance, and Mother Culture Preparation

B. thuringiensis strains could be purchased from various culture collection centers, such as from Microbial Type Culture Collection and Gene Bank (MTCC) (CSIR-Institute of Microbial Technology) in Chandigarh, India, or from the Vector Control Research Center (ICMR) in Puducherry, India, at a minimal cost. As soon as the culture slant arrives, it should be stored at 4 °C and transferred to fresh medium to obtain a fresh mother culture. The Bt strains can be cultured and maintained in nutrient yeast salt medium (NYSM) (9). Plates with well-isolated colonies should be stored at 4 °C until further use. The composition of NYSM is listed in Table 13.3.

Seed Culture Preparation

The seed culture can be prepared in a shaking flask using NYSM broth. Transfer loop-full Bt strain from the mother culture plate into 10 mL NYSM broth. Incubate the flasks at 30 °C on a rotary shaker at 250 rpm for 6–8 h, which will serve as the

Table 13.3 Composition of NYSM media

Ingredients	g/L
Glucose	5.0
Peptone	5.0
Beef extract	3.0
Yeast extract	0.5
NaCl	5.0
$MgCl_2$	0.203
$CaCl_2$	0.102
$MnCl_2$	0.01
pH	7.0

Suspend all the above ingredients in 1 L distilled water and sterilize at 121 °C, 15 Ibs pressure for 15 min. To prepare a plate for subculturing, add agar (15 g/L)

first stage of the seed inoculum. After incubation, inoculate this 10 mL of culture into 500 mL of NYSM broth and incubate further in the same conditions for 6–8 h. This is the second stage seed culture that can be used to inoculate seed fermenters. An inoculum with 2% v/v of second stage seed culture of Bt is used to inoculate a seed fermenter (20 L) filled with sterilized production media. The seed fermenter is maintained under aerobic conditions and incubated for 6–8 h.

Production Media Bt can be easily mass produced in synthetic media. However, the inherent cost of this medium limits the ultimate cost of Bt production. Thus, waste products from the food industry and sewage treatment plants have been suggested as alternatives to synthetic media (Ballardo et al. 2016, 2017). Starch industry wastewater (SIW), a rich source of carbon and nitrogen, has been suggested as a cost-effective replacement for synthetic media (Gnepe et al. 2014; Kumar et al. 2019). Additionally, it was found that Bt produced more lethal crystal proteins when grown on complex substrates than when grown on synthetic media (Azmi et al. 2015). Therefore, the present chapter on calculating cost-benefit ratio of Bt mass production utilizes SIW as an alternative and effective production medium.

Starch Industry Wastewater (SIW) Starch industry wastewater contains glucose, fructose, xylose, and starch as the main carbon sources. It contains total solids, volatile total solids, suspended solids, volatile suspended solids, total carbon and nitrogen, and ammoniacal nitrogen. The pH of SIW is approximately 3.5. It can be sterilized at 121 °C and 15 Ibs pressure before inoculation of seed culture.

Mass Fermentation

For the mass fermentation process in Bt production, a larger fermenter with 1000–2000 L capacity is used. The working volume is 700 L (for a 1000 L reactor capacity) during the pilot-scale production process. SIW should be filled in fermenter and sterilized by autoclaving at 121 °C and 15 Ibs pressure for 30 min. The pH of the SIW should be set to 7.0 using 2.0 N NaOH, and is maintained throughout the fermentation process. The medium should be allowed to cool (45–50 °C) before inoculation with log phase culture obtained from the seed fermenter. The fermenter should be maintained at 30 °C for 48 h at a continuous stirrer speed of 150 rpm. The fermenter can be set automatically to control the fermentation variables. The pH of the medium should be maintained at 7.0 with an automated pH controller, and never allowed to drop down. Dissolved oxygen (DO) should be maintained at 30% by controlling airflow (1 L/L/min). Antifoaming agents such as mineral oil should be added to avoid foam formation. After confirming the spore-crystal complex under a microscope, the fermentation process should be terminated, and broth should be subjected to downstream processes.

Product Recovery through Harvesting

At the end of fermentation, the SIW in the fermenter has mixture of spores, spore-crystal complexes, inclusion bodies, enzymes, other residual solids, and cell debris, which should be recovered efficiently to formulate the product (Rowe and Margaritis 2004). The separation of parasporal crystals from broth is challenging owing to their small size. Various harvesting methods have been used to recover the Bt products such as acid precipitation, alginate immobilization (Prabakaran and Hoti 2012), centrifugation (Brar et al. 2006a), membrane-based filtration-microfiltration (Chang et al. 2012), and ultrafiltration (Adjalle et al. 2007). Generally, the choice of the harvesting technique varies widely depending on the process throughput, scale of production, impurities, desired end-product concentration (Brar et al. 2006a), and the investor's capacity for investment. However, this technique should have minimum number of operations during process, low cost, and ease of use. Overall, this is the key issue that governs the overall cost of commercial Bt production.

Alginate Immobilization The fermented culture broth is mixed with sterile sodium alginate solution (4%) at a rate of 20 mL per liter of culture broth. Then, calcium chloride solution (50%) is added slowly at a rate of 20 mL per liter of culture broth. This mixture should be thoroughly mixed. The mixture should be kept overnight in a cold room. The following day, the supernatant is removed from the culture broth using a peristaltic pump, and the recovered product is placed in an airtight container and stored at 4 °C. The recovered supernatant can be used to recover some important enzymes produced by *B. thuringiensis* during fermentation, which will result in multi-product recovery from a single fermentation process and ultimately reduce the cost of Bt bioinsecticide production.

Acid Precipitation The fermented broth is harvested by acid precipitation by lowering the pH of the broth to 3.0, using 2 N HCl, and leaving it overnight at 4 ° C. At a lower pH, proteins are unable to interact with the medium and precipitate out of the solution. The supernatant is decanted, and the suspended broth containing a mixture of Bt products is collected using a pump.

Centrifugation The sporulated culture broth is centrifuged to harvest the Bt product using a continuous flow centrifuge. The culture broth is fed into the bottom of the separation chamber rotating at 20,000 rpm, where solid components are retained in the form of a cake on the rotor, and the supernatant is expelled from the top of the chamber, which is collected in the collection tray. The deposited biomass is scooped out and formulated for end product. Although the loss of product is more in this process, it remains a widely used method for commercial production.

Membrane-Based Filtration

Membrane-based processes are extensively used in the production, separation, purification, and formulation of biotechnological products (Charcosset 2006). Microfiltration and ultrafiltration are well-known membrane-based techniques used to recover the proteins produced through bioprocesses (Cui 2005). Membranes that are widely used for protein bioseparation are made from various materials; however, cellulose acetate (CA), polyethersulfone (PES), and polyvinylidene fluoride (PVDF) membranes are preferable for Bt filtration because of their low protein-binding characteristics (Rathore and Shirke 2011).

The fermented broth of Bt can be harvested through tangential flow ultrafiltration using membranes having pore size ranging from 10 nm to 0.1 μm or through the microfilters having pore size ranging between 0.1 μm and 5.0 μm. In this system, the culture broth is pumped tangentially along the surface of Pellicon cassette and pressure is applied to drive the culture broth through the membrane to the filtrate side. Spore-crystal complexes that are too large to pass through the membrane is retained on the upstream side and then collected into separate system.

A separate filtration unit is installed. The methodology for operating the system is provided by the installer to operate the unit. The advantages and disadvantages of a particular product recovery process are listed in Table 13.4.

Formulation

Commercial bioinsecticides must be economical to produce, stable during storage, easy to handle during formulation, storage, and application, and consistently effective in controlling pests. The stability of the Bt product during storage is essential because it can affect the effectiveness of the product after application. The appropriate formulation of a product can play a key role, serving as a benchmark for successful stabilization of the product and its entomotoxicity. The ideal formulation should (a) stabilize viable cells during mixing and storage, (b) be easy to handle and apply, (c) protect the product from adverse environmental conditions, and (d) should enhance activity of microbial agents in the field.

Generally, bioinsecticide/biopesticide formulations are classified into two major types: (1) solid-carrier based and (2) liquid formulations. Carrier-based Bt formulations are in the form of wettable powders, granules, or briquettes, and liquid

Table 13.4 Various harvesting methods for B. thuringiensis spores and spore-crystal complex

Method	Advantages	Disadvantages	Material required	References
Alginate immobilization	Minimum wastage, low initial investment, useful for small-scale industry	Time-consuming	Sodium alginate, calcium chloride	Prabakaran and Hoti (2012)
Acid precipitation	Low initial investment	High wastage, corrosive to handling, low recovery efficiency, loss in supernatant	Hydrochloric acid	Prabakaran and Hoti (2012)
Centrifugation	Moderate wastage	High initial investment, recovery efficiency depends upon the relative centrifugal force (RCF), which is difficult to adjust in large amounts of broth, loss in supernatant	Continuous centrifuge	Brar et al. (2006a, 2006b)
Membrane-based filtration	No wastage, high recovery efficiency	High initial investment as well as maintenance charges due to heavy membrane fouling and frequent membrane replacement, time-consuming	Microfiltration or ultrafiltration unit with membrane cassette	Naseri Rad et al. (2016)

formulations are water or oil-based emulsions (Brar et al. 2006b). The final applicable Bt formulations vary with the type of pathogen, insect species, mode of application, application rate, mode of action, and rheology of the technical material. Most importantly, in the case of the Bt product, the final product formulation is determined by the medium used for fermentation, fermentation conditions, methodology used for product recovery, and adjuvants (additives) used during formulation.

Adjuvants (Additives)

Adjuvants are chemically and biologically active compounds that can alter the physical properties of formulations (Brar et al. 2006b). The key factors to be considered while selecting it for Bt formulations are: (a) it must kill the target pest without harming other insects, (b) it must be biodegradable, (c) it must have adhesive properties, and (d) it must be economical.

Liquid Formulations

Liquid formulations of *B. thuringiensis* have replaced wettable powder formulations in the Indian market because of its ease of use and low production cost. As this technology does not require drying and milling, it is industrially favored. These water-based flowables contain carrier liquid (water), 10–40% Bt cells, 1–3% suspender, 1–5% dispersant, 3–8% surfactant. The concentrated slurry obtained after centrifugation recovery (or the concentrated product obtained after other techniques such as filtration) should be mixed with appropriate amount of distilled water after quality control. Different types of adjuvants and additives should be added to the product to maintain stability and entomotoxicity. Some ingredients used to formulate the Bt product with its specific function and concentration (Kumar et al. 2019) are listed in Table 13.5. All ingredients should be added to the final volume of the formulation (concentrated broth + distilled water) and mixed for 1 h in the mixing tank. It is not necessary to follow this list, but it represents more common ingredients used to suspend, stabilize, facilitate mixing and ease of application.

13.3 Packaging

Liquid formulations are generally packed into HDPE bottles of different volumes. The final packaged product should meet international transportation and safety regulations and the requirements specified by the WHO for pesticides used in public health.

13.4 Quality Control

Cell and Spore Count At the fermentation stage, the progress of bioinsecticide production should be monitored, which can be achieved through viable cell and

Table 13.5 Adjuvants and additives added during the formulation of Bt product

Adjuvants/additives	Its function	Approximate concentration in %
Sodium acetate	Act as buffering agent	1.42
Potassium silicate	Act as buffering agent	0.21
Molasses	UV protectant, for adhesion	0.25
Sorbitol	Flavoring agent	0.83
Xanthan gum	For adhesion	0.17
Carboxy-methyl cellulose	Acts as stabilizer or emulsifier	0.13
Sorbic acid	Acts as antimicrobial compound	0.01
Propionic acid	Acts as antimicrobial compound	0.06
Acetic acid	Acts as antimicrobial compound	0.21

spore counts. For this, withdraw the sample from the fermenter through the sampling point, serially diluted in sterile normal saline (0.85% w/v NaCl), spread on nutrient agar, incubate for 24 h at 30 °C, and calculate CFUs/mL next day. To determine the spore count, heat the sample in a hot water-bath at 80 °C for 15 min following the same procedure as described above.

Entomotoxicity Assay Each fermented broth and the final product should be checked for entomotoxocity before packaging. In India, the mosquito larvicidal activity (LC50) of the product is checked against early fourth instar laboratory-reared *Aedes aegypti* L. larvae. Samples should be serially diluted and placed in cups seeded with 50 larvae in 250 mL of chlorine free tap water. Incubate the cups at 30 ° C for 24 h and score the larvae mortality rate. The entomotoxicity of the Bt formulations should be calculated by observing mortality relative to the standard preparation according to the following formula;

Entomotoxicity of sample (IU/mL) = (LC_{50} of standard/LC_{50} of sample) × entomotoxicity of standard (Zhuang et al., 2011).

IU/mL is the International Unit for entomotoxicity per microliter.

13.5 Economical Aspects

13.5.1 Basic Requirements for Setting up of Bioinsecticide Units

Based on the field visits of the biocontrol production unit and to fulfill described methodological aspects, various facilities required to implement successful projects for 100 tons per annum capacity are listed here.

(a) **Land:** Land is the primary requirement for establishing production units. In the present model, a rented building was assumed; hence no land purchase cost has been considered, except for rent, for calculating the overall cost-benefit ratio (Table 13.9).
(b) **Building and infrastructure**: Buildings should be compartmentalized to facilitate overall production in smaller areas and to maintain sterility. The building should be broken up into reception, administration, media preparation room, inoculation and incubation room, area for building a platform for fermenters, formulation and packaging compartment, storage room, quality control laboratory, etc. Because the building is rented here, no infrastructure development cost is applied. However, compartmentalization, lab interiors and furniture purchasing costs must be included to establish the unit (Table 13.9).
(c) **Plant and machinery:** A number of equipments are required for the mass production and product recovery of *B. thuringiensis* bioinsecticides. All machinery required is listed in Table 13.6 with its cost.

Table 13.6 Machineries required for the production of Bt and its cost in INR

Equipment name	Nos	Cost in INR/Nos (in lakhs)	Total cost in INR (in lakhs)
Vertical autoclave (600 × 350 mm)	1	0.8	0.8
Refrigerator (300 L)	1	0.3	0.3
Laminar air flow (3′ × 2′)	1	1.0	1.0
BOD incubator	1	0.6	0.6
Rotary shaker	1	0.25	0.25
Water-bath	1	0.1	0.1
Compound microscope (binocular)	1	0.65	0.65
Electronic weight balance	1	0.2	0.2
pH meter	1	0.3	0.3
Colony counter	1	0.05	0.05
Stainless steel seed fermenters (20 L cap.)	2	1.5	3.0
Stainless steel fermenters (100 L, 70 L working cap.)	2	3.8	7.6
Stainless steel fermenters (1000 L, 700 L working cap.)	1	10.0	10.0
Air compressor oil free type	2	0.25	0.5
Continuous centrifuge machine	1	3.5	3.5
RO plant (200–500 L/h)	1	0.35	0.35
Mixing tanks	2	0.5	1.0
Capping and Labeling machines and other miscellaneous items for packaging	1	1.0	1.0
Air conditioners	4	0.35	1.4
Generator (65 kVA, diesel generator)	1	1.5	1.5
Glassware and plasticwares for microbiological works	–	Approx..	2.5
Other miscellaneous requirements for lab use	–	Approx..	1.4
Grand total			**38.0**

(d) **Raw material:** The media required for seed inoculum preparation and mass fermentation is already described in the mass production technology details. The cost of media is given in the Table 13.7.

(e) **Manpower:** Production of bioinsecticide requires skilled and unskilled labors illustrated in the Table 13.8 along with salary.

(f) **Utilities:** Tap water is required for washing and cleaning, and distilled water is required for media preparation and formulation of the product.

Power: Continuous power supply is essential for bioinsecticide units. The charges are listed under recurring cost Table 13.10.

The ultimate nonrecurring and recurring costs for a production unit of 100 tons per annum capacity are calculated in Tables 13.9 and 13.10, respectively.

Table 13.7 Raw material cost for the production of 100 tons per annum finished product

Medium and specification	Raw material	Quantity	Total cost (INR)/ 100 ton/annum (in lakhs)
NYSM (for subculturing, mother culture and starter culture preparation, quality checking)	Ready to use media from standard company (500 gm) (Himedia)	5 kg	0.15
Production media: Starch Industry wastewater (SIW) (Seed inoculum preparation and fermentation process)	Free of cost Only transportation charge	Approx.. 80,000 L	0.50
Micronutrient (Entomotoxicity checking)	Quality control For insect larvae rearing	Approx..	0.1
Carrier material	Distilled water for liquid formulations	40,000 L	0.25
Adjuvants and additives	Listed in table	Approx..	0.25
Total cost			**1.25**

Table 13.8 Manpower required to run the production unit of 100 TPA capacity

Category	Nos	Salary per head (Rs. In lakhs/month) (INR)	Total cost (Rs. In lakhs/ month) (INR)
Chief microbiologist	1	0.3	0.3
Assistant microbiologist	1	0.15	0.15
Production officer	2	0.20	0.4
Technical staff	1	0.1	0.1
Administrative officer	1	0.15	0.15
Sales officers	5	0.15	0.75
Skilled and unskilled labors	5	0.045 (0.03–0.06 depending upon the work allotment)	0.225
Total cost			**2.075**

Table 13.9 Nonrecurring expenses for establishing 100 TPA unit

Particular	Total cost in INR (in lakhs)
Building rent	3.6
Lab interior and furniture	7.0
Equipment and machinery	38.0
Miscellaneous fixed assets (computer, printer, fax, stationary items)	1.5
Grand total	**50.1**

Table 13.10 Recurring expenses per annum for 100 TPA unit

Particulars/annum	Cost (Rs. in lakhs/annum (INR)
Raw material (cost/100ton/annum)	1.25
Packaging (cost/100 ton/annum) HDPE bottle (1 L) 21 Rs./piece with cap	2.016
Manpower cost	24.9
Utilities – Power (4000 units @ Rs. 5/unit)	2.4
Contingencies	
Marketing and selling expenses	7.2
Repair and maintenance	3.0
Total	**40.766**

Table 13.11 Income calculation

Particulate	Values
Production/annum	100 ton
Selling/annum	90 ton
Price/L	250 INR
Income/annum for selling 90 ton	**225.00 INR (lakhs)**

Table 13.12 Total income and expenditure for model project of 100 TPA bioinsecticide unit (Rs. in lakhs)

Sr. no.	Particulars	Values
1.	Capacity utilized in ton	100
2.	Income (Rs. In lakhs/annum)	225.0
Expenditure		
3.	Technology know how	10.0
4.	Nonrecurring expenses	50.1
5.	Recurring expenses	40.766
6.	Administrative expenses	1.5
7.	Taxes and insurance	3.0
8.	Loss due to contamination low entomotoxicity achievement	5.0
9.	Depreciation for fixed assets @ 5%	2.5
Total expenditure		**112.866**
Net benefit		**112.134**

13.6 Income per 100 Ton of Production

The average cost of *B. thuringiensis* liquid formulation in the current market ranges 250–350 INR/L. Government institutions and organizations may cost less than this, and on the other hand, some private factories may cost double. The calculation for income is presented in Tables 13.11 and 13.12.

13.7 Profitability

Net profit – 112.866 INR in lakhs.
% profit of sale – 50.6%.

References

Adjalle KD, Brar SK, Verma M, Tyagi RD, Valero JR, Surampalli RY (2007) Ultrafiltration recovery of entomotoxicity from supernatant of Bacillus thuringiensis fermented wastewater and wastewater sludge. Process Biochem 42(9):1302–1311

Al-Momani F, Obeidat M (2013) Ecology, toxicity, and hydrolytic activities of Bacillus thuringiensis in forests. Turk J Agric For 37(1):76–82

Azmi NU, Ghafar NSA, Yin CJ, Yakubu S, Adli AA, Aziz NAA, Mustafa M (2015) Toxicity of bacillus thuringiensis biopesticide produced in shrimp pond sludge as alternative culture medium against Bactrocera dorsalis (Hendel). Acta Biol Malays 4(1):5–16

Ballardo C, Abraham J, Barrena R, Artola A, Gea T, Sánchez A (2016) Valorization of soy waste through SSF for the production of compost enriched with Bacillus thuringiensis with biopesticide properties. J Environ Manag 169:126–131

Ballardo C, Barrena R, Artola A, Sánchez A (2017) A novel strategy for producing compost with enhanced biopesticide properties through solid-state fermentation of biowaste and inoculation with Bacillus thuringiensis. Waste Manag 70:53–58

Bizzarri MF, Bishop AH (2008) The ecology of Bacillus thuringiensis on the phylloplane: colonization from soil, plasmid transfer, and interaction with larvae of Pieris brassicae. Microb Ecol 56(1):133–139

Brar SK, Verma M, Tyagi RD, Valéro JR (2006b) Recent advances in downstream processing and formulations of Bacillus thuringiensis based biopesticides. Process Biochem 41(2):323–342

Brar SK, Verma M, Tyagi RD, Valéro JR, Surampalli RY (2006a) Efficient centrifugal recovery of Bacillus thuringiensis biopesticides from fermented wastewater and wastewater sludge. Water Res 40(6):1310–1320

Chang M, Zhou S, Sun Q, Li T, Ni J (2012) Recovery of Bacillus thuringiensis based biopesticides from fermented sludge by cross-flow microfiltration. Desalin Water Treat 43(1–3):17–28

Charcosset C (2006) Membrane processes in biotechnology: an overview. Biotechnol Adv 24(5): 482–492

Cui Z (2005) Protein separation using ultrafiltration—an example of multi-scale complex systems. China Particuol 3(6):343–348

Elleuch J, Jaoua S, Ginibre C, Chandre F, Tounsi S, Zghal RZ (2016) Toxin stability improvement and toxicity increase against dipteran and lepidopteran larvae of Bacillus thuringiensis crystal protein Cry2Aa. Pest Manag Sci 72(12):2240–2246

Glare T, Caradus J, Gelernter W, Jackson T, Keyhani N, Köhl J, Marrone P, Morin L, Stewart A (2012) Have biopesticides come of age? Trends Biotechnol 30(5):250–258

Gnepe JR, Tyagi RD, Brar SK, Valéro JR, Surampalli RY (2014) Corrosion and stability study of Bacillus thuringiensis var. kurstaki starch industry wastewater-derived biopesticide formulation. J Environ Sci Health B 49(11):889–896

Höfte H, Whiteley H (1989) Insecticidal crystal proteins of Bacillus thuringiensis. Microbiol Rev 53(2):242–255

Kashyap L, Goswami TN, Patel VK, Sharma RK (2017) Bacillus thuringiensis and insect pest management. In: Biopesticides and bioagents. Apple Academic Press, pp 331–369

Kumar LR, Ndao A, Valéro J, Tyagi RD (2019) Production of Bacillus thuringiensis based biopesticide formulation using starch industry wastewater (SIW) as substrate: a techno-economic evaluation. Bioresour Technol 294:122144

Kumar V (2015) A review on efficacy of biopesticides to control the agricultural insect's pest. Int J Agric Sci Res 4(9):168–179

Montesinos E (2003) Development, registration and commercialization of microbial pesticides for plant protection. Int Microbiol 6(4):245–252

Naseri Rad S, Shirazi A, Kargari MMA, Marzban R (2016) Application of membrane separation technology in downstream processing of bacillus thuringiensis biopesticide: a review. J Membr Sci Res 2(2):66–77

Osman GEH, Already R, Assaeedi ASA, Organji SR, El-Ghareeb D, Abulreesh HH, Althubiani AS (2015) Bioinsecticide Bacillus thuringiensis a comprehensive review. Egypt J Biol Pest Control 25(1):271

Prabakaran G, Hoti SL (2012) Immobilization in alginate as a new technique for the separation of bacillus thuringiensis var. israelensis spore crystal complex. Biol Control 61(2):128–133

Rathore AS, Shirke A (2011) Recent developments in membrane-based separations in biotechnology processes. Prep Biochem Biotechnol 41(4):398–421

Rodrigo MA, Oturan N, Oturan MA (2014) Electrochemically assisted remediation of pesticides in soils and water: a review. Chem Rev 114(17):8720–8745

Rowe GE, Margaritis A (2004) Bioprocess design and economic analysis for the commercial production of environmentally friendly bioinsecticides from bacillus thuringiensis HD-1 kurstaki. Biotechnol Bioeng 86(4):377–388

Stankus T (2013) Review of biopesticides: pest management and regulation: by Alastair Bailey, David Chandler, WP Grant, Justin Greaves, Gillian Prince, and Mark Tatchell. CAB International, Wallingford, UK, p 2010

Vu KD, Tyagi RD, Surampalli RY, Valéro JR (2012) Mathematical relationships between spore concentrations, delta-endotoxin levels, and entomotoxicity of Bacillus thuringiensis preparations produced in different fermentation media. Bioresour Technol 123:303–311

Zhuang L, Zhou S, Wang Y, Liu Z, Xu R (2011) Cost-effective production of bacillus thuringiensis biopesticides by solid-state fermentation using wastewater sludge: effects of heavy metals. Bioresour Technol 102(7):4820–4826

Chapter 14
Mass Multiplication, Production Cost Analysis, and Marketing of *Trichoderma*

Dhruti Amin

Abstract Due to the increasing demand for chemical-free crop products globally, the Indian market for biopesticide/biofungicides has seen significant growth since 2014. *Trichoderma sp.* are mainly asexual fungi present in all agricultural soil. It is one of the most versatile biocontrol agents used to manage phytopathogens, especially fungi. These fungi can produce many antibiotic substances, and their ability to parasitize other fungi benefits plant growth and development. *Trichoderma virens*, *T. atroviride*, and *T. harzianum* are the most reported *Trichoderma sp.*, which can induce plant resistance against various plant pathogens, and certain strains can also substantially influence plant growth and development. Besides, the Indian Government has initiated awareness programs and various schemes in full swing among the farmers/growers to use biocontrol agents/biopesticides/bi-fungicides. A small effort has been made to fascinate and uplift entrepreneurs in agribusiness, contributing to maintaining sustainability and profit-making. This chapter deals with the critical, technical details of the mass production and cost-benefit ratio to establish the small- and large-scale biocontrol production unit of *Trichoderma.*

Keywords *Trichoderma* spp. · Biocontrol · Plant growth promotor · Mass production · Popularization

14.1 Introduction

Trichoderma is a genus of fungi belonging to the Hypocreaceae family, a globally known fungal biocontrol agent to manage various foliar and soil-borne plant pathogens. It is generally isolated from the rhizosphere soil and decaying plant organic matter (Harman et al. 2004). The difficulties encountered during identifying *Trichoderma* isolates at the species level become more significant because the morphological differences are rare and hard to observe. The concept of "species

D. Amin (✉)
C. G. Bhakta Institute of Biotechnology, Uka Tarsadia University, Surat, Gujarat, India

N. Amaresan et al. (eds.), *Agricultural Microbiology Based Entrepreneurship*, Microorganisms for Sustainability 39,
https://doi.org/10.1007/978-981-19-5747-5_14

aggregate" categorized *Trichoderma* strains into nine aggregates based on morphological features reported by Rifai (1969). Unfortunately, some of the "species aggregate" consists of two or more morphology that is non-differentiable. Later in 1991, Bissett reviewed Rifai's work and attempted to integrate similar forms within the species concept based on morphology, including the characteristics of the conidiophore branching system. As a result, *Trichoderma* was classified into five sections: Saturnisporum, Pachybasium, Longibrahiatum, Trichoderma, and Hypocreanum. The *Trichoderma* genus has about 10,000 species (Waghunde et al. 2016). Morphological and molecular approaches are used for the identification of *Trichoderma* species. In the beginning, *Trichoderma* strains appeared white and cottonish, then developed into yellowish-green to deep green compact tufts, especially at the center of a growing spot or in concentric ring-like zones on the agar surface.

Trichoderma spp. is an effective biocontrol agent, eco-friendly and cheap, nullifying the ill effects of chemicals. It not only kills harmful fungi via mechanisms like mycoparasitism and antibiosis but also plays a vital role in plant growth promotion. *Trichoderma* species promote plant growth, contribute to nutrient utilization, and induce plant defenses against biotic and abiotic stresses. Several formulations contain mixtures of different species that provide a broad range of direct and indirect beneficial effects for the plants. *T. harzianum* and *T. viride* are the two most commonly used species, effective when applied to about 87 different crops in India (Sharma et al. 2014). The usage of this microbial inoculant in *Trichoderma*-based products attracts the attention of researchers to discover more about the other potential benefits of *Trichoderma* spp.

Commercial unit development of *Trichoderma* as a biocontrol agent / biopesticides/biofungicides comprises four essential areas for production, marketing, and selling such as:

(i) Mass production.
(ii) Formulation type.
(iii) Shelf life and storage.
(iv) Application method.

The details are as follows:

14.1.1 *Mass Production Process of* Trichoderma *as a Biocontrol Agent*

Biocontrol agents are produced and marketed by many commercial firms and are available in the global market (Velivelli et al. 2014). In India, more than 250 BCA products are available in the market. The formulation of commercial biocontrol agents for agricultural applications should possess several desirable characteristics and have substantial proof to convince farmers. These include good market potential, easy preparation, unfussy application, high stability during transportation and storage, abundant viable propagules with good shelf life, sustained efficacy, and accepted cost.

The steps involved in mass production are:

1. Selection of potent *Trichoderma* species.
2. Preparation of mother culture: Prepare molasses yeast medium containing 30 g molasses, 5 g yeast, and 1000 ml distilled water and dispense into conical flasks and sterilize at 15 lb. pressure for 15 min in an autoclave. After the medium is cooled, it is inoculated with 10-day-old fungal disc of *Trichoderma viride* and then incubated for 10 days for fungal growth to act as a mother culture.
3. For mass multiplication: Prepare and add molasses yeast medium in the fermenter and sterilize as described earlier. Then after the medium is cooled, the mother culture is added to the fermentor @ 1.5 L/50 L of the medium and incubated at room temperature for 10 days. Then the incubated broth containing the fungal culture is used for commercial formulation preparation using talc powder.

Depending upon the type of *Trichoderma* species, media, substrates, and growing conditions may vary for preparing mother culture and mass multiplication.

There are two methods, such as (i) Solid-state fermentation and (ii) Liquid state fermentation, which are commonly used for inoculum production. The mass production technique should be well-suited to industrial and commercial development methods.

Type of substrates used for *Trichoderma* growth (1). solid based includes (i) grains, (ii) organic matters, (iii) agriculture wastes and (2). liquid based, as mentioned in Table 14.1. Generally, food-based substrates have proved helpful for the preparation of formulation of *Trichoderma*-based biocontrol agents.

14.2 Type of Formulations for *Trichoderma*-Based Biocontrol Agent

Using biocontrol agents at the commercial level has become a reality due to awareness of the negative impact of pesticides. Commercial application of *Trichoderma* either to increase crop health or to manage plant diseases depends on the development of commercial formulations with suitable carriers that support the survival of *Trichoderma* for a considerable length of time. Different technologies become viable only when the research findings are transferred from the lab to the field.

Table 14.1 List of solid and liquid substrates used for different *Trichoderma* spp. production

State of forms	*Trichoderma* spp.	Substrate
1. Solid based:		
(i) Grains	*T. harzianum*	Sorghum, rice, pearl millet, maize
	T. viride	Sorghum, wheat
(ii) Organic matters	*T. harzienum P26*	Neem cake, compost, farmyard manure (FYM), Gliricida leaves
	T. harzianum	FYM, local cow dung, Jersey cow dung, jatropha cake, and neem cake
	T. viride	FYM, vermicompost, poultry manure, goat manure, decomposed coconut, coir pith, peat
(iii) Agriculture waste	*T. harzianum*	Rice bran, paddy straw, groundnut shells, wheat straw, paddy straw, shelled maize cob, paper waste, saw dust, sugarcane bagasse, spent straw, wheat bran, rice bran
	T. viride	Tapioca rind, tapioca refuse, mushroom spent straw, paddy chaff, wheat bran, groundnut shell, rice bran, sugarcane baggase, wheat straw, shelled maize cob, paddy straw, chickpea husk
	T. virens	Spent malt
	T. atroviride	Onion rind (dry onion skin), apple and strawberry pomace, rapeseed meal
	T. hamatum	Tea waste, sorghum straw, wheat straw, wheat bran
2. Liquid based	*T. harzianum strain 1295–22*	Modified RM8
	T. viride	Molasses and Brewer's yeast
	T. harzianum	Local cow urine, Jersey cow urine, butter milk, vermiwash
	T. harzianum strain P1	Defined basal culture medium with mineral solution
	T. harzianum Rifai	Potato dextrose broth, Czapeck's Dox broth and modified Richards' broth

Detail information refer to Waghunde et al. (2016)

14.3 Features of *Trichoderma* for Formulation Development

It is well known that *Trichoderma* spp. can control plant diseases and is eco-friendly compared to agrochemicals. However, the efficacy and consistency of biocontrol agent results were significantly less than commercial agrochemicals. *Trichoderma* should hold features such as high rhizosphere competence, competitive saprophytic ability, plant growth enhancer, ease for mass multiplication, a broad spectrum of action with excellent and reliable control, compatibility with other bioagents, and should tolerate desiccation, heat, oxidizing agents, and UV radiations (Jeyarajan and Nakkeeran 2000). This can enhance the potential of *Trichoderma* formulation, benefit the plants against adverse conditions, lower the usage of agrochemicals, hit the market, and be profitable to the grower.

Several reports of the successful use of formulations of *Trichoderma* in the greenhouse and the field for control of various diseases, especially the soil-borne pathogens, are well known (Cumagun 2014). The yearly requirement of *Trichoderma* has been estimated as 5000 tones to cover 50% of India's area (Jeyarajan 2006). Though *Trichoderma* has excellent potential in managing diseases, it could not be used as spore suspension under field conditions. Thus, the culture of *Trichoderma* should be immobilized in certain carriers and should be prepared as formulations for easy application, storage, commercialization, and field use (Table 14.2).

Among the above-listed carriers, talc-based *Trichoderma* formulation is globally used as a biocontrol agent.

The standards for *Trichoderma* formulations are as follows:

1. Colony Forming Units (CFUs) of *Trichoderma* spp. should be a minimum of 2×10^6 CFU per ml or gm on a selective medium.
2. Free from pathogenic contaminants such as *Salmonella, Shigella*, or *Vibrio*. Other microbial contaminants need not exceed 1×10^4 count ml/gm.
3. Maximum moisture content should not be more than 8% for the dry formulation of fungi.

14.3.1 Shelf Life and Storage of Trichoderma *Formulations*

The shelf life and viability of the formulated product of a biocontrol agent play a significant role in successful marketing. Shelf life mainly depends on carriers, storage material, and culture conditions used for formulation development. Usually, the antagonists multiplied in an organic food base have longer shelf life than the inert or inorganic food bases. The shelf life of *Trichoderma* in coffee husk was more than 18 months, while talc, peat, lignite, and kaolin-based formulations of *Trichoderma* have a shelf life of 3–4 months. The viable propagules of *Trichoderma* in talc formulation were reduced by 50% after 120 days of storage (Sankar and Jeyarajan 1996). Under the ambient condition, the shelf life of talc-based formulation remains up to 180 days. The desiccation rate was less up to the first 75 days of storage than that observed from 75 to 180 days of storage. The formulation retained good numbers of viable propagules (above 106 cfu/g) for more than 150 days of storage reported by Bhat et al. (2009). The storage of *T. viride* formulation in polypropylene bags of various colors revealed that the population of *T. viride* was maximum in milky white bags of 100-micron thickness. In the year 2010–2011, Sriram and team at PDBC, Bangalore, worked on increasing the shelf life of talc formulations of *Trichoderma* using chitin and glycerol in the production medium and heat shock at the end of the log phase of fermentation and found that the shelf life of talc formulation was successfully maintained up to 1 year (Sriram et al. 2010, 2011).

Table 14.2 Various approaches followed in India for *Trichoderma* mass multiplication

Carriers-based formulation	Commercial production	Specification	Developers
Talc	*Trichoderma viride* is grown in the liquid medium mixed with talc powder in the ratio of 1: 2 and dried to 8% moisture under shade	Pulse crops and rice seed treatment at 4–5 g/kg seed to manage several soil-borne diseases. The talc formulations of *Trichoderma* have a shelf life of 3–4 months	It was first developed at the Tamil Nadu Agricultural University by Jeyarajan et al. (1994). Many private industries produce on a large scale for supply to the farmers
Vermiculite-wheat bran	*Trichoderma* multiplied in a molasses-yeast medium for 10 days. 100 g vermiculite, and 33 g wheat bran are sterilized in an oven at 70 °C for 3 days. Then, 20 g of fermentor biomass, 0.05 N HCl, is added, mixed well and dried in the shade	This product is effective in reducing the growth of *Rhizoctonia solani*	Lewis (1991)
Pesta granules	52 ml fermentor biomass is added to wheat flour (100 g) and mixed with gloved hands to form a cohesive dough. The dough is kneaded, pressed and folded by hand several times. Then 1 mm thick sheet (pesta) is prepared and air-dried till it breaks crisply. After drying, the dough sheet is ground, passed through an 18 mesh, and granules are collected	This process appears to be a simple and effective way to formulate	Connick et al. (1991)
Alginate prills	Sodium alginate is dissolved in one portion, and distilled water (25 g/750 ml) and food base is suspended in another portion (50 g/250 ml). These preparations are autoclaved and, when cool, are blended with biomass. The mixture is added dropwise into $CaCl_2$ solution to form spherical beads, which are air-dried and stored at 5 °C	Encapsulation of biocontrol agent and nontoxic to nontarget organisms	Fravel et al. (1999)

(continued)

Table 14.2 (continued)

Carriers-based formulation	Commercial production	Specification	Developers
Press mud	9 days old culture of *T. viride* (prepared in potato dextrose broth) is uniformly mixed with 120 kg press mud. Water is sprinkled intermittently to keep it moist. It is covered with gunny bags to permit air movement and trap moisture under shade. Within 25 days, nucleus culture use for further multiplication becomes ready. The same is added to 8 tons of press mud, mixed thoroughly and incubated for 8 days under shade conditions before field application. Into this is added 8000 times more inoculums in the soil than the recommended doses of biopesticides, which rapidly get established, showing rapid and visible effect. Similarly, other substances could also be effectively used to multiply different bioagents at the mass level	Press mud is a byproduct of the sugar the factory used as a substrate	Sabalpara (2014)
Coffee husk	*Trichoderma* formulation is based on the coffee husk, waste from the coffee curing industry	This product is very effective in managing *Phytophthora* foot rot of black pepper and is widely used in Karnataka and Kerala	Sawant and Sawant (1996)
Oil	Mix the conidia harvested from the solid-state/liquid state fermentation with a combination of vegetable/mineral oils in a stable emulsion formulation. In such formulations, microbial agents are	Used as foliar sprays under dry weather and have a prolonged shelf life. This formulation is effective against soil-borne diseases of groundnut and also control the post-harvest	*T. harzianum* with a shelf life of 8 months has been developed using indigenous constituents at the erstwhile Project Directorate of Biological Control (Batta 2005; Kumar et al. 2014)

(continued)

Table 14.2 (continued)

Carriers-based formulation	Commercial production	Specification	Developers
	suspended in a water-immiscible solvent such as a petroleum fraction (diesel, mineral oils), and vegetable oils (like groundnut.) with a surface-active agent. This can be dispersed in water to form a stable emulsion. Emulsifiable concentrates require a high concentration of an oil-soluble emulsifying agent to form a homogenous emulsion on dilution in water rapidly. The oils used should not have toxicity to the fungal spores, plants, humans, and animals	decay of apple caused by *Botrytis cinerea*	
Banana waste	A pit of various banana waste, sheath pseudo stem and core is chopped 5–8 cm. A pit is prepared, and different ingredients are placed in five layers. Each layer contains 1 ton of banana waste, 5 kg urea, 125 kg rock phosphate, and 1 liter broth culture of *Bacillus polymixa*, *Pleurotus sajor caju*, and *T. viride*. Five different layers are prepared similarly and mixed thoroughly with Banana. Banana waste is decomposed within 45 days, and enriched culture is mass available for field application	Cost effective	Balasubramanian et al. (2008); Kumar et al. (2014)

14.3.2 Application Methods

The most common methods of application of *Trichoderma* are seed treatment, seed biopriming, seedling dip, soil application, and aerial/wound dressing (Ramanujam et al. 2010; Kumar et al. 2014).

14.4 Registration and Quality Control

To prevent the sale of poor quality to the farmers, the Central Insecticide Board (CIB) of the Government of India has made registration of microbial products mandatory before commercial production/selling. Before the registration process, ensure that the products do not adversely affect the environment's biotic and abiotic factors. For detailed information regarding the guidelines including physical, chemical, and biological properties of *Trichoderma* products and their efficacy to the target/nontarget organisms, toxicological reports, packaging, labeling, etc., and registration process refer to the annexure of the Insecticide Act of India.

14.5 Production Cost Analysis for the Establishment of Commercial *Trichoderma*-Based Biocontrol Agents

Trichoderma-based biocontrol agent is one of the most widespread and currently used in sustainable agriculture to control different plant diseases (Hu et al. 2020). Among several beneficial fungi, *Trichoderma* spp. has gained much attention in the commercial agriculture market; more than 60% of registered biological pesticides reported worldwide contain *Trichoderma* spp. (Lorito and Woo 2015). Successful development of products is challenging due to high-cost technological investment (Sachdev and Singh 2016; Singh and Nautiyal 2012). Profitable large-scale production can be achieved by solid-state fermentation (SSF).

Consequently, growing demands for bio-fungicide production to replace excessively used chemical pesticides have recently boosted interest in SSF technology. SSF simulates the natural habitat of fungi and is the preferred choice for these microorganisms to grow and produce useful value-added products (Sadh et al. 2018). SSF is used for the mass production of filamentous fungi, their enzymes, and other metabolites on solid substrates with sufficient moisture but not in the free state (Cavalcante et al. 2008). For biomass production, 35–40% of production cost increases due to using the organic substrate as raw materials (Eltem et al. 2017). Therefore, there is much needed to utilize agro-industrial wastes that are cheap, readily available, and support the extensive growth of *Trichoderma* and contribute to the production of value-added products. It provides avenues for the safe utilization of waste while reducing waste disposal's cost and environmental pollution load. In recent years, the global production, registration, and application of biological pesticides in agriculture as alternatives to chemicals have rapidly increased owing to public concerns about human health, food safety, and the environmental impact (Carvalho 2017; Kumar 2012).

14.6 The Basic Requirements and Project Cost for the Production Unit Are as Follows

1. **Land:** Space required for building culture and rearing rooms, processing room, laboratory, office, and construction of poly house, etc.
2. **Building and civil works:** Area must be far from the industrial unit to avoid pollution problems. About 2400 sq. metre. Area is required to set up a production unit of *Trichoderma*-based biocontrol agent.
3. **Plant and Machinery:** Instruments like laminar flow apparatus and fermentor are required.
4. **Raw material:** Molasses yeast medium for mother culture preparation and various substrates and carrier requirement for formulation development as mentioned in Tables 14.1 and 14.2.
5. **Water:** For washing, cleaning, drinking, and other purposes, water quality should be tested regularly to establish suitability.
6. **Power:** Essential for the production unit.
7. **Manpower:** Labor-intensive project requires a total of 18 members such as three technical staff members, five skilled persons, and ten semi-skilled person.
8. **Unit size:** It must have a 2000 kg/year capacity of biocontrol production from *Trichoderma* fungi.

Outline of expenditure, including contingencies; the project cost works out to Rs. 296.7 lakhs, as shown below.

Project cost	Total cost amount (in lakhs)
Land and site development	19.26
Building	248.85
Plant and machinery	17.07
Miscellaneous fixed assets	4.52
Preoperative cost	2.00
Security deposits	5.00
Total	**296.7**

http://mkuy.apicol.nic.in/Content/ReferenceProject/Bio%20Pesticides%20Bio%20control%20Agent%20producing%20Unit.pdf

14.7 Financial Assistance

The projects on manufacturing biopesticide products would be considered for refinance support by National Bank. All banks may consider financing this project per their technical feasibility, financial viability, and bankability. The means of finance will be:

Means of finance	Amount
Equity (25%)	74.21
Subsidy	50.00
Term loan (14%)	172.64
Total	**296.85**

14.8 Marketing

The Indian government is encouraging research, production, registration, and adoption of biopesticides with an open hand through various rules, regulations, policies, and schemes. The National Farmer Policy (2007) has strongly recommended promoting biopesticides to increase agricultural production and sustain the health of farmers and the environment. It also includes the clause that biopesticides would be treated at par with chemical pesticides in support and promotion. Despite all efforts, the share of biopesticides in India is merely 2% compared to other countries like the USA (40%), Europe, and Oceanic countries (20% each) (Sabalpara 2014). There is difficulty in translating the efficacious biocontrol agents into merchandise due to the unstable and low efficacy of the developed strains under varied environmental conditions and not being able to reach the farmer's field. Also, the proportion of registration of biocontrol agents for commercial availability is very slow, and further bio-products need to be improved to obtain greater levels of disease reduction. The development of formulations with increased shelf life and a broad spectrum of action with consistent performance under field conditions could pave the way for the commercialization of the technology at a faster rate. In this regard, the current chapter mainly focused on mass production, formulation development, quality control, delivery system, and scope to enhance commercialization in India for managing plant diseases. Recent literature surveys have shown that the number of Trichoderma-containing products on the international market has grown exponentially, with more than 300 products now available (Woo et al. 2014; Thakur et al. 2020); in India, there are more than 250 commercial formulations (Singh et al. 2013).

14.9 Concluding Remarks

Trichoderma spp. has potential agriculture applications that help plants improve physiological responses to biotic and abiotic stresses, enhance nutrient uptake, and improve photosynthetic efficiency in different crops. Worldwide, it acts as plant protectants and growth enhancers, in addition to their application in various industrial processes. *Trichoderma* spp. are highly complex, particularly in terms of secondary metabolites production. However, biochemical, molecular, and proteomic approaches made it possible to explore new pathways and novel functions of compounds produced by this genus and their potential applications. *Trichoderma* is most exploited as a biocontrol agent and has many success stories. In India, *Trichoderma viride* and *Trichoderma harzianum* are important biocontrol agents for managing various diseases. Several successful products based on different species of *Trichoderma* have been commercialized in India. The potential *Trichoderma* isolates are formulated using different organic and inorganic carriers through solid or liquid fermentation technologies. Selection of application approaches, such as seed treatment, bio-priming, seedling dip, soil application, and foliar spray of Trichoderma formulation, can vary and depend on the type of plant and phytopathogen. Biocontrol agent formulations can be improved to enhance the performance and consistency of results through molecular approaches, protoplast fusion, genetic recombination, mutation, development of compatible consortia, or addition of water-soluble adjuvants/oils/emulsion.

References

Balasubramanian C, Udaysoorian P, Prabhu C, Kumar GS (2008) Enriched compost for yield and quality enhancement in sugarcane. J Ecobiol 22:173–176

Batta YA (2005) Postharvest biological control of apple gray mold by *Trichoderma harzianum* formulated in an invert emulsion. Crop Protect 23(1):19–26

Bhat KA, Anwar A, Lone GM, Hussain K, Nazir G (2009) Shelf life of liquid fermented product of Trichoderma harzianum in talc. J Mycol Plant Pathol 39(2):263

Bissett J (1991) A revision of the genus Trichoderma. II. Infrageneric classification. Can J Bot 69(11):2357–2372

Carvalho FP (2017) Pesticides, environment, and food safety. Food Energy Sec 6:48–60

Cavalcante RS, Lima HL, Pinto GA, Gava CA, Rodrigues S (2008) Effect of moisture on Trichoderma conidia production on corn and wheat bran by solid state fermentation. Food Bioprocess Technol 1:100–104

Connick W, Daigle D, Quimby P (1991) An improved invert emulsion with high water retention for mycoherbicide delivery. Weed Technol 5:442–444

Cumagun CJR (2014) Advances in formulation of Trichoderma for biocontrol. In: Biotechnology and biology of trichoderma. Elsevier, pp 527–531

Eltem R, Sayit S, Sozer S, Sukan FV (2017) Production of Trichoderma citrinoviride Micropropagules as a biocontrol agent by means of an economical process. U.S. Patent No. 9,551,012

Fravel DR, Rhodes DJ, Larkin RP (1999) Production and commercialization of biocontrol products. In: Albajes R, Lodovica Gullino M, Van Lenteren JC, Elad Y (eds) Integrated pest and disease management in greenhouse crops. Kluwer Academic Publishers, Boston, pp 365–376

Harman GE, Howell CR, Viterbo A, Che I, Lorito M (2004) Trichoderma species—opportunistic, avirulent plant symbionts. Nat Rev Microbiol 2(1):43–56

Hu J, Zhou Y, Chen K, Li J, Wei Y, Wang Y, Wu Y, Ryder MH, Yang H, Denton MD (2020) Large-scale Trichoderma diversity was associated with ecosystem, climate and geographic location. Environ Microbiol 22:1011–1024

Jeyarajan R (2006) Prospects of indigenous mass production and formulation of Trichroderma. In: Rabindra RJ, Ramanujam B (eds) Current status of biological control of plant diseases using antagonistic organisms in India. Project Directorate of Biological Control, Bangalore, pp 74–80. 445

Jeyarajan R, Nakkeeran S (2000) Exploitation of microorganisms and viruses as biocontrol agents for crop disease management. In: Upadhyay et al (eds) Biocontrol potential and their exploitation in sustainable agriculture. Kluwer Academic/Plenum Publishers, USA, pp 95–116

Jeyarajan R, Ramakrishnan G, Dinakaran D, Sridar R (1994) Development of products of *Trichoderma viride* and *Bacillus subtilis* for biocontrol of root rot diseases. In: Dwivedi BK (ed) Biotechnology in India. Bioved Research Society, Allahabad, India, p 25

Kumar S (2012) Biopesticides: a need for food and environmental safety. Biofertil Biopestic 3:1–3

Kumar S, Manibhushan T, Archana R (2014) Trichoderma: mass production, formulation, quality control, delivery and its scope in commercialization in India for the management of plant diseases. Afr J Agric Res 9(53):3838–3852

Lewis JA (1991) Formulation and delivery system of biocontrol agents with emphasis on fungi *Beltsville symposia*. In: Keister DL, Cregan PB (eds) The rhizosphere and plant growth. Agric. Res., vol 14, pp 279–287

Lorito M, Woo SL (2015) Trichoderma: a multi-purpose tool for integrated pest management. In: Principles of plant-microbe interactions. Springer, Berlin/Heidelberg, Germany, pp 345–353

Ramanujam B, Prasad RD, Sriram S, Rangeswaran R (2010) Mass production, formulation, quality control and delivery of Trichoderma for plant disease management. J Plant Prot Sci 2(2):1–8

Rifai MA (1969) Sarawakus Lloyd, a genus of the pyrenomycete family Hypocreaceae. Reinwardtia 7(5):561–578

Sabalpara AN (2014) Mass multiplication of biopesticides at farm level. J Mycol Plant Pathol 44(1): 1–5

Sachdev S, Singh RP (2016) Current challenges, constraints and future strategies for development of successful market for biopesticides. Clim Change Environ Sustain 4:129–136

Sadh PK, Duhan S, Duhan JS (2018) Agro-industrial wastes and their utilization using solid state fermentation: a review. Bioresour Bioprocess 5:1

Sankar P, Jeyarajan R (1996) Biological control of sesamum root rot by seed treatment with *Trichoderma* spp. and Bacillus subtilis. Indian J Mycol Plant Pathol 26:147–153

Sawant IS, Sawant SD (1996) A simple method for achieving high cfu of Trichoderma harzianum on organic wastes for field applications. Indian Phytopathol 49:185–187

Sharma P, Sharma M, Raja M, Shanmugam V (2014) Status of Trichoderma research in India: a review. Indian Phytopathol 67(1):1–19

Singh H, Singh BN, Singh S, Sharma B (2013) Exploring different avenues of Trichoderma as a potent bio-fungicidal and plant growth promoting candidate-an overview. Ann Rev Plant Pathol 5:315–321

Singh P, Nautiyal CA (2012) Novel method to prepare concentrated conidial biomass formulation of *Trichoderma harzianum* for seed application. J Appl Microbiol 113:1442–1450

Sriram S, Palanna KB, Ramanujam B (2010) Effect of chitin on the shelf-life of Trichoderma harzianum in talc formulation. Indian J Agric Sci 80(10):930

Sriram S, Roopa KP, Savitha MJ (2011) Extended shelf-life of liquid fermentation derived talc formulations of Trichoderma harzianum with the addition of glycerol in the production medium. Crop Prot 30(10):1334–1339

Thakur N, Kaur S, Tomar P, Thakur S, Yadav AN (2020) Microbial biopesticides: current status and advancement for sustainable agriculture and environment. In: New and future developments in microbial biotechnology and bioengineering. Elsevier, Amsterdam, The Netherlands, pp 243–282

Velivelli SL, De Vos P, Kromann P, Declerck S, Prestwich BD (2014) Biological control agents: from field to market, problems, and challenges. Trends Biotechnol 32(10):493–496

Waghunde R, Shelake RM, Sabalpara N (2016) Trichoderma: a significant fungus for agriculture and environment. Afr J Agric Res 11(22):1952–1965

Woo SL, Ruocco M, Vinale F, Nigro M, Marra R, Lombardi N, Pascale A, Lanzuise S, Manganiello G, Lorito M (2014) Trichoderma-based products and their widespread use in agriculture. Open Mycol J 8:71–126

Chapter 15
Mass Multiplication, Production Cost Analysis, and Marketing of *Metarhizium*

Sapna Chandwani, Hetvi Y. Naik, Harshida A. Gamit, Komal A. Chandarana, and Natarajan Amaresan

Abstract For plant protection against several insects, biopesticides based on bacteria, viruses, entomopathogenic fungi, and nematodes play a significant role in the regulation of insect population in nature. In this regard, during the last few decades, the use of entomopathogenic fungi as biocontrol agents has received global attention, and the mass production of entomopathogenic fungi, *Metarhizium anisopliae,* is an area of growing interest. It is a fungus that infects insects, particularly, beetle larvae. Generally, *M. anisopliae* is highly virulent, has a broad host range among insects, has transcuticular penetration, and is considered safe to plants, animals, and humans, as well as to the environment. Hence, it can be used commercially for biological control. In several countries, including Australia, USA, New Zealand, Switzerland, and to some extent in Italy, *M. anisopliae* is used as mycoinsecticides for soil insect control under different names. Mass production of *M. anisopliae* is generally carried out in two forms, either as blastospores in liquid media (submerged culture) or as conidia on solid media (surface culture). Solid state fermentation is considered to be a more appropriate technology as it can be easily carried out and the raw materials are available at a lower rate. Moreover, spores produced as living propagules were stable in terms of dry formulations and more tolerant to desiccation. Considering the production cost, benefit cost ratio, and marketing aspect of *M. anisopliae* as the basis of entrepreneurship, this book chapter is designed to involve mass cultivation of the entomopathogenic fungi *M. anisopliae* on locally and easily available agricultural and industrial substrates thereby decreasing the production cost and the benefit cost ratio to trigger the commercialization of *M. anisopliae.*

Keywords *Metarhizium anisopliae* · Mycoinsecticide · Benefit cost ratio · Mass multiplication · Biocontrol agent · Entomopathogenic fungi

S. Chandwani · H. Y. Naik · H. A. Gamit · K. A. Chandarana · N. Amaresan (✉)
C.G. Bhakta Institute of Biotechnology, Uka Tarsadia University, Surat, Gujarat, India

N. Amaresan et al. (eds.), *Agricultural Microbiology Based Entrepreneurship*, Microorganisms for Sustainability 39,
https://doi.org/10.1007/978-981-19-5747-5_15

15.1 Introduction

In several parts of the world, such as Australia, Africa, China, Japan, Brazil, Malaysia, and Mexico, huge losses, including economic and agricultural losses are caused by pests such as grasshoppers, cattle ticks, termites, locusts, etc. (Camargo et al. 2014). In addition, many people are hospitalized, and some are exposed to death due to the vectors of dengue, malaria, and *Bacroftian filariasis,* which are *Anopheles* spp., *Aedes* spp., and *Culex* spp., respectively. Chemical insecticides have been used to eliminate pests and vectors, but the use of chemical insecticides can lead to detrimental effects on pest enemies and can contaminate groundwater. Moreover, agricultural pests, biological vectors, and termites have become resistant to chemical insecticides; therefore, there is an urgent need to substitute chemical insecticides with biocontrol agents, such as bacteria, nematodes, fungi, and viruses (Atheimine et al. 2014). *Metarhizium* spp. are considered a viable option as biocontrol agents because of their safety, narrow host range, eco-friendliness, and ease of mass production. This book chapter deals with the mass production, application, commercialization, and marketing of *Metarhizium* spp. It also addresses the guidelines and instructions for setting up a biopesticide unit in terms of making profits for entrepreneurs.

15.2 Mycoinsecticide: *Metarhizium anisopliae*

The entomopathogenic fungus *M. anisopliae* was first described in 1879 by Metschnikoff. *Metarhizium* is generally found in soil under natural conditions, where infectious spores called conidia are produced and filamentous growth is promoted under moist conditions, infecting soil-dwelling insects on contact (Chetry et al. 2008). These entomopathogenic fungi have been registered and commercialized biocontrol agents for several pests (Chetry et al., 2008). When the spores of the fungus attach to the surface of the insect, they begin to grow and germinate, penetrate the exoskeleton of the insect, and rapidly grow inside the insect, leading the insect to die. Other insects also become infected with the fungus when they come into contact with infected insects (http://www.varshabioscience.com/products/metarhizium.html). *Metarhizium* has been tested in many different ways, including its safety in common people and other (non-insect) animals. *M. anisopliae* is now considered a new generation insecticide against many moths, aphid species, and fly larvae as it produces a number of secondary metabolites such as Destruxint E, which act as mycotoxins (insecticidal toxins) against many pests, and the insect's death is caused by the rapid development of mycotoxins (http://www.varshabioscience.com/products/metarhizium.html).

Biopesticides not only offer an alternative to the indiscriminate use of synthetic pesticides but are also considered to be safe and less toxic than conventional pesticides. In developing countries, mycopesticide use is limited because of the high cost incurred in the mass production of virulent fungi (Sakthivel and Das 2017). Commercial mycopesticides are too expensive for common people; therefore, there is an urgent need to design cost-effective culture techniques in developing countries to produce a higher output of fungal spores and biomass to meet the needs of marginal farmers. Cheap culture media usage of costly conventional media can reduce production costs; therefore, this book chapter is designed on the mass production of *Metarhizium* conidiospores using different cheap substrates and locally available cheap organic materials (Vega et al. 2009; Sakthivel and Das 2017). *Metarhizium* formulations can be stored up to 1 year from the date of manufacturing. Depending on the type of formulation and method of application, requirements of the fungicide *Metarhizium* may vary. For foliar application, 1 ml of *Metarhizium* (liquid) or 10 g of *Metarhizium* (solid/powder) is dissolved in one liter of water and applied on both sides of the leaves. For soil drenching, 3 ml of *Metarhizium* (liquid) or 10 g of *Metarhizium* (powder) is dissolved in 1 l of water and drenched to the base of each plant. In the case of soil application, 250 ml of *Metarhizium* (liquid) or 2 kg of *Metarhizium* (solid) with 100 kg of FYM can be applied to one acre of land.

15.3 Technological Details

Pest management is a major problem in achieving higher yields and agricultural productivity. Therefore, to control termites, biological vectors such as mosquitoes and ticks, agricultural pests, and chemical insecticides are preferred, but the toxic effect of chemical insecticides on the environment and resistance development in pests and vectors toward the use of chemical insecticides as well as public concern have driven the interest of the public toward the use of biological control agents such as fungi and bacteria (Vega et al. 2009). Therefore, this chapter describes the mass production of entomopathogenic fungi, *Metarhizium,* using different cheap substrates and locally available cheap organics for setting up biopesticide units, the methodology and detailed aspects of cost-benefit ratio have also been discussed for mass production of entomopathogenic fungi *Metarhizium*, in order to establish a small-scale or large-scale industry.

The technical details of the project such as the production process and project cost will be discussed. *Metarhizium* species production involves the following core areas of production and selling; (i) mass production (ii) formulation type (solid or liquid) (iii) spore harvesting and drying (iv) moisture content assessment (v) Quality control check (vi) mode of application of product.

15.4 Mass Multiplication/Production Process of *Metarhizium* sp.

The production process of *Metarhizium* sp. involves (I) starter culture/fungus mother culture preparation (II) mass multiplication (III) spore drying and harvesting (IV) product formulation (solid/liquid) (V) packaging, and (VI) quality check. An outline of this process is shown in Fig. 15.1.

15.4.1 Fungal Mother Culture/Starter Culture Preparation

Metarhizium strains could be obtained from any authorized Research Institute, State Agriculture University, from any fungi collection at the Research Institute of Plant Protection, or from the Institute of Microbial Technology (IMTECH) in the form of a slant. To maintain the mother culture, it is necessary to subculture the fungus first under aseptic conditions and to obtain a pure culture through single spore isolation before proceeding toward the mass multiplication of *Metarhizium* (Sakthivel and Das 2017). Potato dextrose broth (PDB) can be used for this purpose. PDB is prepared at a lower rate by the following method as shown in Fig. 15.2.

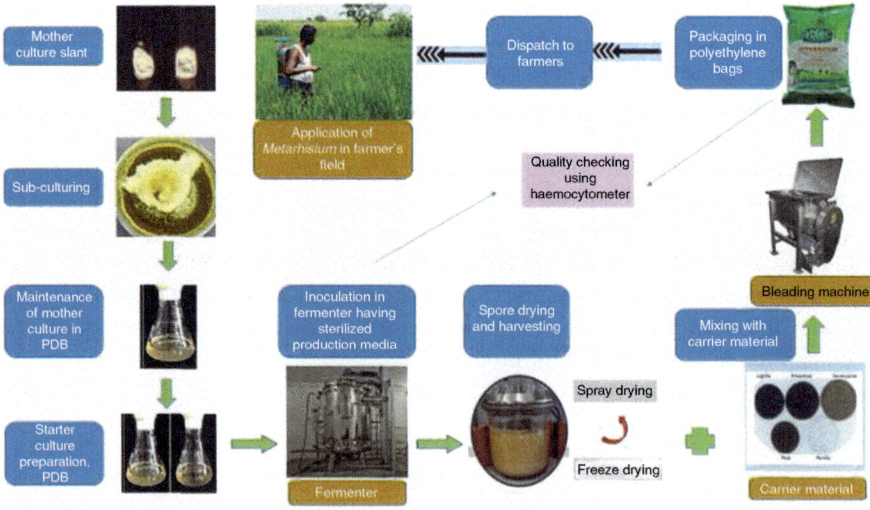

Fig. 15.1 Schematic representation of mass production process of *Metarhizium* sp.

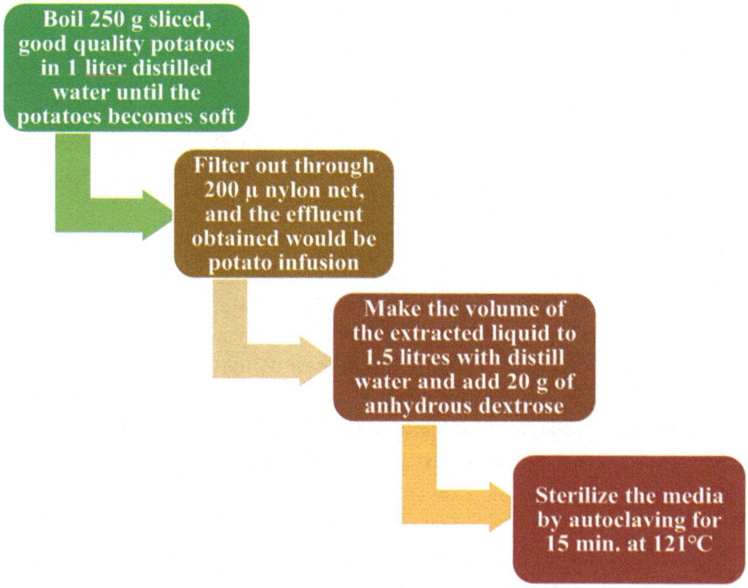

Fig. 15.2 Schematic representation of PDA preparation to use in preparation of starter culture

15.4.2 Mass Multiplication

Preparation of Conidial Suspension

Transfer the spores from pure culture slants to PDA, and then prepare a conidial suspension from the sporulated culture by flushing 10 ml of sterile distilled water on the fungal bed. The suspension is collected in a conical flask in laminar air flow using a microaspirator and gently shaken to make it homogenous; this suspension is then used as the seed material for mass production of *Metarhizium* to carry out further experiments (Habibu et al. 2016).

Subculturing of Spores

First, prepare PDB as described above in Fig. 15.2, after sterilization of the medium, cool and pour into sterile rectangular virgin plastic trays with tight fitting lids. Inoculate the medium with 5 ml of *Metarhizium* culture spore suspension and incubate at 26–28 °C for 8 days. These spores can be used in further experiments (Sakthivel and Das 2017).

Production of Conidiospore on Solid Media

For culturing *M. anisopliae,* different types of solid media such as wheat bran, rice bran, mashed potato dextrose, milled rice, and tamarind can be used to mass multiply

Metarhizium at low rates. Mix 200 g of solid media with 20 g anhydrous dextrose and 30 g agar powder and make the volume to 2 l with distilled water. Boil the media in a large glass vessel and sterilize it in an autoclave. Cool the sterile growth medium and pour it into 500 ml UV-sterilized shallow plastic trays, and allow to solidify. Inoculate the solidified medium with 1 ml of diluted conidial suspension (1×10^7 conidia/ml) and incubate in a cabinet at room temperature for 14 days (Sakthivel and Das 2017).

Production of Conidiospore on Liquid Media

For cost-effective mass production of *Metarhizium* using cheap substrates, five types of broths can be prepared separately using rice bran, wheat bran, potato, milled rice, and tamarind extract. Add 200 g of each substrate and 20 g of anhydrous dextrose to 2 L distilled water. Boil the medium and sterilize using autoclave. In 1 L sterile broth, inoculate 10 ml of diluted conidial suspension in LAF and incubate it in a cabinet at room temperature for 14 days (Sakthivel and Das 2017).

After incubation, harvest the biomass and centrifuge at 10,000 rpm for 5 min at 4 °C. Weigh fresh and dry biomass of the samples. Follow a similar procedure for 3 L capacity fermenter and autoclave it as described above. Inoculate the medium with 3×10^9 submerged spores and incubate at 600 rev / min for 72 h at 30 °C. Harvest the biomass after 72 h and centrifuge at 10,000 rpm for 5 min at 4 °C. Resuspend the pellets in 200 ml deionized water and homogenize it using a magnetic stirrer. Determine the volume, fresh weight, dry weight, and spore concentration of the fresh biomass before proceeding toward the drying process (Kassa et al. 2004).

15.4.3 Spore Drying and Harvesting

Spray-Drying

Formulate submerged conidia produced in the above medium with skimmed milk powder and molasses in a ratio of 1:1:0.08 by weight and homogenize at a low speed. Use a laboratory spray dryer for drying. Adjust the inlet and outlet temperatures to 64 ± 2 and 48 ± 2 °C, respectively. For the spray nozzle, the flow of compressed air should be 600 NL/h during the spray-drying process. Repeat the experiment four times (Stephan and Zimmermann 2001).

Freeze-Drying

For freeze-drying, formulate fresh submerged conidia with skimmed milk powder and glycerol at a ratio of 1:0.4:0.16 by weight and homogenize for spray-drying. Carry out the freeze-drying using a freeze dryer. Pour 100 g of the formulated submerged conidia onto a plate and freeze it using liquid nitrogen. Transfer the frozen material to a freeze-dryer. Adjust the heating temperature of the plate to 40 °C for the first 3 h and then reduce to 30 °C for the next 21 h. After 24 h, terminate the drying process when the temperature of the dried material reaches 25 °C. Pulverize a portion of the freeze-dried submerged conidia through a 120 μm mesh sieve, and repeat the experiment four times (Kassa et al. 2004).

Measure the spore yield, viability, and moisture content of submerged conidia and determine the moisture content using a moisture tester. Suspend 1 g of the respective product in 100 ml water and sonicate for 3 min to count the submerged conidia. Count the spore concentration using a hemocytometer in samples, which are diluted to 1:100 and at each drying process, count three samples for the respective treatments (Kassa et al. 2004). Assess the fresh and dried submerged conidia viability using 1.5% water agar amended with 50 mg/l chloramphenicol, 30 mg/l of streptomycin sulfate, and 0.005% benomyl. Suspend the submerged conidia from the different treatments in deionized water with 0.01% v/v Tween 80 and adjust the concentration to 1×10^6/ml. Put several drops of 2 μl spore suspension onto the agar plate and incubate at 25 °C. Assess the viability by counting 300 spores in each sample using a microscope after 8, 16, and 24 h of incubation. If necessary, staining can be performed using lactophenol cotton blue. Repeat the test four times and three samples can be used to assess viability (Kassa et al. 2004).

15.4.4 Product Formulation

To formulate the products of *Metarhizium,* the entomopathogenic fungi are blended with inert ingredients referred to as carriers. The formulations should have longer shelf life and be safe enough, effective, and easy to use according to CIBRC (Central Insecticides Board and the Registration Committee) guideline for maintaining the quality parameters of the biopesticide before entering the product into the market. Currently, biopesticides are formulated as solid carriers that involve talc, peat, lignite, clay, and a combination of coconut oil and soybean oil (Nithya and Rani 2017). In India, wettable granules (WG), wettable powder (WP), aqueous solutions (AS), and suspension concentrations (SC) are commonly used formulations. For *Metarhizium,* talc and oily formulations are preferred (Nithya and Rani 2017). The formulation of coconut oil and soybean oil at a ratio of 50:50 has been reported to improve viability during storage (Nithya and Rani 2017).

15.5 Packaging

Packaging is considered to be the major constraint driving the biopesticide business market and a vital component as it maintains the quality of the formulated product (Kaur and Joshi 2018). Packaging of a biopesticide should be leak-proof and ensure longer shelf-life of the living ingredients (*Metarhizium*). Packaging should be performed according to the standards and guidelines, but many private firms or companies may perform this task as per their convenience and overlook the quality of the packaging material, which in turn affects the quality of the formulated product as well as farmers or public expectations. For packaging of the formulated products, sterilized polyethylene bags is used, and after packaging, the packed product can be stored at room temperature of 25 ± 2 °C and at refrigeration temperature of 4 ± 2 °C for up to 6 months. At monthly interval up to 6 months, the viability of the formulated products of *M. anisopliae* can be calculated by conidial viability determination (Kaur and Joshi 2018).

15.5.1 Quality Control and Product Specifications

Quality control is considered the essential criterion at each stage of product formulation, as it can lead to the success or failure of the formulated product in the market. Determination of spore concentration and viability is important for ensuring the good quality of *M. anisopliae.*

15.6 Spore Concentration Determination

One day after the formulations obtained, the viability of stored conidia from different formulations can be determined to estimate the initial germination level up to 40 weeks at intervals of 5 weeks. From each stored conidial suspension, 1 g of the respective product is dissolved in 100 ml distilled water and sonicated for 3 min, and at each drying process, as well as after product formulation, spore concentration is determined using a hemocytometer in samples diluted to 1:100 (Kassa et al. 2004).

15.7 Spore Viability Determination

The viability of fresh and dried submerged conidia after packaging is assessed using 1.5% water agar amended with 50 mg/l chloramphenicol, 30 mg/l of streptomycin sulfate, and 0.005% benomyl. Submerged conidia from the different treatments are suspended in deionized water with 0.01% v/v Tween 80, and the concentration is adjusted to $1\text{x}10^6$ / ml. 2 μl spore suspension is dropped onto the agar plate and incubated at 25 °C. The viability is assessed by counting 300 spores in each sample using a microscope after 8, 16, and 24 h of incubation (Kassa et al. 2004).

15.8 Human Risk Assessment

It has been reported that no harm is caused by exposure to the *Metarhizium* strain, as it includes all anticipated exposure reliable information (Strasser et al. 2011).

15.9 Subchronic and Chronic Toxicity

Replication, infectivity, survival, toxicity, or persistence of the microbial agent was not observed in the animals treated during the acute oral infectivity test; hence, subchronic and chronic toxicity were not required (Strasser et al. 2011).

15.10 Dose Response Assessment

Toxicological endpoints were not identified (Strasser et al. 2011).

15.11 Food Exposure and Risk Characterization

For the general population, including infants and children, the risk from the consumption of residues is not expected in the absence of any toxicological endpoints. Furthermore, any dietary exposure can be inadvertent, as this is a non-food use product (Strasser et al. 2011).

15.12 Project Details

Several facilities are required to establish the biopesticide unit and implement a project for the production of *Metarhizium*. The infrastructure and laboratory facilities required for the production of entomopathogenic fungi *Metarhizium* can be applied to the production of other biofertilizers as well as biopesticides. Many private companies and governmental institutes produce both biofertilizers and biopesticides by making use of the common facilities ensuring the economic viability of the project installation. The production of biopesticides based on entomopathogenic fungus *Metarhizium* in India may costs up to 140–145 lakhs. The component-wise budget outlay for mass production of *Metarhizium* of a single unit costs between 50 and 55 lakhs. Therefore, the cost for each particular is counted accordingly for mass production of *Metarhizium*.

Land and Location For establishing any project or manufacturing business, land is considered to be the primary requirement. To set up a laboratory, office, and other facilities for mass production of *Metarhizium* for single unit, 210–350 sq. metereof land is required costing Rs. 11,000–15,000 per sq. meter (Box 15.1).

The location of the biopesticide unit should be far from the residential area of the local people; therefore, the setup of a biopesticide unit should be in such a way that it may not affect the common people. It should not be set up in areas with high temperature and humidity, as it may increase the production and maintenance costs.

Building Infrastructure The building should be divided into different compartments such as quality assurance laboratories; culture preparation and maintenance rooms; fermenter arrangement facilities; labeling, packaging, handling, and dispatching areas; storage facilities; administrative office; staff room; pantry; etc. The infrastructure designed should facilitate the optimum temperature and humidity required for the culture maintenance. A floor plan should facilitate the maximum land usage. The entire site should either be fenced with barbed wire or with compound wall and gates. The required area for buildup of infrastructure for mass production of *Metarhizium* for single unit is 130–150 sq. meter costing between Rs. 20,000 and 24,000/sq. meter (Box 15.1).

Box 15.1 Component—Wise Budget Outlay for Mass Production of *Metarhizium* for Single Unit

Sr. no.	Particulars	Cost/sq. meter (in Rs.)	Required area (in sq. meter)	Total (Rs. in lakh)
1	Land	11,000	210	23.10
2	Infrastructure	20,000	135	27
	Total			50.1

Utilities:

(i) **Power supply:** For mass production of *Metarhizium,* generally, three phase supply is needed. Depending upon the machinery operated, the requirement is nearly about 85–100 KVA. A standby generator is also needed as an option of power supply.

(ii) **Water supply:** At each and every step of *Metarhizium* production, water supply is mandatory whether it is broth preparation, cleaning of glassware, autoclave, hot air oven and fermenter cleaning, etc. The average requirement of water supply will be around 3000–6000 L/day.

Machinery A number of well-established equipments are required to fulfill the standards as well as to achieve the maximum production capacity. The instrumentation for mass production of *Metarhizium* is illustrated in the following table (Tables 15.1, 15.2, 15.3 and 15.4).

Financial Analysis

Benefit Cost Ratio = Total Net Return/Total Cost (Tables 15.5 **and** 15.6)

Table 15.1 Equipments required for mass production of *Metarhizium*

1. Nonrecurring contingencies:

Sr. no.	Specifications/items	Capacity	Rs./unit	Quantity	Rs. (lakh)
1	Laminar air flow	300 lux	0.98	2	1.96
2	Autoclave	22–200 l	0.3	2	0.6
3	Laboratory refrigerator	400 l	0.65	2	1.3
4	RO-based distillation unit	1.5 l/h	0.25	1	0.25
5	Fermentor	500–1000 l	2.5	1	2.5
6	Hot air oven	85 kg	0.9	1	0.9
7	Electronic digital balance	0.1 mg–120 g	0.25	1	0.25
8	Air conditioner with stabilizer	1.5 ton	0.29	4	1.16
9	Microscope	–	0.29	1	0.29
10	Gas stove	–	0.04	1	0.04
11	Drinking water with UV. RO. System	80 l	0.7099	1	0.7099
12	Hemocytometer	–	0.0106	1	0.0106
13	Fire extinguishers for lab safety	2 kg	0.0012	3	0.0036
14	Bottle seller	–	0.36	1	0.36
15	Packet sealer	–	0.053	1	0.053
16	Generator	1KVA	0.5	1	0.5
17	Furniture	–	1	1	1
18	Xerox machine	–	0.6	1	0.6
19	Computer with accessories	–	0.385	1	0.385
20	Miscellaneous	–	0.5	1	0.5
	Total		**10.58**		**13.41**

2. Recurring contingencies:

Sr. no.	Particulars	Number	Rs. (lakh)
1 (i)	Labor: Technical staff @ 22,000/month	2	5.28
(ii)	Skilled labor @ 300/day	5	1.8
(iii)	Accountant @ 23,500	1	2.82
2	Glasswares (Petridish, flask, beaker, etc.)	–	3
3	Stationery items	–	0.5
4	Chemicals	–	7
5	Maintaining mother culture	–	1
6	POL (hiring of vehicles)	–	1
7	LPG gas cylinder	–	0.37
8	Labeling, packaging, and handling of materials	–	1
9	Electricity and water charges	–	1
10	T.A and D.A	–	0.5
11	Miscellaneous	–	0.5
	Total		**25.77**

Table 15.2 Budget requirement for 3 consecutive years

Sr. no.	Particulars	Year			Total (Rs. in lakh)
		I	II	III	
1	Land and infrastructure	50.1	–	–	50.1
2	Nonrecurring items	13.41	–	–	13.41
	Subtotal (A)	**63.51**	–	–	**63.51**
3	Chemicals and stationery	7.5	8.25	9.1	24.85
4	Glasswares	3	–	–	3
5	Salaries for staff	9.90	10.89	11.98	32.77
6	Maintaining mother culture	1	1.10	1.21	3.31
7	POL (hiring of vehicles)	1	1.10	1.21	3.31
8	LPG gas cylinder	0.37	0.41	0.45	1.23
9	Labeling, packaging, and handling of materials	1	1.10	1.21	3.31
10	Electricity and water charges	1	1.10	1.21	3.31
11	T.A and D.A	0.50	0.55	0.60	1.65
12	Miscellaneous	0.50	0.55	0.60	1.65
	Subtotal (B)	**25.77**	**25.05**	**27.57**	**78.39**
	Grand total (A + B)	**89.28**	**25.05**	**27.57**	**141.9**

Table 15.3 Budget requirement (if building is rented) for 3 consecutive years

Sr. no.	Particulars	Year			Total (Rs. in lakh)
		I	II	III	
1	Rent of building	3.65	4.0	4.4	12.05
2	Nonrecurring items	13.41	–	–	13.41
	Subtotal (A)	**17.06**	**4.0**	**4.4**	**25.46**
3	Chemicals and stationery	7.5	8.25	9.1	24.85
4	Glasswares	3	–	–	3
5	Salaries for staff	9.90	10.89	11.98	32.77
6	Maintaining mother culture	1	1.10	1.21	3.31
7	POL (hiring of vehicles)	1	1.10	1.21	3.31
8	LPG gas cylinder	0.37	0.41	0.45	1.23
9	Labeling, packaging, and handling of materials	1	1.10	1.21	3.31
10	Electricity and water charges	1	1.10	1.21	3.31
11	T.A and D.A	0.50	0.55	0.60	1.65
12	Miscellaneous	0.50	0.55	0.60	1.65
	Subtotal (B)	**25.77**	**25.05**	**27.57**	**78.39**
	Grand Total (A + B)	**42.83**	**29.05**	**31.97**	**103.85**

Table 15.4 Project details for mass production of *Metarhizium*

Sr. no.	Production of *Beauveria bassiana*	Production per year in liter or kg	Market value of product (Rs. per liter or kg)	Earning per year (Rs. in lakh)
1	Liquid	9000	490	44.1
2	Solid	18,000	350	63
Total		**27,000**	**840**	**107.1**

Table 15.5 Benefit cost ratio (BCR) of purchased land

Year	Cost (in lakh)	Return (in lakh)	Net Return (in lakh)
1	89.28	107.1	17.82
2	25.05	123.20	98.15
3	27.57	135.52	107.95
Total	**141.9**	**365.82**	**223.92**
BCR			**1.57**

Table 15.6 Benefit cost ratio (BCR) of rented land

Year	Cost (in lakh)	Return (in lakh)	Net return (in lakh)
1	42.83	107.1	64.27
BCR			**1.50**

15.13 Summary

For the mass production of *Metarhizium*, the investment cost required will be Rs. 141.9 lakh and the production per year was found to be 27,000 l/year. The net return obtained from investment will be Rs. 223.92 lakh. It is estimated that an approximately Rs. 82.02 lakh can be gained as a profit and the benefit cost ratio of the project model was found to be 1.57. This indicates the feasibility and profitability in establishment of mass production unit of *Metarhizium*. In the proposed model, B: C ratio is more, therefore, finance may be useful for the implementation of the project.

References

Sakthivel A, Das SM (2017) Laboratory assessment of affordable culture media for the propagation of entomopathogenic fungi, used in mycopesticide production. J Agrobiotechnol 8(1):33–42

Atheimine MO, Bashir MO, Ely SO, Kane CMH, Babah MAO, Benchekroun M (2014) Efficacy and persistence of Metarhizium acridum (Hypocreales: Clavicipitaceae) used against desert locust larvae, Schistocerca gregaria (Orthoptera: Acrididae), under different vegetation cover types. Int J Trop Insect Sci 34(2):106–114

Camargo MG, Marciano AF, Sá FA, Perinotto WM, Quinelato S, Gôlo PS, Angelo IC, Prata MC, Bittencourt VR (2014) Commercial formulation of Metarhizium anisopliae for the control of Rhipicephalus microplus in a pen study. Vet Parasitol 205(1–2):271–276

Chetry YDG, Kelle S, Nagel P, Kafle L (2008) Mass production and application yuiop[]\789+of Metarhizium anisopliae against white grubs in Nepal. J Int Cooperation 3:15–30

Habibu A, Muhammad M, Okpanachi IY, Tahir F and Abdulhameed A (2016) Mass production of entomopathogenic fungi Metarhizium anisopliae from biomass substrates. In *Proceedings of NTBA/NSCB Joint Biodiversity Conference; Unilorin* (Vol. 77, p. 79pp)

Kassa A, Stephan D, Vidal S, Zimmermann G (2004) Production and processing of Metarhizium anisopliae var. acridum submerged conidia for locust and grasshopper control. Mycol Res 108(1):93–100

Kaur N, Joshi N (2018) Talc formulation of Metarhizium anisopliae (Metsch.) Sorokin and its viability on storage. J Entomol Zool Stud 6(4):853–856

Nithya PR, Rani OR (2017) Evaluation of carrier materials for formulating entomopathogenic fungus Lecanicillium Lecanii (Zimmermann) Zare and Gams. J Biol Control 31(1):50–55

Stephan D and Zimmermann G (2001) Locust control with Metarhizium flavoviride: drying and formulation of submerged spores. *EUR 19692–COST action 830–Formulation of microbial inoculants*, pp.27–34

Strasser H, Hutwimmer S, Burgstaller W (2011) Metabolite toxicology of fungal biocontrol agents. In: *Regulation of biological control agents*. Springer, Dordrecht, pp 191–213

Vega FE, Goettel MS, Blackwell M, Chandler D, Jackson MA, Keller S, Koike M, Maniania NK, Monzon A, Ownley BH, Pell JK (2009) Fungal entomopathogens: new insights on their ecology. Fungal Ecol 2(4):149–159

Chapter 16
Mass Multiplication, Production Cost Analysis, and Marketing of *Beauveria*

Sapna Chandwani, Hetvi Y. Naik, Harshida A. Gamit, Komal A. Chandarana, and Natarajan Amaresan

Abstract Recently, the excessive use of chemical pesticides for pest management has led to environmental contamination, and the interest of consumers has diverted towards the use of biopesticides to control plant pests and plant diseases. The entomopahogenic fungus *Beauveria bassiana* plays a key role in the management of various veterinary, forestry, and agricultural arthropod pests. However, the use of entomopathogenic fungi as biocontrol agents is less expensive and safer. *Beauveria* formulated in liquid or dry formulations with large numbers of aerial conidia are typically deployed in a wide range of applications in a chemical paradigm. Solid-state fermentation has emerged as an appropriate technology for the mass production of *Beauveria* to yield hydrophobic aerial conidia, the principal active ingredients of mycoinsecticides. An effective production methodology that can be easily adopted involves the use of different media such as wheat bran, rice bran, Sabouraud dextrose agar (SDA), and sorghum, which were used as substrates for mass multiplication of *B. bassiana*. Organic media, nonsynthetic media, and the biomass of fungal grain media can also be used for production. The economics of *B. bassiana* were evaluated based on the final yield. However, worldwide production of fungal spores that demand low inputs was carried out using simple technologies while considering the aspect of production and cost-benefit analysis as well as marketing of *B. bassiana*, targeting the basis of entrepreneurship. Hence, this book chapter was designed to provide an overview of the mass cultivation of the entomopathogenic fungus *B. bassiana* on locally available agricultural and industrial waste to decrease the production cost and to increase the cost-benefit ratio for the production of virulent spores, i.e., solid and liquid media, and to accelerate the commercialization of *B. bassiana*.

Keywords *Beauveria bassiana* · Mycoinsecticide · Cost-benefit analysis · Biocontrol · Entomopathogenic fungi

---

S. Chandwani · H. Y. Naik · H. A. Gamit · K. A. Chandarana · N. Amaresan (✉)
C.G. Bhakta Institute of Biotechnology, Uka Tarsadia University, Surat, Gujarat, India

N. Amaresan et al. (eds.), *Agricultural Microbiology Based Entrepreneurship*, Microorganisms for Sustainability 39,
https://doi.org/10.1007/978-981-19-5747-5_16

16.1 Introduction

The introduction of chemical pesticides is now open to serious questions owing to increased resistance of insects, vertebrates, pathogens, and weeds (National Research Council 1986). In addition, for companies, it is very costly and difficult to discover new synthetic pesticides because to find a new, commercially acceptable, synthetic pesticide, companies had to screen at least 150,000 chemicals, which requires an investment of more than $250 million and takes around 10 years time to launch into the market; on the other hand, to discover microbial biopesticide, investment of $1–2 million and 3–5 years time is required to reach the market (Marrone, 2014). Moreover, societal, governmental, and market-driven demands for chemical-free residue in the food industry have decreased the use of chemical pesticides and increased the growth of organic agriculture. Various integrated pest management (IPM) programs and environmentally friendly approaches to crop protection from various insect pests have become attractive (Ravensberg 2015). Recently, interest in formulating microbial biopesticides has increased as an emerging tool, even among large chemical companies, to expand their portfolios for the management of pests and diseases in vegetable crops to attain sustainable agriculture. However, several constraints and technological challenges in the mass production, application, commercialization, and marketing of *B. bassiana*-based biopesticides are addressed in this chapter. This chapter serves as guideline for entrepreneurs interested in setting up biopesticide units to extend their financial assistance.

16.2 Biopesticide: *Beauveria bassiana*

Agostino Bassi di Lodi in 1834 discovered *B. bassiana*, also known as the white muscardine disease-causing pathogen in silkworm (*Bombyx mori*) (Deba et al. 2017), which is a widely distributed soil-inhabiting fungus throughout the world formerly known as *Botrytis bassiana*, belonging to class Deuteromycetes (Khachatourians 1991). This fungus can be isolated from insects, soil, and mites; infects a wide range of hosts; and is a pathogen for various insect orders such as Hemiptera, adult sawtoothed grain beetles, Lepidoptera, Homoptera, Hymenoptera, Orthoptera, and Coleoptera (Dannon et al. 2020). *Beauveria* species suppress plant diseases via various mechanisms such as antibiosis, competition, mycoparasitism, induced systemic resistance, and endophytism. In addition to these mechanisms, various secondary metabolites (bassinolide, beauvericin, oosporein, beauvolide, beauviroloide, tenellin, and bassianin), volatile organic compounds, and hydrolytic enzymes (lipases, chitinases, cellulases, amylases, proteases, and caseinases) also have antimicrobial and antifungal properties against various plant pathogens (Ownley et al. 2010). The infection pathway comprises the following steps when *Beauveria* species attack their host insects percutaneously: (1) spore attachment to the insect cuticle by electrostatic and chemical forces, (2) germination of the spore on the cuticle where the fungi excretes enzymes like chitinases, lipases, proteases, and lipoxygenases, (3) its penetration through the cuticle by mechanical pressure initiated by a specialized structure formed in the germinative tube, the appressorium, (4) overcoming the host immune response and its proliferation within the host cell and (5) saprophytic outgrowth from the dead host where new conidia are produced (Keswani et al. 2013).

In general, conidial germination of *B. bassiana* takes about 10 h and is completed in 20 h at 25 °C. These germinated spores allow hyphal penetration through non-sclerotised areas of the cuticle such as mouthparts, joints, and between segments, which produce chitinases and extracellular proteases that degrade chitinous and proteinaceous components. Subsequently, via extensive vegetative growth and production of toxic secondary metabolites, the fungus invades other tissues of the host insect, leading to the death of the host (Logrieco et al. 2002).

The entomopathogenic fungus *B. bassiana* is the most effective biological control agent and is environmentally friendly than chemical pesticides. According to literature, the consumption of *B. bassiana* extract has positive impact on the immune system and is not harmful to human health (Faria and Wraight 2001). No adverse effects or toxicities were observed in the nontarget organisms. Moreover, it is relatively cheap and easy to maintain and culture several *B. bassiana* strains under laboratory conditions, compared to the production of chemical pesticides, and allergic reactions have rarely been reported in people manipulating the fungus (Lambert 2010). *B. bassiana* formulations can be stored up to 1 year from the date of manufacturing. Depending on the type of formulation whether liquid or solid based, 3 L or 4 kg *B. bassiana* formulation is required per hectare of land in 500 litres of water (i.e., 6 mL or 8 g/L of water) which is approximately Rs.1500–1800/hectare. Furthermore, *B. bassiana* conidial solutions can be easily applied using various equipment and as application methods of synthetic insecticides (Dannon et al. 2020).

16.3 Technological Details

To achieve higher crop yields and agricultural productivity, pest problems are considered a major problem. The loss of agricultural productivity of crops due to insect pests was estimated to be 15.7% (Dhaliwal et al. 2015). Recently, a valuable tool for crop protection or pest management has been based on the biological control of pests via microbial pathogens, such as bacteria, nematodes, viruses, and fungi (Anand et al. 2009). The essential component for utility in the IPM program involves the mass production of the entomopathogenic fungus *B. bassiana.* To disseminate the technology broadly and to establish the bankability of *B. bassiana* mass multiplication, as well as to extend financial assistance to entrepreneurs who are interested in setting up biopesticide units, the methodological aspects for mass production of *B. bassiana* and detailed aspects of cost-benefit ratio have been discussed.

To establish a small-scale or large-scale industry, the production process, and project cost, along with other technical details, will be discussed to make use of *B. bassiana* in the local market to attain sustainable agriculture. *B. bassiana* production mainly involves the following core areas of its production and selling; (i) mass production, (ii) type of formulation (solid or liquid), (iii) spore harvesting and drying, (iv) culture moisture content assessment, (v) Quality control of spores and (vi) mode of application.

16.4 Mass Multiplication/Production Process of *Beauveria Bassiana*

The production process of *Beauveria bassiana* involves; (I) preparation of starter culture/fungus mother culture (II) mass multiplication, (III) spore drying and harvesting, (IV) product formulation, (V) packaging and (VI) quality check. An outline of this process is shown in Fig. 16.1.

16.4.1 Preparation of Fungi Mother Culture/Starter Culture

The *B. bassiana* strain could be obtained from an authorized research institute, the State Agriculture University, or from any fungi collection of the Research Institute of Plant Protection in the form of a slant. To maintain the mother culture, it is necessary to subculture the fungus first under aseptic conditions and obtain a pure culture

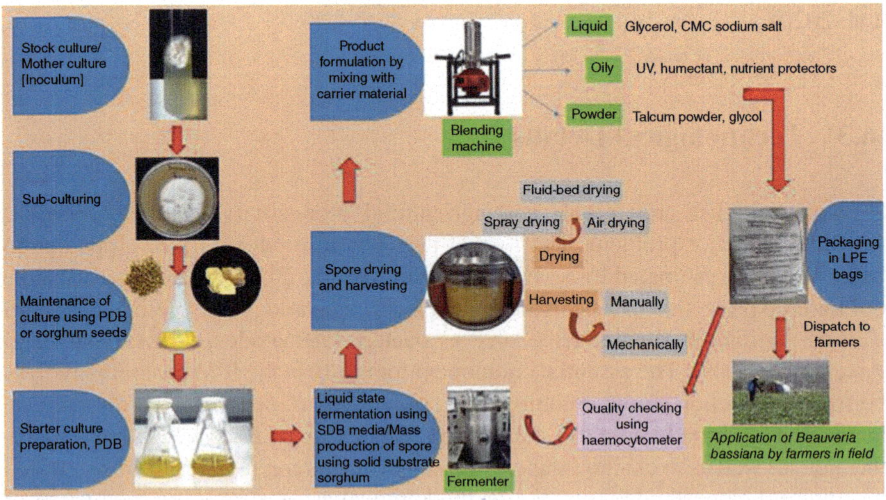

Fig. 16.1 Schematic representation of mass production process of *Beauveria bassiana*

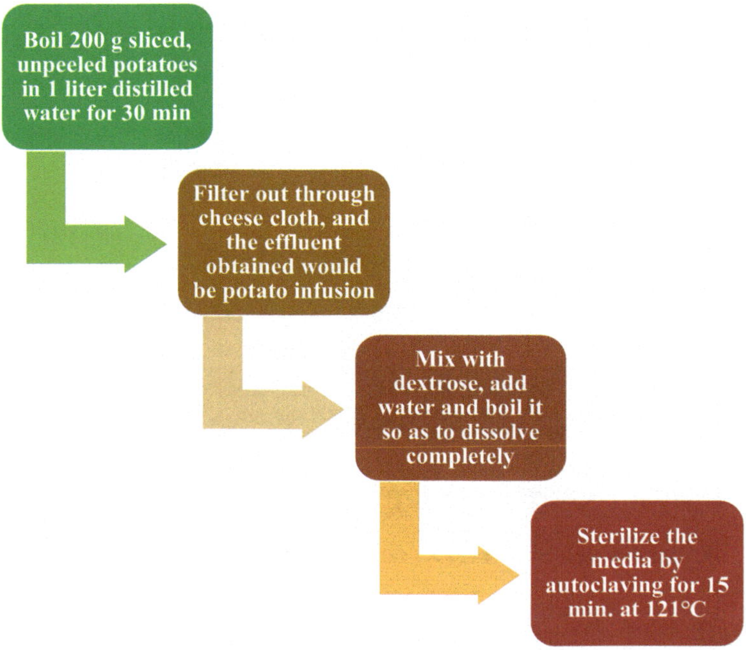

Fig. 16.2 Schematic representation of preparation of PDA to use in preparation of starter culture

through single spore isolation before proceeding towards the mass multiplication of *B. bassiana* (Swathi et al. 2017). For this, potato dextrose broth or sorghum seeds are utilized. Potato dextrose broth is prepared at a lower rate by the method shown in Fig. 16.2.

The materials required for maintaining culture in sorghum seeds are sorghum and water and the methodology is shown in Fig. 16.3.

16.4.2 Mass Multiplication

Preparation of Liquid State Fermentation

Inoculate 1 month old sporulated culture of *B. bassiana* into a sterilized 5 L capacity Sabouraud dextrose broth (SDB) and keep on an orbital shaker or in a BOD incubator at 28 °C for a couple of weeks for the growth of blastospores. This culture can be used as inoculum for fermentor for mass production (Swathi et al. 2017). The medium composition of the SDB is shown in the Box 16.1.

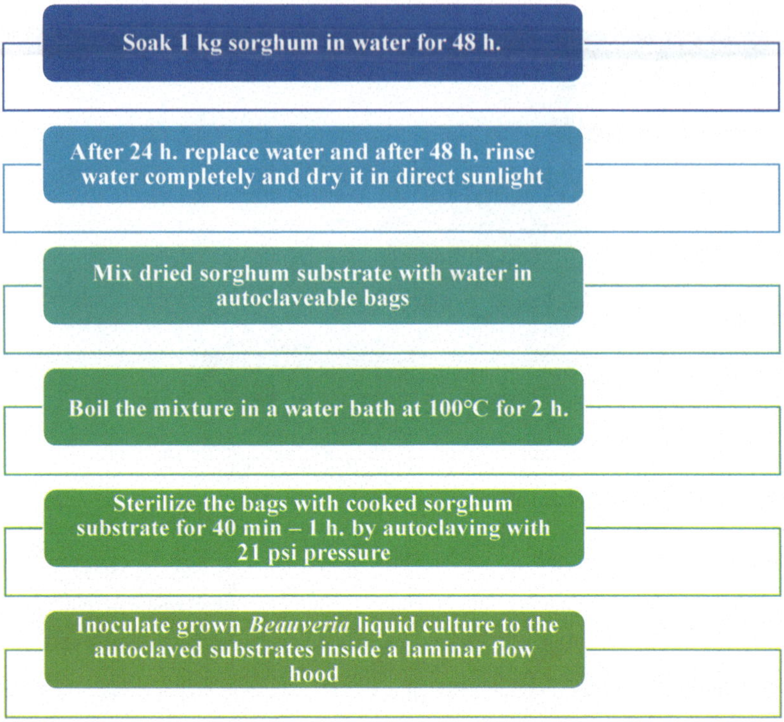

Fig. 16.3 Step by step procedure for maintaining culture in sorghum seeds

Box 16.1 Composition of Sabouraud Dextrose Broth (SDB)

Ingredients	g/L
Dextrose (glucose)	20.0
Peptone	10.0
pH	5.6 ± 0.2

Suspend total 30.0 g of the above ingredients in 1000 ml distilled water. Heat if necessary to dissolve the medium completely. Sterilize it by autoclaving at 15 Ibs pressure at 121 °C for 15 min. Mix well and dispense as desired.

Add SDB as a substrate into the fermenter with 1% mother culture. Five litre of mother culture is required for 500 L of SDB for one batch harvesting. After 6–7 days *Beauveria* can be harvested and packed (Belay and Tenkegna 2017).

Mass Production of Spore on Solid Substrate

After preparing the bags with the cooked sorghum substrate as mentioned in Fig. 16.3, follow the proper procedure for mass multiplication of *B. bassiana*. The mouth of the bags is inserted into a cylindrical autoclaveable PVC pipe during the time of inoculation out. The mouth of the bag is pulled out through the PVC pipe and the PVC pipe is wrapped in such a way that the mouth remains open. The mouth of the bags is covered with sterile cheesecloth, tissue paper, and aluminium foil to cover from inside to outside, and a rubber band is placed around the PVC pipe opening. The inoculated bags is then placed inside a growth chamber. To allow air to circulate freely around the bags, open shelves are used to prevent heating of the substrate as the fungus grew. Inoculated substrate is evenly distributed inside the bag as a thin layer to maximize the surface area for gas exchange and growth. After 3–4 days of inoculation, the substrate is gently mixed without opening the bags. Repeat the mixing process 1–2 more times, 3–4 days apart. To ensure a better oxygen supply for fungal mycelial growth and sporulation, remove aluminium cover after 1 week of inoculation. Fungal mycelia completely cover the substrate within 15 days and become sporulated within 20 days (Belay and Tenkegna 2017).

16.4.3 Spore Drying and Harvesting

Spore Drying

Transfer the fungal biomass plus the substrate from the bags onto trays wrapped in sterile paper. To enhance hyphal maturation, sporulation, and eventual drying, two trays are used for each bag so that the substrate could be spread into a thin layer. For complete drying, the sample cultures are kept in a chamber with an average relative humidity of 22 ± 6% and an average temperature of 12 ± 4 °C. Meanwhile, when the cultures are exposed to an open tray, 200 g of subsample culture (sorghum plus conidia) from each treatment is collected in duplicate for moisture content assessment (Belay and Tenkegna 2017).

During the air-drying process, the moisture content of sorghum and culture is assessed at every 24 h interval for 10 days until and unless the moisture content was found to be stable. To monitor the air drying efficiency, the same amount of subsamples from each treatment is taken and oven-dried at 105 °C. The final moisture content of the harvested spore is calculated using the following formula (Rao et al. 2006).

$$\%\text{of moisture} = \frac{\text{MLod} - \text{MLad}}{\text{MLod}} \times 100$$

Where,

MLod = Moisture loss due to oven dry.
MLad = Moisture loss due to air dry.

Spore Harvesting

Spore harvesting can be done both manually and mechanically for 20 min. First, the large clumps are broken manually, and then the contents gently mashed and shaken. The dried culture is placed into a fine mesh bag (25 × 50 cm) through a 35-mesh sieve (279 μm pore size), which is held inside a plastic bag (35 × 60 cm or larger) to separate the spore powder from the dried substrate. The aerial conidia is harvested manually by shaking the fungus colonized substrate. The separated conidia is passed from the grains through the mesh, weighed and stored in a plastic bag (Gouli et al. 2005).

16.4.4 Product Formulation

To formulate products of *B. bassiana,* entomopathogenic fungi should be blended with ingredients referred to as carriers. To maintain the quality parameters before launching the formulated product into the market, the product should be effective, safe, easy to handle, and use, and should have longer shelf life as per the guidelines prescribed by Central Insecticides Board and Registeration Committee (Dannon et al. 2020). Formulations of biopesticides in liquid or wettable powder form include the following.

Liquid Formulation

Liquid formulations can be prepared using carboxymethyl cellulose sodium salt and glycerol. To 1 L of fresh fermented *Beauveria* culture, 10 mL of carboxymethyl cellulose sodium salt and 10 mL of glycerol is added to formulate the product (Dannon et al. 2020).

Wettable Powder

To 1 kg of talc powder, 10 mL fermented fresh *Beauveria* culture and 10 mL of glycerol is mixed to formulate the product in a wettable powder form (Dannon et al. 2020).

16.5 Constrains Related to Biopesticide Formulation Based on *B. bassiana*

There are many constraints related to biopesticide formulation based on *B. bassiana,* which include the following:

1. Difficulty in finding materials that need to be blended or combined.
2. Maintaining the viability of conidia and also their stability such as desiccation tolerance, shelf life, UV protection, etc.
3. Maintaining the effectiveness of the formulated product.
4. Making the dosage more precise (Kpindou et al. 2012).

To obtain the product in granulated forms, a previously harvested spore coating or the growth and sporulation of the fungus on the surface of a granular nutrient carrier is preferred, and this method has already been applied to easily industrialized auxiliary microorganisms, but for entomopathogenic fungi, this method is still insufficiently tested in the field. On the other hand, for better results, oily formulations containing UV, humectant, and nutrient protectors are preferred for spore growth and germination, as oils provide better adhesion properties and efficient spore application to hydrophobic cuticles and insects. According to Croda's technical expertise, the use of mild and biocompatible surfactants such as sorbitan esters, polysorbates, and low risk non-ionic polymers is considered the best approach for choosing a formulation adjuvant (Dannon et al. 2020).

16.5.1 Packaging

Collect the spore powder after sieving, then weigh and store in separate sterile vials for packaging in low density polyethylene (LDPE) packets. Furthermore, to extend the shelf life of aerial conidia, active packaging is preferred through the use of oxygen and moisture scavengers. This type of active packaging helps to improve the shelf life of aerial conidia of *B. bassiana,* and thus can be stored at temperatures as high as 50 °C. In conventional systems such as vacuum packaging or replacement of internal storage atmosphere with an inert gas, the oxygen level is found to be 0.3–3% whereas, oxygen absorbing technology reduces the oxygen level below 0.01%, which is lower than the oxygen level mentioned in conventional systems (Cruz et al. 2007). However, very little information is known about the use of an active packaging approach for liquid formulations that produce blastospores.

Several studies have revealed that air -and spray-drying processes can be used to dry liquid cultures producing blastospores. Once the formulations are dried, they were packaged in sealed Mylar bags with combinations of oxygen and moisture scavangers and then stored at 4 or 28 °C. After drying and storage, blastospore viability can be evaluated by assessing the insecticidal efficacy (Mascarin et al. 2016).

16.6 Quality Control and Product Specification

Quality control is considered the main consideration for microbiologists at each stage of product formulation, influencing the success or failure of the product in the market. To ensure good quality of the *B. bassiana* formulation, it is important to determine the spore concentration as well as its viability.

16.7 Determination of Spore Concentration

To determine spore concentration, 1 g of harvested spores is air dried for 10 days, transferred into 10 mL of sterile distilled water containing 0.1% triton-X-100 solution in a 100 ml conical flask and shaken for 10 min. Subsequently, the suspension is filtered through a double layer muslin cloth. Perform the serial dilution of the suspension and count the spores using a haemocytometer to determine the concentration, i.e., the number. of spores per gram suspension (Belay and Tenkegna 2017).

16.8 Determination of Spore Viability

To assess the viability of the spores, a sample of 1 g of conidia is taken after harvesting the spore powder and suspending it in 0.01% triton-X-100. The suspension is filtered through sterile cheesecloth and the spore concenteation is adjusted to 1×10^7 spore/mL to determine spore viability (Belay and Tenkegna 2017).

16.9 Food Quality Clearance

Formulated *B. bassiana* products were registered in 1995 after satisfying all the criteria of the Food Quality Protection Act (FQPA). The registrant provided information regarding production analyses determining adequate quality control measures of nominal limits, contaminants and metabolites for consideration in the exemption from tolerance on 21 January 1999; the certainty for registering the product was that no harm would result in cumulative and aggregated exposure to *B. bassiana,* including all potential exposure via drinking water as well as all anticipated dietary exposures (Längle 2006).

16.10 Risks Posed Due to Drinking Water Exposure

Generally, *B. bassiana* is found in soil, as it is not an aquatic microorganism; therefore, *B. bassiana* is not screened for drinking water as an indicator of microbial contamination. The possibility of *B. bassiana* exposure to drinking water is reduced by both percolation through soil and municipal treatment of drinking water; thus, the possibility of *B. bassiana* transfer to drinking water is minimal to negligible. The agency concluded that although the presence of *B. bassiana* is negligible, if oral exposure through drinking water occurs, there would be no risk due to the lack of toxicity and ubiquitous nature of *B. bassiana* (Längle 2006).

16.11 Toxicology

Toxicity studies were found to be acceptable and no incidents of hypersensitivity have been reported for *B. bassiana* microbial pesticide active ingredient. Because *B. bassiana* satisfied the Tier I test guideline requirements in compliance with the FQPA of 1996, Tier II tests and Tier III toxicology tests were not required (Längle 2006).

16.12 Risks Posed by Multiple Routes Including Dermal, Oral and Inhalation

After considering all the various routes of exposure, such as drinking water, dietary, dermal, and inhalation, the agency considered that *B. bassiana* is not known to produce metabolites that are dermally absorbed or that is a human pathogen. In addition, the risk assessment of dietary exposure was found to be minimal or almost negligible for all segments of the population, including children and infants (Längle 2006).

16.13 Project Details

Certain facilities are required to successfully establish a biopesticide unit and implement production project. The infrastructure and laboratory facilities required for the production of entomopathogenic fungi *B. bassiana* can also be applied to the production of other biofertilizers as well as biopesticides. Many private companies and governmental institutes produce both biofertilizers and biopesticides by making use of the common facilities ensuring the economic viability of the project installation. The projects set up for the production of biopesticides based on entomopathogenic fungi *B. bassiana* in India vary from 140 lakhs to 150 lakhs. The component-wise budget outlay for the mass production of *B. bassiana* for single unit costs between 45 lakhs and 55 lakhs. Therefore, the cost of each particular is counted accordingly for the mass production of *B. bassiana*.

Land and Location Land is prerequisite for establishing a project or a manufacturing business. To set up a laboratory, office, and other facilities for mass production of *B. bassiana* for a single unit, 200–300 sq. metre of land is required costing Rs. 10,000–11,000 per sq. metre (Box 16.2).

The location of the biopesticide unit should be in such a way that it should not disturb the common people, and therefore a biopesticide unit should be set up far from local people residences. It should not be set up in areas with high temperature and humidity, as it may increase the production and maintenance costs.

Building Infrastructure The building should be divided into different compartments, such as quality assurance laboratories; mother culture preparation and maintenance rooms; fermenter arrangement facilities; labelling, packaging, handling, and dispatching areas; storage facilities; administrative offices; staff rooms and pantries. The designed infrastructure should facilitate the optimum temperature and humidity conditions required for culture maintenance. The entire site should be either fenced with a barbed wire or with a compound wall and gates. The required area for buildup of infrastructure for mass production of *B. bassiana* for single unit is 130–140 sq. metre costing between Rs. 20,000 and 25,000 / sq. metre (Box 16.2).

Box 16.2 Component-wise budget outlay for mass production of *B. bassiana* for single unit

Sr. no.	Particulars	Cost/sq. metre (in Rs.)	Required area (in sq. metre)	Total (Rs. In lakhs)
1	Land	10,764	210	22.61
2	Infrastructure	20,000	130	26
	Total			48.61

Utilities

(i) **Power supply:** For mass production of *B. bassiana,* generally, three-phase supply is needed. Depending upon the machinery operated, the requirement is nearly about 75–100 KVA. A standby generator is also needed as an option of power supply.
(ii) **Water supply:** Water supply is mandatory at each and every step of *B. bassiana* production whether it is broth preparation, cleaning of glassware, autoclave, hot air oven, fermenter cleaning, etc. The average requirement of water supply will be around 4000–6000 liters/day.

Machinery A number of well-established equipments are required to fulfill the standards as well as to achieve the maximum production capacity. The instrumentation for mass production of *Beauveria bassiana* is illustrated in the following table (Tables 16.1, 16.2, 16.3 and 16.4).

Financial Analysis
Benefit-Cost Ratio = Total Net Return/Total Cost (Tables 16.5 **and** 16.6).

Table 16.1 Equipments required for mass production of *Beauveria bassiana*

1. Nonrecurring contingencies

Sr. no.	Specifications/items	Capacity	Rs./ unit	Quantity	Rs. (lakhs)
1	Laminar air flow	300 lux	0.98	2	1.96
2	Autoclave	22–200 L	0.3	2	0.6
3	Laboratory refrigerator	400 L	0.65	2	1.3
4	RO-based distillation unit	1.5 l/h	0.25	1	0.25
5	Fermentor	500–1000 L	2.5	1	2.5
6	Hot air oven	85 kg	0.9	1	0.9
7	Electronic digital balance	0.1 mg-120 g	0.25	1	0.25
8	Air conditioner with stabilizer	1.5 ton	0.29	4	1.16
9	Microscope	–	0.29	1	0.29
10	Gas stove	–	0.04	1	0.04
11	Drinking water with UV. RO. System	80 L	0.7099	1	0.7099
12	Haemocytometer	–	0.0106	1	0.0106
13	Fire extinguishers for lab safety	2 kg	0.0012	3	0.0036
14	Bottle seller	–	0.36	1	0.36
15	Packet sealer	–	0.053	1	0.053
16	Generator	1 KVA	0.5	1	0.5
17	Furniture	–	1	1	1
18	Xerox machine	–	0.6	1	0.6
19	Computer with accessories	–	0.385	1	0.385
20	Miscellaneous	–	0.5	1	0.5
	Total		**10.58**		**13.41**

2. Recurring contingencies

Sr. no.	Particulars	Number	Rs. (lakh)
1 (i)	Labour: Technical staff @ 22,000 / month	2	5.28
(ii)	Skilled labour @ 300 / day	5	1.8
(iii)	Accountant @ 23,500	1	2.82
2	Glasswares (Petridish, flask, beaker, etc.)	–	3
3	Stationery items	–	0.5
4	Chemicals	–	7
5	Maintaining mother culture	–	1
6	POL (hiring of vehicles)	–	1
7	LPG gas cylinder	–	0.37
8	Labelling, packaging and handling of materials	–	1
9	Electricity and water charges	–	1
10	T.A and D.A	–	0.5
11	Miscellaneous	–	0.5
	Total		**25.77**

Table 16.2 Budget requirement for 3 consecutive years

Sr. no.	Particulars	Year			Total (Rs. in lakhs)
		I	II	III	
1	Land and infrastructure	48.61	–	–	48.61
2	Nonrecurring items	13.41	–	–	13.41
	Subtotal (A)	**62.02**	–	–	**62.02**
3	Chemicals and stationery	7.5	8.25	9.1	24.85
4	Glasswares	3	–	–	3
5	Salaries for staff	9.90	10.89	11.98	32.77
6	Maintaining mother culture	1	1.10	1.21	3.31
7	POL (hiring of vehicles)	1	1.10	1.21	3.31
8	LPG gas cylinder	0.37	0.41	0.45	1.23
9	Labelling, packaging and handling of materials	1	1.10	1.21	3.31
10	Electricity and water charges	1	1.10	1.21	3.31
11	T.A and D.A	0.50	0.55	0.60	1.65
12	Miscellaneous	0.50	0.55	0.60	1.65
	Subtotal (B)	**25.77**	**25.05**	**27.57**	**78.39**
	Grand Total (A + B)	**87.79**	**25.05**	**27.57**	**140.41**

Table 16.3 Budget requirement (if building is rented) for 3 consecutive years

Sr. no.	Particulars	Year			Total (Rs. in lakhs)
		I	II	III	
1	Rent of building	3.6	3.96	4.356	11.916
2	Nonrecurring items	13.41	–	–	13.41
	Subtotal (A)	**17.01**	**3.96**	**4.356**	**25.326**
3	Chemicals and stationery	7.5	8.25	9.1	24.85
4	Glasswares	3	–	–	3
5	Salaries for staff	9.90	10.89	11.98	32.77
6	Maintaining mother culture	1	1.10	1.21	3.31
7	POL (hiring of vehicles)	1	1.10	1.21	3.31
8	LPG gas cylinder	0.37	0.41	0.45	1.23
9	Labelling, packaging and handling of materials	1	1.10	1.21	3.31
10	Electricity and water charges	1	1.10	1.21	3.31
11	T.A and D.A	0.50	0.55	0.60	1.65
12	Miscellaneous	0.50	0.55	0.60	1.65
	Subtotal (B)	**25.77**	**25.05**	**27.57**	**78.39**
	Grand Total (A + B)	**42.78**	**29.01**	**31.926**	**103.716**

Table 16.4 Project details for mass production of *B. bassiana*

Sr. no.	Production of *B. bassiana*	Production per year in L or kg	Market value of product (Rs. per L or kg)	Earning per year (Rs. in lakhs)
1	Liquid	8000	500	40
2	Solid	16,000	450	72
Total		**24,000**	**950**	**112**

Table 16.5 Benefit-cost ratio (BCR) of purchased land

Year	Cost (in lakhs)	Return (in lakhs)	Net return (in lakhs)
1	87.79	112.00	24.21
2	25.05	123.20	98.15
3	27.57	135.52	107.95
Total	**140.41**	**370.72**	**230.31**
BCR			**1.64**

Table 16.6 Benefit-cost ratio (BCR) of rented land

Year	Cost (in lakhs)	Return (in lakhs)	Net return (in lakhs)
1	42.78	112.00	69.22
BCR			**1.61**

16.14 Summary

For the mass production of *B. bassiana,* the investment cost required will be Rs. 140.41 lakh and the production per year was found to be 24,000 L per year. The net return obtained from investment will be Rs. 230.31 lakh. It is estimated that Rs. 89.9 lakh can be gained as a profit, and the benefit-cost ratio of the project model was found to be 1.64. This indicates the feasibility and profitability of the mass production unit of *B. bassiana.* In the proposed model, the B:C ratio is higher, therefore, financing may be useful in the implementation of the project.

References

Anand R, Prasad B, Tiwary BN (2009) Relative susceptibility of Spodoptera litura pupae to selected entomopathogenic fungi. BioControl 54(1):85–92

Belay YC, Tenkegna TA (2017) Bioassay and pilot mass production of entomopathogenic fungus, Beauveria bassiana for the control of coffee berry borer (Hypothenemus hampei: Scolytidae), Ferrari. J Appl Biosci 117:11669–11683

Cruz RS, Soares NDFF, Andrade NJD (2007) Efficiency of oxygen: absorbing sachets in different relative humidities and temperatures. Ciência e Agrotecnologia 31(6):1800–1804

Dannon HF, Dannon AE, Douro-Kpindou OK, Zinsou AV, Houndete AT, Toffa-Mehinto J, Elegbede I, Olou BD, Tamò M (2020) Toward the efficient use of Beauveria bassiana in integrated cotton insect pest management. J Cotton Res 3(1):1–21

Deba L, Rajesha T, Tombisanaa RK, Majumdera D (2017) Antagonistic potential of Beauveria sp. against phytopathogens. Bull Environ Pharmacol Life Sci 6(3):207–212

Dhaliwal GS, Jindal V, Mohindru B (2015) Crop losses due to insect pests: global and Indian scenario. Indian J Entomol 77(2):165–168

Faria M, Wraight SP (2001) Biological control of Bemisia tabaci with fungi. Crop Prot 20(9): 767–778

Gouli V, Gouli S, Brownbridge M, Skinner M and Parker BL (2005) Manual for Mass Production of Entomopathogenic fungi in developing countries with particular reference to Beauveria bassiana and Metarhizium anisopaliae. *University of Vermont, Entomology Research Laboratory, UK*

Keswani C, Singh SP, Singh HB (2013) Beauveria bassiana: status, mode of action, applications and safety issues. Biotech Today 3(1):16–20

Khachatourians GG (1991) Physiology and genetics of entomopathogenic fungi. Handbook Appl Mycol 2:613–663

Kpindou OD, Djegui DA, Glitho IA, Tamò M (2012) Sensitivity of Helicoverpa armigera (Hübner) (lepidoptera: Noctuidea) to the entomopathogenic fungi, Metarhizium anisopliae and Beauveria bassiana in laboratory. J Agric Biol Sci 7(12):1007–1015

Lambert N (2010) *Lutte Biologique Aux Ravageurs: Applicabilité Au Québec* (Doctoral dissertation, Université de Sherbrooke.)

Längle T (2006) September. Beauveria bassiana (Bals.-Criv.) Vuill.–A biocontrol agent with more than 100 years of history of safe use. In *Workshop on Current Risk Assessment and Regulation Practice Salzau, Germany*

Logrieco A, Moretti A, Ritieni A, Caiaffa MF, Macchia L (2002) Beauvericin: chemistry, biology and significance. In: Advances in microbial toxin research and its biotechnological exploitation. Springer, Boston, MA, pp 23–30

Marrone PG (2014) The market and potential for biopesticides. In: Biopesticides: state of the art and future opportunities. American Chemical Society, pp 245–258

Mascarin GM, Jackson MA, Behle RW, Kobori NN, Júnior ÍD (2016) Improved shelf life of dried Beauveria bassiana blastospores using convective drying and active packaging processes. Appl Microbiol Biotechnol 100(19):8359–8370

National Research Council (1986) Pesticide resistance: strategies and tactics for management. National Academies Press

Ownley BH, Gwinn KD, Vega FE (2010) Endophytic fungal entomopathogens with activity against plant pathogens: ecology and evolution. BioControl 55(1):113–128

Rao NK, Hanson J, Dulloo ME, Ghosh K and Nowell A (2006) *Manual of seed handling in genebanks* (No. 8). Bioversity International

Ravensberg W (2015) September. Crop protection in 2030: towards a natural, efficient, safe and sustainable approach. In *International Symposium Swansea University* (pp. 7–9)

Swathi P, Das SB, Visalakshy PNG (2017) Mass multiplication of Beauveria bassiana on substrates. Agric Update 12(TECHSEAR-3):794–798

Chapter 17
Chaetomium sp.: An Insight into its Antagonistic Mechanisms, Mass Multiplication, and Production Cost Analysis

Parthasarathy Seethapathy, Harish Sankarasubramanian, Rajendran Lingan, and Raguchander Thiruvengadam

Abstract Abiotic and biotic factors conspire to keep crop output at a bare minimum. In agricultural production, to combat these factors, beneficial microorganisms are used to boost yields and as a substitute for synthetic fertilizers and fungicides. The genus, *Chaetomium* is well-known for its fast, saprophytic colonization, similar to that of *Trichoderma. Chaetomium* spp. produce a variety of compounds that stimulate plant growth and inhibit pathogen infestation. There is very little research attempting to understand the metabolic and physiological changes that occur in crops following exposure to *Chaetomium* spp., which might be beneficial in protecting them against harmful environmental conditions. This chapter summarizes current knowledge regarding the antagonistic activity of *Chaetomium* spp. against a variety of oomycetes and fungal diseases of crop plants. During antagonism, *Chaetomium* spp. produce an array of antibiotics and secondary metabolites in their associated environment, thereby preventing pathogen entry and colonization of plant tissues and reducing the harmful microbial population. This leads to the development of biopesticide formulations, which improves crop productivity and disease management strategy. The goal of this chapter is to keep readers up to date on current breakthroughs in the scope, biopesticide production, production economics, and market potential of the genus *Chaetomium*, as well as the issues that researchers and entrepreneurs will face in the coming years.

Keywords Antagonist · *Chaetomium globosum* · Formulation · Production technology · Benefit-cost analysis

P. Seethapathy (✉)
Department of Plant Pathology, Amrita School of Agricultural Sciences, Amrita Vishwa Vidyapeetham, Coimbatore, India

H. Sankarasubramanian · R. Lingan · R. Thiruvengadam
Department of Plant Pathology, Centre for Plant Protection Studies, Tamil Nadu Agricultural University, Coimbatore, India

N. Amaresan et al. (eds.), *Agricultural Microbiology Based Entrepreneurship*, Microorganisms for Sustainability 39,
https://doi.org/10.1007/978-981-19-5747-5_17

17.1 Introduction

The impact of the green revolution on avoiding death and hunger in major countries is beginning to wane, as crop output declines significantly due to poor growth of crop plants because of elevated diseases and pests. Plant disease is diminishing the quantity and quality of food, fibre, and fuel crops at a time when farming is struggling to feed the world's largest and fastest burgeoning population. Regardless of the cause, either catastrophic or chronic, fatalities account for an average of 42% of the annual production of the six major food crops. Farmers have traditionally relied on chemical management using synthetic pesticides to suppress these plant pathogens, but these chemicals have been shown to be harmful to human health and livestock as well as to the environment. Biological control has developed into a viable method of controlling the pests and pathogens associated with diversified crops in the twenty-first century while maintaining ecological equilibrium. As an alternative to chemical fungicides, numerous fungi and bacteria have been effectively employed as antagonistic microorganisms against a variety of plant pathogens (Aggarwal 2015). Additionally, they are resistant to adverse climatic circumstances and do not cause adverse effects associated with chemicals, such as pathogenic resistance to fungicides. Biopesticides represent a tiny proportion of the total pesticide market in the world, but their growth is predicted to be higher than that of synthetic insecticides in the coming decades. Some beneficial strains of microorganisms like *Agrobacterium* spp., *Ampelomyces quisqualis*, *Bacillus* spp., *Chaetomium* spp., *Coniothyrium minitans*, *Eudarluca filum*, *Gliocladium* spp., *Paenibacillus* spp., *Pantoea* spp., *Pseudomonas* spp., *Pythium oligandrum*, *Serratia* spp., *Streptomyces* spp., *Trichoderma* spp. and yeasts represent the major biopesticide products on the commercial scale against several plant pathogens across the world (Parthasarathy et al. 2017). A resurgence of interest in all facets of biocontrol agents and microbial consortia may also facilitate the delivery of a "one-pot" microbial cell factory to meet the growing demands of farmers and the environment in the twenty-first century. To fulfil this opportunity and to acquire a complete understanding of these microbial biopesticide applications and commercialization, it is critical to continue developing economically viable products and processes. The preponderance of biological control research is focused on soil-borne pathogens. For both foliar and root-infecting diseases, there have been several recent publications on the effective antagonistic behaviour of fungi. In this context, *Chaetomium* spp. and *Trichoderma* spp. have been explored for the control of several aerial and soil-borne pathogens (Soytong et al. 2001; Aggarwal et al. 2014). Based on the habitat and conditions, *Chaetomium* behaves as both friends and foes. Many species of *Chaetomium* are beneficial to plants, bio-diversity, environment, human health and animals because of their pharmacological (Fatima et al. 2016), industrial, and agricultural applications (Soytong et al. 2021). *Chaetomium*-based biopesticides degrade and recycle spontaneously. These products are also considered eco-friendly due to their low impact on nontarget creatures. The use of *Chaetomium* will also reduce agricultural and biological residues in the environment, making the product potentially safer for consumers. Certain species, on the other hand, are detrimental due to their potential to invade and degrade surfaces, as well as their aptitude to behave as pathogens and cause disease in plants, animals and human populations.

17.2 Significance

Chaetomium species are well-known for their antagonistic potential towards a wide range of plant diseases, and the genus is estimated to have around 400 species (Abdel-Azeem 2020). The synthesis of lytic enzymes and secondary metabolites is critical for their antagonistic activity. The genus has been widely documented to exhibit a broad spectrum of biological activities, including assisting in the biodegradation of municipal trash, acting as an efficient antioxidant, and producing pharmaceutical and industrial enzymes. Additionally, it has been implicated as a contaminant in human illnesses such as type 1 allergic reactions and nonspecific infections (Fatima et al. 2016). Despite seeming to be saprophytic ascomycetes with only sporadic rendezvous in human and animal disease occurrences, *Chaetomium* spp. are capable of causing a wide variety of mycoses in humans, including corneal ulcers, empyema, keratitis, onychomycosis, sinusitis, pneumonia, and fatal disseminated cerebral disease, particularly in immunocompromised patients and intravenous drug users. The genus *Chaetomium* in animals can complete their life cycle as fimicolous, resulting in infections such as depilation, erythematous epilation, scales, and dermatitis with mild itchiness on their skin (Sugiyama et al. 2008). On the other hand, ***Chaetomium*** *globosum* and *C. cochliodes* are potential antagonists in plant disease management by controlling a wide-range of seed, foliar, and soil-borne pathogenic fungi (Soytong et al. 2001). It is now used commercially as a biopesticide against a variety of fungal and bacterial diseases, insect pests, and nematodes and also induces plant growth promotion. Recent emphasis on the cellulolytic character of this organism opens up numerous opportunities in food production, reforestation, life sciences, textiles, and health care industries.

17.3 Taxonomy

Chaetomium is one of the most intimate genera of Pyrenomycetes, and it is confronted by a variety of biological and derived commodities. Numerous researchers have discussed the taxonomy of *Chaetomium*. According to http://www.indexfungorum.org, around 444 species have been described in the genus since its inception. Since, Gustav Kunze introduced the genus *Chaetomium* in 1817, with *C. globosum* Kunze as its generic type, and the largest family Chaetomiaceae in 1885, numerous taxonomic studies have been conducted to classify this group of fungi. The centum structure and perithecial habit are the

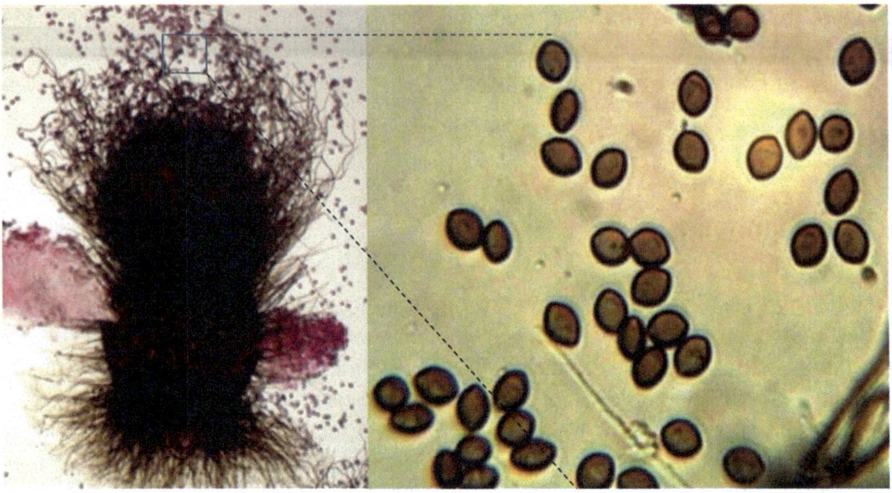

Fig. 17.1 Morphological features of *C. globosum*

primary characters used to identify ascomata, asci, and ascospores, but they have not been sufficient to resolve relationships and determine the taxonomic position of *Chaetomium*. Alexopoulos (1962) classified *Chaetomium* and its related genera in the order Chaetomiales. This decision was drawn on the basis of the presence of perithecial hairs, the absence of paraphyses, and the evanescent asci. During these changes, the genus has remained within the Chaetomiaceae family, with *C. globosum* Kunze & Schmidt serving as the basal taxon in the family. *Chaetomium* is a genus of the class Sordariomycetes, the order Sordariales, and the family Chaetomiaceae. This ascomycetous genus is distinguished by ostiolate ascomata that are typically covered in setae, clavate, fusiform or cylindrical, fasciculate, evanescent asci, and dark to grey-brown, single-celled ascospores with one or two germ pores (Fig. 17.1). On culture media, the genus can produce an *Acremonium*-like state (anamorphic stage), which is described by flask-shaped perithecia bounded by dark, stiff hairs. Several researchers reported that no anamorph of *C. globosum* has been formally identified for *Chaetomium* spp. (Abdel-Azeem 2020). On the other hand, it has been associated with several anamorphic genera, including *Acremonium*, *Botryotrichum*, *Chrysosporium*, *Histoplasma*, *Humicola*, *Phialophora*, *Scopulariopsis*, and *Scytalidium*. With the exception of *Acremonium*, all of these genera were described after *Chaetomium*. Phylogenetic investigations conducted by various researchers revealed that the genus *Chaetomium* was not monophyletic in nature (Zhang et al. 2017). Chemotaxonomy based on secondary metabolites is an influential taxonomic tool that can be used as a supporting relation to resolve classification problems in *Chaetomium*. Among the secondary metabolites produced by *Chaetomium*, cytochalasans and azaphilones demonstrated a distribution pattern and frequency coherent in their use as chemotaxonomic markers for the *Chaetomium* species (Yang et al. 2021). In addition, since the majority of these anamorphic genera looked to be polyphyletic, a phylogenetic assessment and taxonomy revision were required for these genera. Therefore, molecular advancements for *Chaetomium* species identification are crucial, as seems to be the history of DNA barcoding development for all economically efficient fungi.

17.4 Ecology

Due to its potential biocontrol activity and great adaptation to a variety of ecological situations, the fungus *C. globosum* has gained international notice (Darshan et al. 2021a). The majority of species in this genus are cosmopolitan, common in soil and air, and capable of establishing themselves in a wide range of organic and inorganic substrates. The genus *Chaetomium* is a saprobic fungus that is usually coprophilous, also known as dung fungi, mesophilic, and thermophilic in nature. It is agriculturally associated with plant surfaces, seeds, soil on cellulose-rich substrates, or on dung, wood, and organic composts through the formation of symbiotic, endophytic, and ectophytic relationships without causing harm to their hosts. *Chaetomium* species might well be degraders of cellulosic and other biological resources because they rapidly degrade cellulose and produce thermostable cellulases. Additionally, it can act as an antagonist against a variety of soil and foliar pathogenic microorganisms. The increased prevalence of *Chaetomium* species in a diverse selection of ecological environments, including deserts, polar regions, salterns, mangrove forests, indoor environments, water bodies, food materials, ruminant dung, living plants, lichens, decaying wood, human systems, and animals, shows the versatility of the genus *Chaetomium* in a wide range of ecological environments. The natural habitats of numerous species in this clade are even in arid climates, and their survival at elevated temperatures likely contributes to their survival in mammalian tissue. As a result of their widespread distribution, these species have a profound effect on ecosystems, agriculture, food production, biotechnology, and human and animal health. Due to the diversity of its species and habitats, *Chaetomium* spp. may develop diverse biosynthetic gene clusters that turn into a variety of secondary metabolites in order to adapt to changing ecological conditions. More than 200 compounds with a diverse spectrum of bioactive properties have been identified in *Chaetomium* spp. to date, but given the diversity of species, it is possible that more bioactive secondary metabolites will be discovered in this fungus (Hung et al. 2015). The amount of information known regarding the ecological distribution of *Chaetomium* taxa is constantly increasing. Hence, it is reasonable to anticipate that the genus's biogeography will become more fully understood in the near future.

17.5 Biological Control Potential of *Chaetomium* spp.

Understanding the antagonistic potential and mode of action of some biocontrol agents, such as *Chaetomium* spp., is crucial to becoming more effective against certain pathogens in a range of ways. Numerous *Chaetomium* species are adapted to operate as biological control agents by inhibiting the growth of fungus, bacteria, insect pests, and nematodes through mycoparasitism, competition, antibiosis, lysis, induced resistance, and plant growth promotion (Fig. 17.2), or a combination of these mechanisms (Aggarwal 2015; Soytong et al. 2021; Thiruvengadam et al. 2020). *C. globosum* is one of the most common species growing as a saprophyte in the rhizosphere, phyllosphere, and as a normal colonizer of the soil. *Chaetomium* spp. has been exploited in various studies as a biocontrol agent (Table 17.1). *C. globosum* has been reported to be a potential antagonist of various plant pathogens, especially foliar, soil- and seed-borne pathogens, such as *Alternaria alternata, Alternaria triticina, Bipolaris oryzae, Bipolaris sorokiniana, Colletotrichum gloeosporioides, Cochliobolus sativus, Fusarium graminearum, Fusarium fujikuroi, Fusarium oxysporum, Helminthosporium victoriae, Macrophomina phaseolina, Phytophthora* spp.*, Pyricularia oryzae, Pythium* spp., *Sclerotium rolfsii, Sclerotinia sclerotiorum, Rhizoctonia solani, Setosphaeria turcica, Tilletia indica,* and *Venturia inaequalis* (Soytong et al. 2001; Aggarwal et al. 2014; Soytong et al. 2021; Shanthiyaa et al. 2013; Zhang et al. 2013; Kumar et al. 2021; Darshan et al. 2020, 2021b; Aggarwal et al. 2004). *Chaetomium cupreum* can solely control pathogenic species, which includes *Fusarium, Phytophthora, Rhizoctonia, Erwinia, Xanthomonas* etc. (Soytong et al. 2001). *C. cupreum*, and *C. globosum* in particular have been studied extensively to manage root diseases of citrus and strawberry and have been reported to reduce damping-off of sugar beet (Soytong et al. 2001; Tomilova and Shternshis 2006). The mechanism of *Chaetomium* spp. and reaction

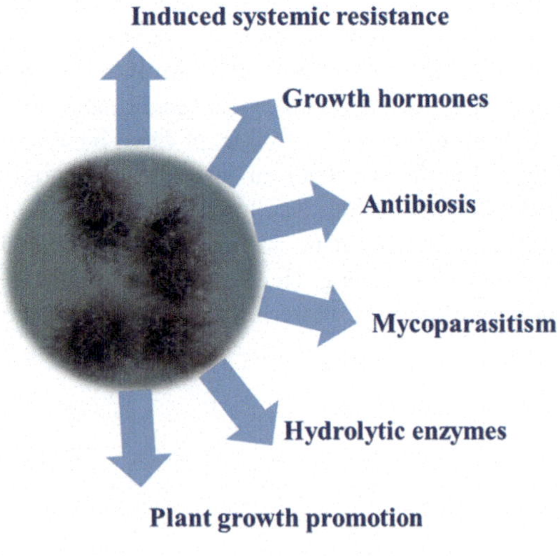

Fig. 17.2 Mechanism of *Chaetomium* spp. towards targeted pathogens and its host

Table 17.1 Antagonistic activity of *Chaetomium* spp. against various plant pathogens

Species	Pathogen	Crop	References
Chaetomium spp.	*Pyricularia oryzae*, *Rhizoctonia oryzae*	Rice	Soytong et al. (2021)
	Curvularia lunata	Corn	Soytong et al. (2021)
	Fusarium oxysporum f. sp. *lycopersici*	Tomato	Pothiraj et al. (2021)
	Sclerotium rolfsii	Corn	Soytong et al. (2021)
	Pyrenophora graminea	Barley	Linkies et al. (2021)
C. cupreum	*Colletotrichum gloeosporioides*	Coffee	Vilavong and Soytong (2017)
	Rigidoporus microporus	Para rubber	Soytong et al. (2021)
C. cochlioides	*Phytophthora* sp.	Durain	Rujira (2017)
C. spirale	*Valsa ceratosperma*	Apple	Ya-fen and Jin-jie (2005)
C. elatum	*Pyricularia oryzae*	Rice	Song et al. (2016)
	F. Oxysporum f.sp. *lycopersici*	Tomato	Soytong (2015)
C. globosum	*Fusarium roseum* f. sp. *cerealis*	Corn	Soytong (2010)
	Venturia inaequalis, *Zygophiala jamaicensis*; *Gloeodes pomigena*	Apple	Davis et al. (1992)
	Pythium ultimum	Sugarbeet, citrus	Soytong et al. (2021); Park et al. (2005); Pietro (1992)
	Verticillium dahliae	Cotton	Zhang et al. (2021)
	Erysiphe graminis f. sp. *hordei*	Barley	Park et al. (2005)
	Pyricularia oryzae	Rice	Gandalera et al. (2013)
	Cochliobolus sativus	Wheat	Aggarwal et al. (2004)
	Sclerotium cepivorum	Onion	McLean and Stewart (2000)
	Rhizoctonia solani	Potato	Arunkumar et al. (2021)
	Phytophthora infestans	Potato	Soytong et al. (2021); Shanthiyaa et al. (2013)
	Alternaria brassicicola	Sugarbeet	Abdel-Azeem (2020)
	Sclerotinia sclerotiorum	Rapeseed	Zhao et al. (2017)
	Botrytis cinerea	Lily	Soytong et al. (2001)
	Diaporthe phaseolorum f. sp. *meridionalis*	Soybean	Dhingra et al. (2003)
	C. cladosporioides, *C. herbarum* and *C. sphaerospermum*	Faba bean	Gamal et al. (2021)
	Fusarium lycopersici	Tomato	Rajendran et al. (2021)
C. cupreum and *C. globosum*	*Phytophthora parasitica*	Tangarine	Soytong et al. (2001); Soytong et al. (2021)
	Phytophthora palmivora	Black pepper, Durian	
	Phytophthora cactorum	Strawberry	

with these pathogens are many and complex, which are influenced by soil type, temperature, pH, moisture, soil environment, and also by the presence of other microorganisms. In bio-management of plant diseases, *C. globosum* is largely exploited as a single strain rather than as component of a consortium (Aggarwal 2015). However, microbial consortia composed of antagonistic bacteria and fungi already have the additional advantage of protecting plants under a variety of conditions and seasons by occupying complementary or distinct niches.

As a bioagent, the endophytic *C. aureum* HP047 significantly reduced the Pitch canker disease on *Pinus radiata* seedlings caused by *F. circinatum* (Martínez-Álvarez et al. 2016). In addition, *C. aureum* also effectively controlled the pathogenic *Magnaporthe grisea* causing rice blast disease in vitro and in vivo, and the sheath blight disease incited by *R. solani*. Based on in vitro interactions (Moya et al. 2016), two endophytic *Chaetomium* (C2 and C5), were isolated from barley seedlings and were identified as potential bioagents against fungal pathogens, *Drechslera teres* (net blotch) and *B. sorokiniana* (spot blotch). The nanoparticles derived from three different species such as *C. elatum*, *C. lucknowense*, and *C. brasiliense* showed antifungal potential against *Magnaporthe oryzae*, incitant of rice blast (Song et al. 2020). Chaetoglobosin A from *C. globosum* inhibited root-knot nematode (*Meloidogyne incognita*) egg hatch and juvenile mobility (Hu et al. 2013), and its culture broth inhibited the hatching of soybean cyst nematode (*H. glycines*) (Nitao et al. 1999). Zhou et al. (Moya et al. 2016) recorded that *C. globosum* TAMU 520 is an endophyte of cotton and systemically colonizes the cotton plants through seed treatment. The endophytic *C. globosum* suppressed the infection of root-knot nematode, which reduced female reproduction. Kooliyattil et al. (Hu et al. 2013) found *C. globosum* as a fungal parasite which was isolated from the eggs of *Globodera pallida* (cyst nematode) in the USA. It showed the greatest reduction of the infection (76%) by *G. pallida* in potato. Zhou et al. (Moya et al. 2016) stated that endophytic *C. globosum* showed a negative effect on the fecundity of cotton aphids (*Aphis gossypii*) and beet armyworms (*Spodoptera exigua*).

17.5.1 Mycoparasitism

Biocontrol agents are capable of parasitizing the intended host fungal hyphae mechanically and occasionally with the aid of lytic enzymes, and this is regarded as a critical strategy for plant pathogen control. Dingle and McGee (Dingle and McGee 2003) documented that, *Chaetomium* spp. isolated from healthy wheat leaves inhibited the growth and spread of *Puccinia recondita* f. sp. *tritici* rust pustules on these leaves. Later, the study showed that (Larran et al. 2016) *Chaetomium* spp. reduced the growth of *Pyrenophora tritici-repentis* and average disease severity on wheat leaves in greenhouse experiment. *Drechslera oryzae*, which causes rice leaf spot, was effectively controlled in Thailand by *C. cupreum*, *C. brasiliense*, *C. cochliodes*, *C. globosum*, and *C. elatum* (Tathan et al. 2012). *C. globosum* strain no.05 strongly inhibited the in vitro mycelial growth of several phytopathogenic fungi on potato dextrose medium including; *S. turcica*, *Verticillium mali*, *C. diplodiella*, *C. glocosporioides*, *C. fimbriata*, and *S. sclerotiorum* (Zhang et al. 2013). The antagonistic strains, viz., *C. lucknowense*, and *C. cochiliodes*, showed a clear zone of inhibition and were grown over *Ganoderma boninense*, the incitant of oil palm basal stem rot disease (Soytong 2014). In the dual culture assay, strains of *C. globosum*, *C. cupreum*, and *C. lucknowense* caused noticeable inhibition of *Phytophthora palmivora* (root rot of pomelo) and *P. nicotianae* mycelial growth, and reduced 92–99% of its sporangial production (Hung et al. 2015); these species were found to inhibit the vegetative and reproductive structures of *F. oxysporum*, the causal agent of wilt and root-rot disease of tea (Huu Phong et al. 2016). In the study (Van Thiep and Kasem 2015), *Chaetomium* spp. included *C. cochliodes*, *C. bostrychodes*, and *C. gracile*, which inhibited the growth and spore production of *F. roseum,* causing coffee and tea wilts, and *C. gloeosporioides,* causing leaf anthracnose of coffee in the dual culture assay. In another study, strains such as *C. globosum*, *C. cochliodes*, *C. aureum*, *C. nozdrenkoae*, *C. ramosissimum*, and *C. elatum* were shown to inhibit the mycelial growth of *Pyrenophora graminea,* causing leaf stripe of barley, and *B. sorokiniana,* causing spot blotch of wheat (Linkies et al. 2021).

17.5.2 Antibiosis

Antibiosis is defined as the destruction or inhibition of harmful microorganisms by substances such as specific or nonspecific secondary metabolites or by the production of antibiotics that inhibit the growth of another microorganism. Chaetomin mycotoxin and microbial volatile organic compounds was recorded for the first time from *Chaetomium* spp. (Abdel-Azeem 2020); however, a wide variety of bioactive compounds were reported later in *Chaetomium* spp. such as azaphilones, and cytochalasans (Yang et al. 2021), chaetomin, chaetocin, chaetominine, chaetomugilin I, J, chaetoglocin A, globosuxanthone, globosumone C, parietin, prenisatin, prochaetoglobosin I, II (Kumar et al. 2021), armochaetoglobins A–J, chaetochromins, chaetoglobosin A, B, C, D, J, N, Q, R, T, W, chaetosin, cochliodinol, chaetochalasin A, mollicellin, oosporein, polyhydroxylated steroids, sterigmatosystin and xanthenone (Aggarwal 2015; Kumar et al. 2021; Park et al. 2005; Mao et al. 2010), cheatospirolactone and orsellides (Xu et al. 2018), chaetomugilin A, D, E, F, chaetoviridin A, B, E, chaetoglobin B, chaetoquadrin A, chaetocochin B and F, chaetoglobosin Q and N, and globoxanthone A (Kumar et al. 2021; Darshan et al. 2021b). Chemically, these metabolites are classified as anthraquinones, chaetoglobosins, chromones, depsidones, steroids, terpenoids, and xanthones. Some of these compounds presented significant biological activities including; antifungal, anti-inflammatory, anti-tumour, cytotoxic, and enzyme inhibition (Fatima et al. 2016). *C. cupreum* delivered bioactive substances that significantly inhibited *F. oxysporum* f. sp. *lycopersici* causing tomato wilt (Soytong et al. 2001), and it was later discovered that this isolate of *C. cupreum* produced rotiorin and rotiorinols A to C, which were also antifungal against *Candida albicans* (Huu Phong et al. 2016). Chaetoviridins A and B isolated from *C. globosum*, showed potent disease control efficacy against *M. grisea* (rice blast) and *Puccinia recondita* (wheat leaf rust) (Park et al. 2005). Moreover, application of the conidial suspension, nano-rotiorinol, nano-trichotoxin, and bioformulation of the *C. cupreum* reduced coffee anthracnose incidence (Vilavong and Soytong 2017). Zhao et al. (Zhao et al. 2017) evaluated chaetoglobosin A and D from *C. globosum* strain CDW7 as an endophyte of *Ginkgo biloba* for its biocontrol potential against rape rot caused by *S. sclerotiorum*. The compounds cladosporin, chaetoatrosin A, and chaetoviridin A derived from endophytic *C. globosum* were shown to be pathogenic to *F. oxysporum* in dual culture assay (Fierro-Cruz et al. 2017). *Chaetomium* is a prolific producer of species-specific metabolites in the environment, and thus mapping these enzymes in terms of their multimode actions with commercial implications would be extremely beneficial for designing the future formulations (Aggarwal 2015; Arunkumar et al. 2021; Song et al. 2020). Recently, Darshan et al. (Darshan et al. 2021b) identified key regulator genes responsible for biosynthesis of secondary metabolites, polyketide synthase, antibiotics, hydrolytic enzymes, and putative fungistatic metabolites against *B. sorokiniana*.

17.5.3 Plant Growth Promotion

Specific strains of *Chaetomium* spp. have been shown to promote plant growth and induce better crop yields both in greenhouse and field conditions, following their application as biological products. In greenhouse and field conditions, crops such as rice, pepper, citrus, tomato, corn, durian, birds of paradise, and carnations treated with Ketomium® have better plant growth and yields than untreated plants. Ketomium, a broad spectrum mycofungicide, has been registered as a biofertilizer for degrading organic matter and for inducing plant immunity and stimulating plant growth (Soytong et al. 2001). The role of endophyte *C. globosum* in the growth of *Capsicum annuum* by the production of gibberellins and indole acetic acid was studied. In that, the culture filtrate of *C. globosum*, when applied to the pepper plants, revealed that there was increased shoot growth, chlorophyll content, plant biomass, and leaf area as compared to fungal-free medium and water applied to pepper plants (Khan et al. 2012). They reported that the culture filtrate analysis of *C. globosum* showed the presence of gibberlic acid and IAA compared to the culture filtrate of the pathogen *F. fujikuroi*. Symbiosis of such endophytic fungi offers advantages to host plants in transport and assimilation of biochemical compounds necessary for plant growth and to counteract biotic and abiotic stresses. The sequential application of *C. globosum* Cg-6 as a liquid formulation (Shanthiyaa et al. 2013) or isotonic formulation (Arunkumar et al. 2021) through tuber treatment, soil application, and foliar spray enhanced the growth, number of leaves, and branches of potato crop compared to untreated control.

17.5.4 Induced Systemic Resistance

Induced systemic resistance (ISR) occurs in host plants as a localized or systematic response to pathogen infection. ISR induced by antagonistic agents uses the same set of genes and gene products as systemic acquired resistance (SAR). *C. globosum* produces Chaetoglobosin C, which can induce a localised and systemic oxidative burst in carrots, potatoes, sweet potato, tomato, and tobacco and this substance acts to trigger plant immunity for disease resistance (Soytong et al. 2001). Specifically, several researchers reported the induction of defence enzymes by *C. globosum*, especially in wheat against *B. sorokiniana* and *Puccinia triticina* (Aggarwal 2015). We investigated the induction of different defence genes, viz., peroxidase (PO), polyphenol oxidase (PPO), β-1, 3 glucanase, catalase, and superoxide dismutuase, PR-proteins and NPR1 in potato against *P. infestans*, chilli against *C. gloeosporioides*, and turmeric against *P. aphanidermatum* using the biocontrol agent *C. globosum* (Thiruvengadam et al. 2020; Arunkumar et al. 2021; Shanthiyaa et al. 2014). The reduced disease incidence by *C. globosum* may be a result of cell wall strengthening through deposition of lignin monomer and induction of defence-related enzymes. It is determined that *C. globosum* may be used as an excellent resistance inducer against fungal pathogens. Recently, a study revealed the involvement of *C. globosum* Cg-2 in the active participation of jasmonic acid (JA) and salicylic acid (SA) signalling transduction pathways, which further indicated the involvement of induced systemic resistance (ISR) and systemic acquired resistance (SAR) in the systemic resistance against spot blotch of wheat (Darshan et al. 2020), and early blight disease of tomato (Singh et al. 2021).

17.5.5 Extracellular Enzymes Production

C. globosum isolates are potent sources of secreting fungal wall lytic enzymes such as amylase, proteases, cellulase, chitinase, laccase, pectinase, and lipase for their antagonistic actions (Abdel-Azeem et al. 2016; Darwish and Abdel-Azeem 2020). Microbial antagonists like *Chaetomium* spp. can degrade cellulose-rich organic residues in the environment and specific isolates can inhibit several plant pathogens, viz., *Alternaria brassicola, A. raphani, B. sorokiniana, Fusarium* spp., and *P. ultimum* (Aggarwal et al. 2004). In our study, *C. globosum* Cg-6 isolate produced cellulolytic enzymes (exoglucanase, endoglucanase, cellobiase and β 1,3 glucanase), pectinolytic enzymes (pectin methyl esterase), and laccase in vitro (Shanthiyaa et al. 2014), with greater biocontrol potency against *P. infestans* as their cell walls are composed mainly of glucan and cellulose constituents. *Chaetomium* species may deteriorate cellulosic and other organic compounds and might even be used as antagonistic potential against a range of microbial pathogens. Several studies on *C. cellulolyticum*, *C. erraticum*, *C. fusisporale*, *C. globosum*, and *C. thermophilum* species have previously been conducted to determine their cellulolytic potential, distribution, complexities, and cellulose constituent properties (Darwish and Abdel-Azeem 2020). *C. globosum* produced higher proportion of laccase among the ascomycete fungi during the solid state formation of coffee pulp (Parani and Eyini 2012). Endo-1,3-glucanases (EC 3.2.1.39) and exo-1,3-glucanases (EC 3.2.1.58) from *Chaetomium* sp. have properties for degrading cell walls, inhibiting mycelium growth, and spore germination of plant pathogenic fungi (Jiang et al. 2017). Exo-1,3-glucanases produced by *C. globosum* can degrade the cell walls of plant pathogens, including *R. solani*, *Gibberella zeae*, *Fusarium* spp., *C. gloeosporioides*, *P. infestans*, and *Phoma* spp. (Shanthiyaa et al. 2013; Shanthiyaa et al. 2014). *C. globosum* produced various pectinolytic enzymes such as polygalacturonate trans-eliminase (PGTE), pectin trans-eliminase (PTE), polygalacturonase (PG), pectin methyl esterase (PME), protopectinase (PP), and cellulolytic (C_1 and C_x) enzymes (Upadhyay et al. 2020). Production, partial purification, and characterization of extracelluar xylanase from *C. globosum* were studied. Seeds of some legumes were amended with xylanase produced by *C. globosum*, which showed positive inhibition of seed borne mycoflora such as *Fusarium solani*, *R. solani*, and *M. phaseolina* (Atalla and El Gamal 2020).

17.6 Commercial Formulations

The future of agriculture is heavily reliant on bio-based farm inputs. This is the first option in agro-input industries. A wide range of biological control agents have been developed as commercial mycofungicide products. Ecologically safer plant disease management products require biofungicide formulations with high efficacy and long stability. Future biofungicides should have a better cost-efficiency balance than current biopesticides. Also, new formulations like nano-emulsion, nano-suspension, nano-capsule suspension, etc., will arise from newly created nanobiotechnology. Choosing mycotoxin-free, ecologically safe, and nontoxic *Chaetomium* spp. strains is critical in the creation of biofungicides. *C. globosum* formulation composed of colloidal cellulose and a vegetable oil-based spreader-sticker (Soy-Dex) has exhibited long-term foliar tenacity and survival and growth of *C. globosum (*Davis et al. 1992*)*. In recent years, several mycofungicides have been patented and registered for plant disease control. Mycofungicide formulations include wettable powders and granules based on *C. globosum*, *C. cupreum*, *Trichoderma harzianum*, and others (Kaewchai et al. 2009). The application of liquid formulation (Shanthiyaa et al. 2013) or isotonic formulation (Arunkumar et al. 2021) of *C. globosum* as tuber treatment and soil application resulted in a reduction of late blight disease incidence (72%) and black scurf disease (71%), respectively, on potato. These findings were in accordance with previous results (Park et al. 2005; Soytong and Ratanacherdchai 2005). Ketomium, a broad spectrum mycofungicide, which is a formulated form of powder and pellets of *C. globosum*, was most efficient in suppressing raspberry spur blight caused by *Didymella applanata* and could also reduce potato disease caused by *R. solani* (Shternshis et al. 2005). Earlier studies demonstrated that this species has the ability to suppress several soil-borne pathogens including, *Sclerotium cepivorum*, *P. ultimum*, and *Diapothe phaseolorum* f. sp. *meridionalis* (Pietro 1992; McLean and Stewart 2000; Dhingra et al. 2003). The *Chaetomium* mycofungicide is a biological fungicide developed from *C. cupreum,* CC01-CC10 and *C. globosum,* Cg1-Cg12 that constitute 1,500,000 CFU/g. It is supposed to be applied at 10 g/20 L of water sprayed onto the soil to control pathogens, particularly *P. infestans*, following soil adjustment with biofertilizer. This experiment demonstrated that after spraying the plants on a regular basis, *Chaetomium* mycofungicide can influence potato dry rot and minimize late blight symptoms on the leaves and stems (Soytong and Ratanacherdchai 2005). In our study, we found the sterile distilled water was amended with sources, viz., trehalose, poly-vinyl pyrrolidone and glycerol, the results of which revealed that glycerol supported a greater number of *C. globosum* (TNAU-Cg6) spores throughout the observation period up to 6 months of storage, followed by trehalose and poly-vinyl pyrrolidone. *C. globosum* significantly inhibited *F. oxysporum* f. sp. *lycopersici* conidial production. Similarly, *C. lucknowense* demonstrated antimicrobial activity against the same pathogen with an ED50 of 188 g/mL. Separately, the authors prepared oil formulations from *C. globosum* and *C. lucknowense* conidia. The inoculum was adjusted to 2.5 10^6 conidia/mL and then introduced to sterilized palm oil individually. Additionally, the formulations significantly increased tomato plant yield when compared to the artificial fungicide and the untreated control (Charoenporn et al. 2010). The soil application of the *C. globosum* talc formulation had a maximum population count of 9.0×10^4 (cfu/10 g of soil) compared to the untreated control (Shanthiyaa et al. 2013). The nano-particles from *C. globosum* and *C. cupreum* were able to control the growth of *C. lunata* causing leaf spot on rice (Tann and Soytong 2016). It was demonstrated that the isotonic aqueous formulation of *C. globosum,* TNAU-Cg6 treated plants showed resistance against *P. infestans (*Arunkumar et al. 2021*).*

17.7 Mass Production Technology

Recently, *C. brasiliense* mass multiplication on a wide range of substrates was revealed. In that, the substrate, 100 g FYM, was ground and sieved through a 1 mm mesh, while 50 g each of cotton stalk, sorghum stalk, pigeonpea stalk, sugarcane stalk, wheat straw, and paddy straw was presoaked for 2 h in sterile water. For mass multiplication, each substrate was placed in autoclavable high density polythene bags with a 0.2 mm thickness and a 20 × 25 cm size with moisture adjusted to 60%, then sealed tightly, and sterilized twice for 30 min at 1.05 kg/cm^2 pressure. Two agar discs (9 mm) were inoculated into each bag from the periphery of a previously grown culture of *C. brasiliense*. The bags were incubated at room temperature for 30 days. Observations were made at various intervals of time. The number of colony forming units was counted using the serial dilution plate technique (Vaja and John 2020).

Biopesticides are being desiccated and preserved in a dry environment, or they are being processed as oil suspensions in preparation for commercial use. A lack of survivability, severe contamination, and poor field performance are the three most significant limitations of carrier-based bioproducts. In relation, the use of liquid inoculants provides significant advantages, such as a high cell count, zero contamination, longer shelf life, greater protection against environmental stress, and ease of handling, and this technology is considered a breakthrough in the field of biopesticide technology. In our study, *C. globosum* Cg-6 was grown in oatmeal broth for 3 weeks and centrifuged at 10,000 rpm for 15 min to pellet the ascospores. Using a thermocouple psychrometer, the water potential of the spores was determined to be −2.52 MPa. One litre of sterile distilled water, 0.15 M trehalose (51.86 g), 2.4% polyvinylpyrrolidone (24 g), 0.19 M glycine (14.27 g), and 0.12 M glycerol (11.09 ml) were added, and flasks with no chemical additions served as controls. At 15-day intervals up to 180 days during the storage period, flasks were incubated at room temperature and the formulation was analyzed for spore viability and contamination using the serial dilution method (Arunkumar et al. 2021).

17.8 Production Cost Analysis

Sr.	Particulars of item	Amount Year 1	Amount Year 2
Cost			
1	Total capital cost	1,600,000	–
2	Total registration cost (cost includes multilocational, multi-seasonal field trials, toxicological tests, biosafety tests, registration for product and label crops, etc.)	1,500,000	–
3	Total operational cost	500,000	550,000
Total cost		3,600,000	550,000
Benefit			
4	Sale of bioproducts (10 tonnes/annum) – talc @ 300/kg (first year); 350/ kg (second year).	2,500,000	2,500,000
5	Sale of bioproducts (10,000 L/annum) – liquid @ 500/ lit (first year); 550/ lit (second year).	3,300,000	3,300,000
6	Sale of efficacy data, toxicological data and strain – 10 lakhs/unit (5 units/annum approx.)	500,000	500,000
Total benefit		6,300,000	6,300,000
Net benefit		2,700,000	5,750,000

Net benefit (first year): Rs. 2,700,000.
Net benefit (second year): Rs. 5,750,000.
Benefit-cost ratio = Present value of expected benefits/present value of expected costs.

B:C ratio (first year) = 1.75.
B:C ratio (second year) = 11.45.

The proposed cost of production demonstrates that the better the benefit-cost ratio, the greater the profit on *Chaetomium* cultivation. From the second year on, the financial benefits will be greatly multiplied.

17.9 Market Trend

Over the last four decades, research on biofungicides has exploded, and the literature on biofungicides and their opportunities as microbial control agents is enormous. Throughout this time period, numerous businesses have engaged in biopesticide-related activities, and various products have been produced, registered, and introduced to the market. The use of biopesticides is increasing at a rapid rate. However, their total use accounts for only a fraction of the total global use of agrochemicals. Chemical pesticides currently dominate the market and have a substantial impact on product manufacturing due to their broad range of potential applications and high inhibitory efficacy against pests and diseases. Export requirements have expanded to include agro-inputs as importer countries' pesticide residue standards on produce have become increasingly stringent. As a result, microbial pesticides have emerged as a critical tool for meeting the requirement for residue minimization. According to Business Communications Company (BCC), Inc. research, the global microbial biopesticide and chemical pesticide market was valued at USD 61.2 billion in 2017 and is predicted to increase to roughly USD 79.3 billion by 2022. Globally, microbial biopesticide use has been rising at a rate of above 10% annually. However, these biopesticides will significantly contribute to their global market consumption requirements, which are expected to expand further in the future, as they will be substituted for chemical pesticides, thus reducing the over-reliance on chemical pesticides. In the European Union, biopesticides are evaluated using the same laws as synthetic active chemicals (Organization for Economic Co-operation and Development-OECD), which necessitates the addition of various new provisions to existing legislation, and the development of new guidelines facilitates the registration of potential biopesticide products. It is thought that the European Union has fewer registered active ingredients in biopesticides than the USA, India, Brazil, or China. By the early 2050s, it is projected that, biopesticide use would be comparable to synthetics, although significant uncertainties in acceptance rates, particularly in countries such as Africa and Southeast Asia, account for the majority of the uncertainty in such forecasts. Microbial pesticides, which include bacteria, fungi, viruses, protozoa, and nematodes, are the most frequent source of biopesticides and account for the majority of the world's production and use of biocontrol agents. Considering its potential, large agro-industrial companies are entering the biopesticide business through technology and product in-licensing, strategic partnerships, and procurement (Marrone 2014). Major players in the microbial biopesticide market at a world level include: AG Biotech Australia Pty Ltd., AgraQuest Inc., AgroGreen, BASF SE, Bayer Crop Protection, BioWorks Inc., BionTech Inc., Certis USA LLC, Chr, Hansen, Devgan, EMD, FMC, GmBh, Greeneem, Isagro SpA, Kumiai Chemical Industry Co. Ltd., Marrone Bio Innovations, Novozymes AS, Nufarm Ltd., Koppert B.V., Pasturia, Prophyta Biologischer Pflanzenschutz, San Jacinto Environmental and the Dow Chemical Company, Valent BioSciences Corporation, etc. According to numerous studies, the global market for biopesticides will be worth USD 5084.9 million in 2020, representing a lower compounded annual growth rate (CAGR). According to predictions, the global synthetic pesticide market will be worth USD 11,438.1 million in 2026, a CAGR of 14.5–17% during the forecast period (2021–2026). From manufacture to consumption, COVID-19 affects the entire process of the biopesticide industry. COVID-19 restricted labour travel, reduced farmer demand, and closed production facilities. Repeated lockdowns hampered production and trade, causing serious labour shortages. Demand from a few large markets has decreased, putting pressure on the profitability of biopesticide manufacturers and distributors (Marrone 2014).

Farmers have been successfully introduced to the integrated biological control of plant pathogens. Disease control technology can be demonstrated independently or in conjunction with other control measures. The products have been scientifically proven to have not only protective properties, but also curative and growth promoting properties. The first biological product of *Chaetomium* is a novel broad-spectrum fungicide from *Chaetomium* (Thailand Patent No. 6266, International Code: AO 1 N 25/12, and registered as Ketomium® biofungicide for plant disease control), which was developed and improved from 22 strains of *C. cupreum* CC01-CC10 and *C. globosum* CG01-CG12 (Soytong and Ratanacherdchai 2005). The formulation has been successfully used in infested field soils in combination with cultural control to provide long-term protection against several fungal and oomycete diseases. Additional research is being conducted to develop bioactive metabolites from active *Chaetomium* spp. for disease control and immunity in plants. Recent research widely focuses on diversified formulations with extended shelf life and viability. In this context, several powder- and liquid-based formulations of antagonistic *Chaetomium* spp. and their consortia were available on the market. The antagonistic activity of *Chaetomium* spp. in the formulated product was retained at high levels even after storage for 2 years (Tomilova and Shternshis 2006). *Chaetomium* is a unique board-spectrum biofungicide that is registered in Cambodia, China, Laos, India, Thailand, Vietnam, and BioAgriCert, International Federation of Organic Agriculture Movements (Table 17.2). These countries contributed and tested *Chaetomium* biofungicide. Biodegradable nano-elicitors constructed from active metabolites from *Chaetomium* species are the new unique science for plant immunity that has been contributed in Cambodia, China, Finland, Indonesia, India, Laos, Myanmar, Thailand, and Vietnam. Our experience in research investigations in this area of specialization would contribute to sustainable development goals in agriculture. Commercialization is the ultimate and most arduous stage of microbial product development. The most crucial considerations are the cost of development and the time required to bring a product to market (Marrone 2014). Thus, it is vital to

Table 17.2 *Chaetomium* spp.-based products in the market

S. no.	Product	Active ingredient	Company	Formulation
1.	Con-Blight	*C. globosum*	T-Stanes & Company Ltd., Coimbatore, India	Talc and liquid
2.	Biokuprum	*C. cupreum*	Agri Life, Medak, India	Soluble powder, liquid, lyophilized powder
3.	*Chaetomium globosum*	*C. globosum*	Krishi Rashayan exports pvt. Ltd., Kolkotta, India	Wettable powder
4.	Ketomium	*C. cupreum* and *C. globosum*	Gungxi Giulin Green Harvest Fertilizer Factory, China and Vien Di Truyen Nong Nghiep, Vietnam	Pellet and powder
5.	Biocin	*C. cupreum* and *C. globosum*	SP Inter, Vientiane, Lao PDR, Vietnam	Liquid
6.	Novacide	*C. cupreum* and *C. globosum*	Nova Science Co. Ltd., Thailand	Powder
7.	Ketocin	*C. cupreum*	Neoworld Ltd., Thailand	Powder
8.	Alglow	*C. globosum*	Amruth Organic Fertilizers, India	Powder
9.	Super *Chaetomium*	*C. globosum*	Champ Agro Advance, Malaysia	Powder

investigate all of these critical elements in the successful commercialization of microbial pest control products during the product's development phase, and these critical factors have been described.

17.10 Opportunities

The existing system, which relies almost entirely on synthetic chemicals, is undergoing change. Biopesticides employ novel and sophisticated modes of action and agronomic methods. The majority of agricultural biopesticides are well-suited for use in conjunction with conventional pesticides, which contributes to the rapid growth of biopesticide use. Today, there are an increasing number of biological active components and products that can compete with and complement conventional chemical pesticides. Demand is also being fuelled by our increasing understanding of the mode of action of biopesticides and the refinement of application rates and methodologies. The reasons for consideration during biopesticide development have been reported numerous times, but the predominant issues are as follows: variable quality and efficacy of biocontrol products, their costs relative to effectiveness, competitive pressure with pesticides, licensing, exaggeration of the market growth, undervaluation of the presumption that market implementation will be easy, miscalculation of the minimum total time and budget to market, and frequently suboptimal collaboration.

Technically, the current trend in formulation technology of chemical pesticides and biopesticides is away from wettable powders and suspension concentrates and towards water-dispersible granules and isotonic liquids. Controlled release formulations are being created for maximum effectiveness, and nanotechnology is expected to enable the development of novel formulations such as nano-emulsions, nano-suspensions, and nano-capsules (Soytong et al. 2021; Tann and Soytong 2016). Additionally, it is critical to prioritize the selection of appropriate adjuvants in order to maximize biopesticide activity. Only through collaboration between biologists and chemists can novel formulations and applications be produced and offered to the market, decreasing field performance inconsistency and motivating producers to adopt this new technology. The majority of current bioformulation research efforts are directed towards distribution and application methods, more precisely the effect of plant defense induction kinetics on the timing and location of product application.

17.11 Future Potential

Farming and the community as a whole should profit from the balanced and prudent use of conventional chemical pesticides and biopesticides, but it is critical to prioritize biopesticide research in order to gain more benefits in the future. The *Chaetomium* biofungicide was effective in preventing diseases and encouraging plant development. It has been successfully applied to infected soils for long-term plant disease protection. In particular, *Chaetomium* and its metabolites paired with emerging technologies like nanotechnology are showing promising outcomes. Except for downy mildew, powdery mildew, rust, and smut, which are obligate parasites, it can thus be used to control plant diseases in good agricultural practices, pesticide-free, non-agrochemical, and organic agriculture. According to the research findings, using formulations containing *Chaetomium* results in improved crop yields by leveraging physiological activities or mechanisms, which may help prevent significant losses due to pest and disease incidence. Integrated plant disease control strategies utilizing *Chaetomium*-based biofungicides and biostimulant formulations have been successfully unified with other control strategies to achieve sustainability in plant protection. When formulated products containing *Chaetomium* are used in soils and plants, a variety of beneficial mechanisms are considered. Thus, several methodological steps are required to develop effective microbial formulations that perform consistently in field conditions in the near future under a climate-changing scenario.

Acknowledgement We acknowledge the scientific information from the project DST-SERB No. EMR/2017/003860 for inclusion in this review article.

References

Abdel-Azeem AM (2020) Recent developments on genus *Chaetomium*. Springer

Abdel-Azeem AM, Gherbawy YA, Sabry AM (2016) Enzyme profiles and genotyping of *Chaetomium globosum* isolates from various substrates. Plant Biosyst 150:420–428. https://doi.org/10.1080/11263504.2014.984791

Aggarwal R (2015) *Chaetomium globosum* : a potential biocontrol. Indian Phytopathol 68:8–24

Aggarwal R, Sharma S, Gupta S, Shukla R (2014) Development of conventional and real time Pcr assay for the rapid detection and quantification of a biocontrol agent, *Chaetomium globosum*. J Plant Pathol 96:477–485. https://doi.org/10.4454/JPP.V96I3.023

Aggarwal R, Tewari AK, Srivastava KD, Singh DV (2004) Role of antibiosis in the biological control of spot blotch (*Cochliobolus sativus*) of wheat by *Chaetomium globosum*. Mycopathologia 157:369–377. https://doi.org/10.1023/B:MYCO.0000030446.86370.14

Arunkumar K, Parthasarathy S, Lingan SH, Raguchander RT (2021) Evaluating an isotonic aqueous formulation of *Chaetomium globosum* Kunze for the management of potato black scurf disease caused by *Rhizoctonia solani* Kuhn in India. J Plant Pathol. https://doi.org/10.1007/s42161-021-00971-6

Atalla SMM, El Gamal NG (2020) Production and characterization of xylanase from pomegranate peel by *Chaetomium globosum* and its application on bean under greenhouse condition. Bull Natl Res Cent 44. https://doi.org/10.1186/s42269-020-00361-5

Charoenporn C, Kanokmedhakul S, Lin FC et al (2010) Evaluation of bio-agent formulations to control *Fusarium* wilt of tomato. African J Biotechnol 9:5836–5844. https://doi.org/10.4314/ajb.v9i36

Darshan K, Aggarwal R, Bashyal B, Singh J (2021b) Antifungal metabolite profiling in *Chaetomium globosum* potential strain Cg2 effective against *Bipolaris sorokiniana*. Indian J Agric Sci 91(5):776–782

Darshan K, Aggarwal R, Bashyal BM, Haritha Mohan M (2021a) Deciphering the network of interconnected pathways of *Chaetomium globosum* antagonistic related genes against *Bipolaris sorokiniana* using rna seq approach. J Biol Control 34:258–269. https://doi.org/10.18311/jbc/2020/26736

Darshan K, Aggarwal R, Bashyal BM et al (2020) Transcriptome profiling provides insights into potential antagonistic mechanisms involved in *Chaetomium globosum* against *Bipolaris sorokiniana*. Front Microbiol 11:1–19. https://doi.org/10.3389/fmicb.2020.578115

Darwish AMG, Abdel-Azeem AM (2020) *Chaetomium* enzymes and their applications. In: Abdel-Azeem AM (ed) Recent developments on genus *Chaetomium*. Springer Nature Switzerland AG, pp 241–249

Davis RF, Backman PA, Rodriguez-Kabana R, Kokalis-Burelle N (1992) Biological control of apple fruit diseases by *Chaetomium globosum* formulations containing cellulose. Biol Control 2: 118–123. https://doi.org/10.1016/1049-9644(92)90035-C

Dhingra OD, Mizubuti ESG, Santana FM (2003) *Chaetomium globosum* for reducing primary inoculum of *Diaporthe phaseolorum* f. sp. *meridionalis* in soil-surface soybean stubble in field conditions. Biol Control 26:302–310. https://doi.org/10.1016/S1049-9644(02)00167-6

Dingle J, McGee PA (2003) Some endophytic fungi reduce the density of pustules of *Puccinia recondita* f. sp. *tritici* in wheat. Mycol Res 107:310–316. https://doi.org/10.1017/S0953756203007512

Fatima N, Muhammad SA, Khan I et al (2016) *Chaetomium* endophytes: a repository of pharmacologically active metabolites. Acta Physiol Plant:38

Fierro-Cruz JE, Jiménez P, Coy-Barrera E (2017) Fungal endophytes isolated from *Protium heptaphyllum* and *Trattinnickia rhoifolia* as antagonists of *Fusarium oxysporum*. Rev Argent Microbiol 49:255–263

Gamal E, Elnaser A, El M (2021) Morphological, molecular characterization, plant pathogenicity and biocontrol of *Cladosporium* complex groups associated with faba beans. Sci Rep:1–12. https://doi.org/10.1038/s41598-021-93123-w

Gandalera EE, Divina CC, Dar JD (2013) Inhibitory activity of *Chaetomium globosum* Kunze extract against Philippine strain of *Pyricularia oryzae* Cavara. J Agric Technol 9:333–348

Hu Y, Zhang W, Zhang P et al (2013) Nematicidal activity of chaetoglobosin a poduced by *Chaetomium globosum* NK102 against *Meloidogyne incognita*. J Agric Food Chem 61:41–46. https://doi.org/10.1021/jf304314g

Hung PM, Wattanachai P, Kasem S, Poaim S (2015) Biological control of *Phytophthora palmivora* causing root rot of pomelo using *Chaetomium* spp. Mycobiology 43:63–70. https://doi.org/10.5941/MYCO.2015.43.1.63

Huu Phong N, Pongnak W, Soytong K (2016) Antifungal activities of *Chaetomium* spp. against *Fusarium* wilt of tea. Plant Prot Sci 52:10–17. https://doi.org/10.17221/34/2015-PPS

Jiang C, Song J, Cong H et al (2017) Expression and characterization of a novel antifungal Exo-β-1,3-glucanase from *Chaetomium cupreum*. Appl Biochem Biotechnol 182:261–275. https://doi.org/10.1007/s12010-016-2325-z

Kaewchai S, Soytong K, Hyde KD (2009) Mycofungicides and fungal biofertilizers. Fungal Divers 38:25–50

Khan AL, Shinwari ZK, Kim YH et al (2012) Role of endophyte *Chaetomium globosum* lk4 in growth of *Capsicum annuum* by producion of gibberellins and indole acetic acid. Pak J Bot 44:1601–1607

Kumar R, Kundu A, Dutta A et al (2021) Chemo-profiling of bioactive metabolites from *Chaetomium globosum* for biocontrol of sclerotinia rot and plant growth promotion. Fungal Biol 125:167–176. https://doi.org/10.1016/j.funbio.2020.07.009

Larran S, Simón MR, Moreno MV et al (2016) Endophytes from wheat as biocontrol agents against tan spot disease. Biol Control 92:17–23. https://doi.org/10.1016/j.biocontrol.2015.09.002

Linkies A, Jacob S, Zink P et al (2021) Characterization of cultural traits and fungicidal activity of strains belonging to the fungal genus *Chaetomium*. J Appl Microbiol 131:375–391. https://doi.org/10.1111/jam.14946

Mao BZ, Huang C, Yang GM et al (2010) Separation and determination of the bioactivity of oosporein from *Chaetomium cupreum*. Afr J Biotechnol 9:5955–5961. https://doi.org/10.4314/ajb.v9i36

Marrone PG (2014) The market and potential for biopesticides. In: Coats et al (eds) Biopesticides: state of the art and future opportunities, ACS Sympos. American Chemical Society, pp 245–258

Martínez-Álvarez P, Fernández-González RA, Sanz-Ros AV et al (2016) Two fungal endophytes reduce the severity of pitch canker disease in *Pinus radiata* seedlings. Biol Control 94:1–10. https://doi.org/10.1016/j.biocontrol.2015.11.011

McLean KL, Stewart A (2000) Application strategies for control of onion white rot by fungal antagonists. N Z J Crop Hortic Sci 28:115–122. https://doi.org/10.1080/01140671.2000.9514131

Moya P, Pedemonte D, Amengual S et al (2016) Antagonism and modes of action of *Chaetomium globosum* species group, potential biocontrol agent of barley foliar diseases. Bol la Soc Argentina Bot 51:569–578. https://doi.org/10.31055/1851.2372.v51.n4.16334

Nitao JK, Meyer SLF, Chitwood DJ (1999) In-vitro assays of *Meloidogyne incognita* and *Heterodera glycines* for detection of nematode-antagonistic fungal compounds. J Nematol 31:172–183

Parani K, Eyini M (2012) Biodegradation of coffee pulp waste by different fungal associations. Biosci Discov J 3:222–228

Park JH, Gyung JC, Kyoung SJ et al (2005) Antifungal activity against plant pathogenic fungi of chaetoviridins isolated from *Chaetomium globosum*. FEMS Microbiol Lett 252:309–313. https://doi.org/10.1016/j.femsle.2005.09.013

Parthasarathy S, Rajalakshmi J, Narayanan P et al (2017) Bio-control potential of microbial antagonists against post-harvest diseases of fruit crops: a review. Res Rev J Bot Sci 6:17–23

Pietro AD (1992) Role of antibiotics produced by *Chaetomium globosum* in biocontrol of *Pythium ultimum*, a causal agent of damping-off. Phytopathology 82:131

Pothiraj G, Hussain Z, Singh AK et al (2021) Characterization of *Fusarium* spp. inciting vascular wilt of tomato and its management by a *Chaetomium*-based biocontrol consortium. Front Plant Sci 12:1–15. https://doi.org/10.3389/fpls.2021.748013

Rajendran L, Datchinamoorthy D, Raja K, Gandhi K, Thiruvengadam R (2021) Identification of genes responsible for Chaetoglobosin in *Chaetomium globosum* for managing *Fusarium* wilt in tomato. In: International Conference on Emerging Trends in Plant Protection for Sustainable Vegetable Cultivation. pp 454–456

Rujira T (2017) Bioactive test of metabolites from *Chaetomium cochliodes* against *Phytophthora* sp. Int J Agric Technol 13:1669–1673

Shanthiyaa V, Karthikeyan G, Raguchander T (2014) Production of extracellular proteins, cellulases and antifungal metabolites by *Chaetomium globosum* Kunze ex. Fr Arch Phytopathol Plant Prot 47:517–528. https://doi.org/10.1080/03235408.2013.813124

Shanthiyaa V, Saravanakumar D, Rajendran L et al (2013) Use of *Chaetomium globosum* for biocontrol of potato late blight disease. Crop Prot 52:33–38. https://doi.org/10.1016/j.cropro.2013.05.006

Shternshis M, Tomilova O, Shpatova T, Soytong K (2005) Evaluation of Ketomium-mycofungicide on Siberian isolates of phytopathogenic fungi. J Agric Technol 1:247–254

Singh J, Aggarwal R, Bashyal BM et al (2021) Transcriptome reprogramming of tomato orchestrate the hormone Signaling network of systemic resistance induced by *Chaetomium globosum*. Front Plant Sci 12. https://doi.org/10.3389/fpls.2021.721193

Song J, Soytong K, Kanokmedhakul S (2016) Antifungal activity of *Chaetomium elatum* against *Pyricularia oryzae* causing Rice blast. Int J Agric Technol 12:1437–1447

Song JJ, Soytong K, Kanokmedhakul S et al (2020) Antifungal activity of microbial nanoparticles derived from *Chaetomium* spp. against *Magnaporthe oryzae* causing rice blast. Plant Prot Sci 56:180–190. https://doi.org/10.17221/41/2019-PPS

Soytong K (2010) Evaluation of *Chaetomium*—biological fungicide to control *phytophthora* stem and root rot of durian. Res J 3:115–124

Soytong K (2014) Biological control of stem rots of oil palms caused by *Ganoderma boninense* using *Chaetomium lucknowense* and *Chaetomium cochiliodes*. J Agric Technol 10:87–92

Soytong K (2015) Testing bioformulation of *Chaetomium elatum* ChE01to control *Fusarium* wilt of tomato. J Agric Technol 11:975–996

Soytong K, Kahonokmedhakul S, Song J, Tongon R (2021) *Chaetomium* application in agriculture. In: Technology in Agriculture. pp. 1–18

Soytong K, Kanokmedhakul S, Kukongviriyapa V, Isobe M (2001) Application of *Chaetomium* species spectrum biological review article fungicide. Fungal Divers 7:1–15

Soytong K, Ratanacherdchai K (2005) Application of mycofungicide to control late blight of potato. J Agric Technol:19–32

Sugiyama K, Sano A, Murakami M et al (2008) Three isolations of *Chaetomium globosum* from erythematous epilation of canine skin. Med Mycol 46:505–510. https://doi.org/10.1080/13693780801968555

Tann H, Soytong K (2016) Bioformulations and nano product from *Chaetomium cupreum* CC3003 to control leaf spot of rice var. Sen Pidoa in Cambodia. Int J Plant Biol 7:59–63. https://doi.org/10.4081/pb.2016.6413

Tathan S, Sibounnavong P, Sibounnavong P et al (2012) Biological metabolites from *Chaetomium* spp to inhibit *Drechslera oryzae* causing leaf spot of rice. J Agric Technol 8:1691–1701

Thiruvengadam R, Johnson I, Rajalakshmi J (2020) *Chaetomium* mediated resistance in the management of pythiaceous fungi. In: National Symposium on "Challenges and Management Approaches for Crop Diseases of National Importance - Status and Prospects." 36th Annual Conference of ISMPP, TNAU, Coimbatore, India Feb 12–14, 2015, pp 182–190

Tomilova OG, Shternshis MV (2006) The effect of a preparation from *Chaetomium* fungi on the growth of phytopathogenic fungi. Appl Biochem Microbiol 42:67–71. https://doi.org/10.1134/S0003683806010108

Upadhyay H, Kumar V, Kumar S (2020) The effect of fungal metabolites on soil-borne pathogenic fungi. New Future Dev Microb Biotechnol Bioeng:59–67. https://doi.org/10.1016/b978-0-12-821007-9.00006-1

Vaja S, John P (2020) Mass multiplication of biocontrol agent *Chaetomium brasiliense* using different substrates under South Gujarat condition sugarcane trash. J Pharmacogn Phytochem 9: 2256–2257

Van Thiep N, Kasem S (2015) *Chaetomium* spp. as biocontrol potential to control tea and coffee pathogens in Vietnam. J Agric Technol 11:1381–1392

Vilavong S, Soytong K (2017) Application of a new bio-formulation of *Chaetomium cupreum* for biocontrol of *Colletotrichum gloeosporioides* causing coffee anthracnose on arabica variety in Laos. Agri 39:303–310. https://doi.org/10.17503/agrivita.v39i3.1070

Xu QL, Xiao YS, Shen Y et al (2018) Novel chaetospirolactone and orsellide F from an endophytic fungus *Chaetomium* sp. J Asian Nat Prod Res 20:234–241. https://doi.org/10.1080/10286020.2017.1320548

Ya-fen X, Jin-jie S (2005) Bio-control trials of *Chaetomium spirale* ND35 against apple canker. J For Res 16:121–124. https://doi.org/10.1007/bf02857904

Yang MY, Wang YX, Chang QH et al (2021) Cytochalasans and azaphilones: suitable chemotaxonomic markers for the *Chaetomium* species. Appl Microbiol Biotechnol 105:8139–8155. https://doi.org/10.1007/s00253-021-11630-2

Zhang G, Wang F, Qin J et al (2013) Efficacy assessment of antifungal metabolites from *Chaetomium globosum* no. 05, a new biocontrol agent, against *Setosphaeria turcica*. Biol Control 64:90–98. https://doi.org/10.1016/j.biocontrol.2012.10.005

Zhang Y, Wu W, Cai L (2017) Polyphasic characterisation of *Chaetomium* species from soil and compost revealed high number of undescribed species. Fungal Biol 121:21–43. https://doi.org/10.1016/j.funbio.2016.08.012

Zhang Y, Zhu H, Ye Y, Tang C (2021) Antifungal activity of chaetoviridin a from *Chaetomium globosum* CEF-082 metabolites against *verticillium dahliae* in cotton. Mol Plant-Microbe Interact 34:758–769. https://doi.org/10.1094/MPMI-02-21-0032-R

Zhao SS, Zhang YY, Yan W et al (2017) *Chaetomium globosum* CDW7, a potential biological control strain and its antifungal metabolites. FEMS Microbiol Lett 364:1–6

Chapter 18
Vermicomposting: A Leading Feasible Entrepreneurship

P. Kavitha

Abstract Vermicomposting is the best approach of converting organic wastes into valuable compost by utilizing the earthworms. Earthworms are extraordinary at consuming almost all sorts of organic material. Its gut possesses cellulase enzyme activity, which converts waste into resourceful manure. Vermicompost acts as an efficient manure in the cultivation of crops and good protein-rich feed. Apart from this, it is economically valuable as its demand has increased rapidly, as an alternative to chemical fertilizers. Hence commercialization of vermicomposting is on the verge of tackling the demand and supply gap. Moreover, it has emerged as an enterprise among farmers and small-scale entrepreneurs in villages and rural India. Yet, commercialization of vermicomposting is not achieved due to lack of awareness among farmers regarding its economic and environmental significance. Therefore, there is need to address these issues and scale up initiatives in order to increase the revenue of vermicomposting. This study discusses its significance, mass production, and economical status thereby, by which large-scale units could be attained. In addition, the business development plan, benefit-cost ratio, and their significance gains importance to establish standard vermicomposting technology and its implementation worldwide. These units would not only promote the quality of environment but also enhance the vermicomposting entrepreneurship, especially in countries like India due to their warmer climate.

Keywords Vermicompost · Organic wastes · Earthworms · Compost · Mass production

P. Kavitha (✉)
Department of Microbiology, Srimad Andavan Arts and Science College (Autonomous), Tiruchirappalli, Tamil Nadu, India
e-mail: kavitha@andavancollege.ac.in

N. Amaresan et al. (eds.), *Agricultural Microbiology Based Entrepreneurship*, Microorganisms for Sustainability 39,
https://doi.org/10.1007/978-981-19-5747-5_18

18.1 Introduction

Vermicomposting refers to the method of converting organic wastes into fertilizers using earthworms, which is most effective. On other hand, vermiculture refers to the culturing of worms by which their reproductive habitat is maintained so as to increase the number of worms to attain economic sustainability. Vermicompost are the digested materials that are excreted by earthworms as vermi-casts upon consumption of any organic material such as agricultural wastes, kitchen leftover, animal excreta, forest litter, farm residues, and so on (Glenn 2007). It is also known as "Black gold", whereas vermicomposting technology is referred as "gold from garbage" (Tara 2003). As the name suggest, vermicomposting has gained importance as a foremost source of entrepreneurship among farmers especially in rural India, due to its significance in environmental enhancement and economic growth. It took a stand, as the most important source of organic fertilizer when the chemical fertilizers, pesticides, insecticides, etc. resulted in the environmental deterioration. Simultaneously, its easy production method and ecological and economic significance were an added advantage (Business model 2019). Furthermore, commercial vermicompost production is not achieved to a greater extent and measures need to be taken to bridge the gap between farmers' unawareness and its significance, when implemented as per the required conditions.

18.2 Role of Earthworms in Sustainable Environment

The history of earthworms dates back to 20 million years and is identified as nature's way of recycling organic nutrients. From ancient times many have recognized the value of these worms where ancient people, including Greece and Egypt, appreciated the role of earthworms played in the soil. The Egyptian Pharaoh, Cleopatra said, "Earthworms are sacred". Charles Darwin being fascinated by the earthworms in 1881, wrote a book named "The formation of vegetable mould through the action of worms, with observations on their habits" and in which he stated that "It may be doubted whether there are many other animals in the world which have played such an important part in the history of the world" (Darwin 1892). The great philosopher Aristotle called earthworms "the intestines of the soil" (Eisenhauer and Eisenhauer 2020) due to their effects on sustainable agriculture.

As per Earles and Williams (2005), sustainable agriculture is the one that produces abundant food without depleting the earth's resources or polluting its environment. In the first decade of the twenty-first century, sustainable agriculture was a set of commonly accepted practices and a model farm economy, but it is still in its infancy. The main challenge for sustainable agricultural practices is the use of chemical fertilizers. Chemically produced plant will accumulate in the human body as toxic chemicals, which are very dangerous. The deleterious effect of the chemical fertilizers starts right from the manufacturing stage, and their products and by-products are more toxic and produce gases like NH_4, CO_2, CH_4, etc. Hence its continuous use without taking any remedial measure or judicious approach to reduce their use will deplete all the natural resources one day and will also threaten the life forms of the earth. The adverse effect of these chemicals on human health and environment can be changed only by the alternate fertilizers like vermicomposting.

18.3 Valuable Compost

Ingested organic matter upon entering into the gut of the earthworm, the cellulase enzyme, and its gut microbe acts in combinations on decomposing it; as a result the decomposition is attained at a faster rate. Vermicompost is a granular organic material that acts as manure by enhancing the physiochemical and biological properties of the soil. Also, it is of nutrient-rich content (Table 18.1) with increased amount of carbon, nitrogen phosphorous, etc.; it is greatly important for the growth of the plants, enhancement of plant hormones, soil enrichment, increasing water and nutrient holding ability and pathogen-inhibiting activity. Based on these observations, Government sectors and non-government organizations (NGO) have been involved in setting up vermicomposting units to the entrepreneurs of southern and central India. Apart from this, community-based organizations, self-help groups, trusts etc., also show interest in developing vermicomposting technology due to its positive effects upon the environment and economic wellness. Though large-scale units have become communal in states like Karnataka, Tamil Nadu, Kerala, Maharashtra, Madhya Pradesh, Gujarat and Rajasthan (Business model 2019), there are certain constrains in implementing a large-scale vermicomposting unit. Further discussion involves current challenges and initiatives required to broaden the technological approach in constructing large-scale units. The major components of

Table 18.1 Nutritional content of the good quality vermicompost

S. no.	Parameters	Concentration	Unit
1.	Carbon	9.15 to 17.98	%
2.	Nitrogen	1.5 to 2.10	%
3.	Phosphorus	1.0 to 1.50	%
4.	Potassium	0.60	%
5.	Calcium	0.17	%
6.	Magnesium	0.06	%
7.	Sulphur	128-548	Ppm
8.	Copper	100	Ppm
9.	Iron	1800	Ppm
10.	Zinc	50	Ppm
11.	C:N ratio	11.64	–
12.	pH	6.8	–

Sources: Business model (2019) and ICAR report

constructing a vermicomposting unit commercially involve mass production, quality, and properties of vermicompost; business plan; marketing and cost analysis.

18.4 Mass Production of Vermicompost

Mass production of vermicompost includes four major components such as:

(i) Selection of worms and its optimal parameters.
(ii) Selection of organic wastes.
(iii) Components for installation of commercial units.
(iv) Method of vermicomposting and its processing.

Mass production and their components involved are pictorially represented in Fig. 18.1.

18.4.1 Worm Selection and its Significant Factors

Among 1800 species (approx.) of earthworms worldwide (Edwards and Lofty 1972) and 350 species (approx.) in India possessing various burrowing and food behaviours, *Eisenia foetida, Eudrilus eugeniae* and *Perionyx excavatus* are the efficient

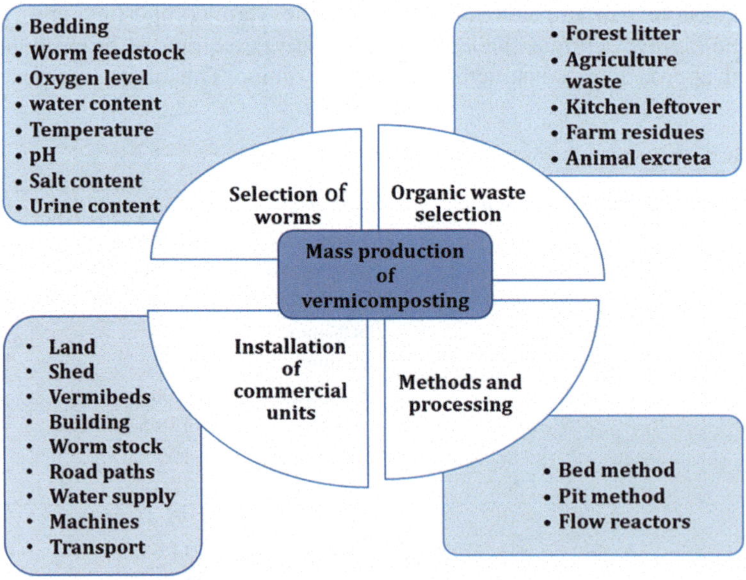

Fig. 18.1 Mass production and their components

species in producing vermicompost. More specifically, *Eisenia foetida* (Red worm) is best suited for vermicomposting due to its rapid multiplication rate, adaptability and worldwide prevalence. Though, a combination of worms with different burrowing habits such as Anecic (deep burrowers), Endogeic (shallow burrowers) and Epigeic (surface dwellers) can be preferred, *E. foetida*, an Epigeic outstands in the efficiency (Card et al. 2004). Moreover, it possesses certain extraordinary characteristics such as active mobility, protective functions and ability to feed on varying organic matter thereby attracting vermicompost producers (Elena et al. 2019). This selection of worms plays a vital role in the quality and time consumption of vermicompost production. Apart from worm selection, its maintenance and providing the required parameter during the process of vermicomposting are also equally important to obtain a quality product. They include:

Bedding

Bedding, the basic habitat of worms in vermicomposting, needs to be selected perfectly to keep the worms active and produce vermicompost in an efficient and eco-friendly manner. It includes three major parameters such as:

(i) Absorbency, i.e. the moisture content should be absorbed by the bedding to prevent worms' death due to dryness.
(ii) Porosity of bedding, i.e. the oxygen flow, should be optimal; it should neither be too loosely packed nor too dense. Therefore, the materials should be chosen wisely upon which the porosity depends.
(iii) Increased Carbon: Nitrogen ratio is considered best for worms' survival, as the increased protein or nitrogen ratio results in production of excess heat due to rapid degradation, which is a threat to the worms. Increased protein content leads to protein poisoning of worms.

Worm Feedstock

Earthworms consume ravenously, almost half of their body weight of organic matter in a day. Especially when it comes to manure, worms feed on dairy and beef manures extremely.

Oxygen Level and Water Content

Moisture content should not be less than 50% as it is extremely dangerous to the worms. On an average analysis, moisture content above 75% is considered ideal for the worms' growth and production of quality vermicompost (Georg 2004). Worms require oxygen for their survival and the O_2 deprivation due to increased oils and fats or other contaminants such as microbial blooms lead to the mortality rate of worms.

Increased water content with respect to decreased aeration is another challenge for O_2 management, as it blocks the oxygen supply completely.

Temperature

Compost worms usually survive at 0 °C and further reducing it to freezing temperatures will affect the feeding ability of worms, by which it dies. Approximately 10–15 °C is recommended for the efficient vermicomposting process and 15–20 °C for vermiculture production. Temperature above 20 °C is appreciable for reproduction. Worms don't prefer more than 35 °C and it may eventually lead to mortality.

pH

Survival rate of worms is observed from pH 5–9 whereas, 7 and above being preferable. Georg (2004) considered 7.5–8 pH to be the most optimum. In some cases, the pH may vary greatly due to its materials, which need to be either increased using calcium carbonate or reduced using acidic bedding.

Salt Content

Manures and bedding should be washed thoroughly to reduce salt content as the worms prefer salt below 0.5%. Habitually, manures contain 8% salt concentration, which is a great threat when used as bedding because worms are unable to escape bedding. Rather when used as feed, the worms eliminate the feed with high salt concentration, which is not a big deal.

Urine Content

Increased urine concentration in the manure will lead to the production of toxic gases in the bedding, which would kill worms. This problem usually arises in the case of cattle cultured in concrete environment, as the urine is stagnated along with manure without being drained through soil (Gaddie and Douglas 1975).

These factors play a major role in the efficiency of vermicomposting production. Furthermore, toxic factors such as deworming medicines, cleaning solutions, detergents, insecticides, tannins from some plant sources, etc., may get mixed up in the materials and cause concerns thereby, requires excess consciousness.

Apart from maintaining these factors the vermicompost units should be prevented from other sources of damage, such as moles and birds that feed on earthworms and centipedes that consumes worms and cocoons. Ants consume the feed of the worms due to its increased sugar content. Among mites, red mites are parasitic (blood sucking insect) to earthworms, which could be avoided at 7 pH and above (Glenn 2007).

18.4.2 Organic Waste Selection

Organic wastes are the basic raw materials in case of vermicomposting, which need to be selected appropriately to avoid the loss of worms or inefficient product. Common raw materials include forest litter, agriculture waste, kitchen leftover, farm residues, animal excreta and so on. But still certain points should be noted while selecting the raw materials as they are the vital components of vermicomposting. Typically, cow dung is most preferred, whereas the residues of leguminous and non-leguminous plant combinations nourish the vermicompost quality further (ICAR report n.d.).

Certain wastes suits well for the bedding purpose, whereas some others work better as feed. For example, Rabbit manure contains increased nitrogen content, unsuitable for bedding, which on other hand contains increased amounts of vitamins and minerals and is considered the better feed. Hog manure is the best feed that produces high quality vermicompost, in contrast sheep or goat manure shows good efficiency as both bedding and feed. Excellent bedding properties are proposed by shredded paper and cardboards (Georg 2004). Other materials such as Straw, bark, wood shavings, etc. show increased C:N ratio, which is highly recommended in case of bedding. In short, carbon sources are better bedding materials whereas nitrogen sources work better as feedstock. On choosing all these materials as per requirement and proper maintenance of the composting process, an efficient vermicompost could be produced (Glenn 2007).

18.4.3 Commercial Installation of Vermicomposting Unit

Basic components for the commercial installation include certain specification at which the process could be carried out much resourcefully. Such specifications include:

(i) Land of about 0.5–0.6 acre would be more convenient and provide easy access of environment for workers and transportation facility.
(ii) Sheds need to be constructed to prevent the area from influence of outer climatic change such as rainfall, etc. It could be made of any material such as wooden, steel, HDPE sheets, etc.
(iii) Vermi-beds of standard sizes are used to produce vermicompost efficiently. Uniform size is preferred throughout the site, to eliminate the loss of efficiency due to size variations.

(iv) It entails construction of a building for the proper management of office work, storage facility of raw material and to produce quality compost.
(v) Worm stock for the initial processing of materials is obligatory. Though worms increase in number by 6 months of inoculation, there is a need to start the procedure with required stock to prevent delay of vermicompost production.
(vi) Road paths are the basic necessity for the accessibility of trolleys, finished products, machineries, etc.
(vii) Water supply is the foremost requirement because worms need proper moisture content and above all, it provides the continuous water facility by which the whole procedure could be carried out much magnificently.
(viii) Machines involve in the procedure right from organic manure collection to the packaging of finished product.
(ix) Transportation is done through trucks at a rate of 3 tonnes per unit, which produces 1000 tonnes of product per year (NABARD project report 2019).

18.4.4 Vermicomposting Technology

Processing of the vermicompost involves standard procedures whereas the prominent methods of vermicomposting are discussed below. The major difference between the methods is that:

(i) Bed method is carried out in the floor with heaps of organic materials and worms, where its size ranges from 6 × 6 × 2 feet. Usually, the length could be varied but the width and height are invariable in order to access the interior portions comfortably. The organic materials are mixed with the worms and allowed to decompose, which produce nutrient-rich vermicompost over a period of time.
(ii) Pit method is the construction of a pit with cement, at a size of 5 × 5 × 3 feet. This method does not gain importance due to certain issues such as water logging, expensive, and poor O_2 supply (ICAR report).
(iii) Flow reactors is a method where the worms are kept undisturbed by adding the organic materials to the top and collecting the vermicompost from bottom. It consists of a hydraulically powered breaker bar that separates the vermicompost from the worms and other undecomposed materials (Glenn 2007).

Processing of Vermicompost

(i) Cool and moist places are best suited for this procedure.
(ii) An initial step of partial decomposition is obtained by mixing cow dung and dried organic material in the ratio of 3:1 and allowing it to decompose for 15–20 days.

(iii) Bedding of 15–20 cm is prepared upon which the earthworms (1500–2000) are allowed to decompose the wastes from topmost layer.
(iv) Required conditions are provided and water is sprinkled at regular intervals to obtain high yields.
(v) After 30 days the bed is mixed or turned up to increase and facilitate aeration.
(vi) Approximately after 45–50 days the vermicompost is harvested, which is approximately 3/4th of the raw materials provided (ICAR report).

Production Cycle of Earthworms

Earthworms produce the valuable compost by their production cycle that impacts the quantity and quality of the vermicompost product readily (Elena et al. 2019). The ideal population for vermicomposting is considered to be 1500 individuals, including both mature and cocoons. Each cocoon is identified to produce 21 eggs whereas the mature worm consists of 10 larvae. Thus, within 2–3 weeks, 10 mother families could be developed in the environment. Moreover, the young worms also become sexually active within 12–20 weeks. This, active multiplication rate possesses the characteristics of increased vermicompost within the stipulated time. Moreover, the worms count can also be raised and simultaneously the production time will decrease. The production cycle includes four phases such as:

(i) Breeding phase, when the young cocoons are hatched out after the mating process. These new cocoons are extremely sensitive to external impact and handling. Therefore, immense care is needed during the sampling of product at this phase.
(ii) Processing phase, when the organic wastes are processed at a higher rate with 1500 worms in 1 m^3 size, within a week.
(iii) Sampling phase, when the degraded product is separated from worms though sieving or a mesh.
(iv) Wintering phase, showing decreased mobility of worms due to their shift in environmental conditions.

18.5 Harvesting

Once the organic matter is completely decomposed, black and granular vermicompost is produced, which could be harvested further. At this stage, watering should be eliminated and the vermicompost is separated in heaps allowing the worms to migrate towards the bottom. After a period of time the vermicompost is sieved, which gives a good quality vermicompost (Dheeraj n.d.).

18.6 Points to be Noted to Produce Quality Product

(i) Handling should not hinder the growth or damage worms.
(ii) Inoculation of some content from the former location is important as the worms require some time to adapt to the current environment.
(iii) *Trichoderma* or *Pseudomonas* addition on vermicompost increases the quality and repels pathogen.
(iv) Nondegradable products such as plastics, etc. should be completely eliminated from the environment.
(v) In some cases, the neem cake when in small amounts increases the worm growth (Srivastava and Zamir Ahmed 2008).

18.7 Business Development Plan

Current scenario of vermicomposting is limited to the individual farmers, who produce vermicompost for their own utilization. Implementation of large-scale units, need to be initiated to commercialize vermicompost production.

18.7.1 Constraints on Setting up Large-Scale Units

Certain constraints are found in a commercial vermicompost venture (Business model 2019). They include:

(i) Lack of awareness and knowledge of vermicomposting technology among farmers. Most of the farmers utilize cow dung as manure at inappropriate methods rather than vermicompost.
(ii) Standard procedures for earthworm management, their growth and multiplication, nurturing of cocoons and tiny worms are essential for the sustainable development of vermicomposting. Moreover, farmers and interested entrepreneurs should be provided with adequate knowledge on technical approach, vermicomposting enterprise and management in order to take it to next level.
(iii) Hands-on experience should be introduced among farmers to attain a well-developed vermicomposting enterprise.
(iv) Scaling up of this enterprise is highly recommended by the introduction of resourceful raw materials, good quality worms, site and accessible environment and more notably man power.
(v) Marketing ground and the retail network of vermicomposting is not much promising, which need to be addressed as an approach of lifting vermicomposting initiatives.

(vi) Financial assistance is essential for the farmers and entrepreneurs to obtain required construction sites, organic matter, worms, etc. Financial assessment from recognized financial institutions remains a great challenge for them.

(vii) Earthworm's unsustainability hinders the vermicomposting enterprise to a greater extent. Even farmers with much experience struggle to nurture and maintain quality worms. Therefore, a proper channel for supply of worms should be initiated.

(viii) Above all these, quality of the vermicompost speaks as an end result. There are no measures in India to monitor the vermicomposting activities. Thus, there is a need for quality testing laboratories and certification systems among entrepreneurs for standardization of quality compost product.

Upon focussing on all these issues, certain plans for the business development of vermicomposting are proposed under Business model (2019).

18.7.2 Initiatives for Surpassing these Constraints

Farmer interest groups (FIG) in the rural environment should be supported for enhancing the vermicomposting enterprise. They can be further merged with farmer producer organisations (FPO) to tackle these issues. FIGs represent the individual farmers and small groups of entrepreneurs, whereas FPOs include cooperatives, production companies and so on. The foremost aim is to initiate the vermicomposting enterprise of FIGs among rural areas by providing technical, financial and logistical ideas and thereby setting up a commercial production unit. On integrating it along with the FPOs, the FIGs could gain importance in various fields as such as grants and subsidies.

The FIGs and FPOs can be unified on the aspect of developing a highly efficient supply chain, a major concern in the current scenario.

FIGs Support through NGO and FPOs

(i) Increasing the mobilisation and circulation of products through grants and subsidy.

(ii) Quality product producing entrepreneurs can be selected to increase the supply of vermicompost.

(iii) Extended trainings and awareness related to vermicomposting and its packaging can be executed.

(iv) Familiarizing the benefits of different schemes such as Paramparagat Krishi Vikas Yojana (PKVY) and Rashtriya Krishi Vikas Yojana (RKVY) to the farmers.

(v) Arrangements can be made to increase the supply of raw materials and other products for vermicomposting.

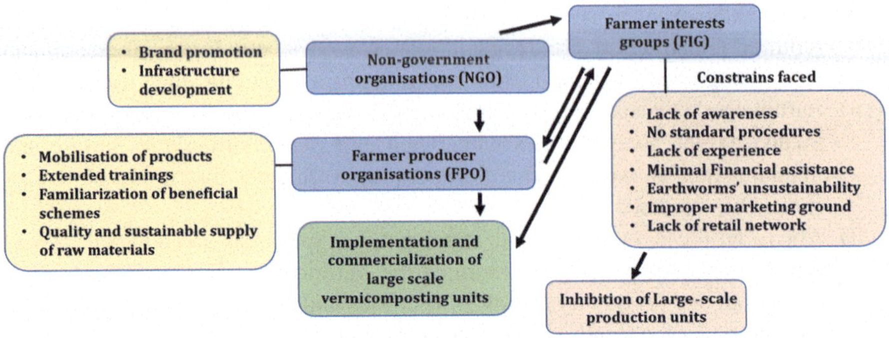

Fig. 18.2 Flow chart representing business plan development

(vi) Imports and exports of products/material can be initiated among FIGs and FPOs.

FPOs Support

(i) Introducing standard procedures for vermicompost production and packing.
(ii) Intense supervision on quality of the product and its commercial distribution can be followed.
(iii) Extension of facilities such as infrastructure, weighing balances, packing bags, storage and transportation could be introduced.
(iv) Brand promotion could be an outstanding idea, when it comes to sales and marketing.

Introducing these measures will definitely be a stepping stone in the field of vermicompost entrepreneurship. Moreover, in today's situation where the emergence of organic farming, natural fertilizers and zero budget manures are intensely focussed, vermicomposting leads in the role. It could be scaled up in locations where crop cultivation, commercial dairy units, cattle population, agriculture and animal husbandry are more popular as they would be an enormous source of raw materials. By following all these norms vermicomposting business would reach its sustainability in 2–3 years. Flow chart regarding the business plan development is depicted in Fig. 18.2.

18.8 Marketing of Vermicompost

Once the business model is set up, the next challenge arises due to its marketing inefficiency. Mostly vermicompost is used only by the farmers for their own purpose but when it comes to commercialization, marketing plays a key role. In some cases, the sale of the vermicompost among other farmers and companies has been reported, but still a commercial rate is undefined. Most often, consumers receive the product directly from the producers, yet to a certain extent sales are also observed through cooperatives (88%) and traders (71%) comparatively, but are too minimal in count. Thus, the analysis of marketing proceduces will definitely pave way in overcoming all these hindrances.

18.8.1 Marketing Analysis

Marketing systems have to be chosen appropriately in order to gain more profit. A producer could attain more revenue when the marketing margin is maintained low and the producer's share is high.

Marketing margin refers to the amount starting from farm gate price to retailer's price. Accordingly, producer's share represents the ratio of money that the producer receives to the money that the consumer pays (Deepa Devkota et al. 2014). It is represented in percentage. Both the values are calculated using the formulae:

$$\text{Marketing margin} = \text{Retailer price (Rs.)} - \text{Farm gate price (Rs.)}$$

$$\text{Producer's share} = [\text{Amount that producer receives}/\text{Amount that consumer pays}] \times 100\%$$

Acharya and Agrawal (1999) calculated the efficiency of marketing using the formula,

$$\text{Marketing efficiency} = \frac{\text{Amount that producer receives (Rs.)}}{\text{Marketing cost (Rs.)} + \text{Marketing margin (Rs.)}}$$

18.8.2 Marketing Systems

Approximately, 85% of the vermicompost are sold directly to the consumers, which include farmers (70%), researchers (5%), government organizations (15%) and NGOs/INGOs (10%). The remaining 15% are sold via cooperatives (10%) and traders (5%). Another important point to be noted is that the marketing was carried out only among their own district and marketing; outside of their district was extremely rare.

Marketing cost on the whole includes package and transportation that varies accordingly (Deepa Devkota et al. (2014).

(i) On transporting the product directly to the consumer, the marketing cost was Rs. 1.5/kg (Rs. 0.5/kg for packaging and Rs. 1/kg for transportation). It is the most predominant method of marketing.
(ii) Marketing via cooperatives showed a marketing rate of Rs. 2/kg, which was shared equally, i.e. Rs.1/kg by both producer and cooperatives. This method is considered more effective as its marketing efficiency and producer's share are higher compared to other methods.
(iii) Trader's marketing cost was identified to be Rs. 2.5/kg, whereas the producer's share was only Rs. 0.5/kg. On the whole, marketing cost via trader is Rs. 3/kg.

Marketing also faces some issues such as increased transportation costs, low demand, seasonal demand, storage problem and low marketable surplus (Shivakumar et al. 2009). Additionally, the lack of marketing facility affects the contribution of organizations towards its development, which is a major concern.

18.9 Benefit and Cost Analysis

Vermicomposting is a much profitable enterprise, as the benefit/cost ratio is always high compared to the capital cost invested. Even though any issues such as increase in earthworm rate, labour and raw material or decrease of vermicompost are faced, the profit would not be lower than 2.7 (benefit/cost ratio), moreover the loss can be easily attained back within 2 years (Shivakumar et al. 2009). Apart from the capital cost of about Rs. 5000–6000/tonne of compost, the cost of vermicompost production is interpretated to be Rs. 2/kg (approx.), which would be highly profitable if sold even at a low cost of Rs. 4–4.5/kg (NABARD project report 2019).

As an approach towards analysing the cost and its profit through vermicomposting, certain tools and statistical analysis are used, which is depicted in detail below (Siddharth et al. 2021).

(i) Fixed cost, are those cost of the components that does not vary, on further processing and product production.
(ii) Variable cost, includes the amount that varies from time to time depending upon the availability, etc. It comprises the rate of earthworm, cow dung, electricity, labour and so on.
(iii) Cultivation cost is the sum of fixed cost and variable cost, which is represented as.

Cultivation cost = Fixed cost +Variable cost.

Apart from the capital cost invested that involves cultivation cost, the profitable amount needs to be calculated to analyse the economic efficiency of vermicomposting entrepreneurship. It is termed as follows:

(i) Gross income is the cost of main product produced. It is calculated as,

Gross income = Product quantity x Price/100 kg.

Table 18.2 Comparison of benefit/cost analysis among three different sites provided by business model (2019), NABARD report and project report by Dheeraj

S. no.	Description	One unit production by FPOs (Rs.)	Production of 200 tonnes per annum (Rs.)	Production of 200 tonnes per annum from individual proprietor (Rs.)
1.	Total capital cost (fixed cost)	90,000 (initial amount)	11,83,300 (initial amount)	3,14,375 (initial amount)
		Average cost of 5 years	**Average cost of 2 years**	**Average cost of 5 years**
2.	Total operational cost	2,36,451	2,64,060	6,33,750
3.	Total cost	2,54,451	8,55,710	6,96,625
4.	Total benefits	3,48,114.8	7,42,500	8,81,498.2
5.	Net benefits	94,783.6	7,61,130	1,84,873.2

(ii) Net income refers to the specific income of the product. It is represented as,

Net Income = Gross Income – Total Cost (all expenses).

(iii) Input/Output Ratio = Gross Income/Total Cost.
(iv) Benefit/Cost Ratio = Net Income/Total Cost.

Deepa Devkota et al. (2014) has stated that each rupee of investment in vermicomposting resulted in Rs. 1.55 revenue using average earthworm population. Furthermore, Heap method of vermicomposting showed increased benefit/cost ratio (3.01) when compared to vat method (2.04). Based on the analysis, it is projected that even when the cost of labour, waste and worms increases by 10%, the benefit/cost ratio will be 3.2. On other hand, if the cost of vermicompost is reduced 10% annually the benefit/cost ratio will not be less than 2.7. This more evidently states that vermicomposting is an outstanding commercial venture (Shivakumar et al. 2009). Table 18.2 shows the profitable venture of vermicomposting by comparing the benefit/cost analysis of individual farmers, production of 200 tonnes per annum and also an individual production of 200 tonnes per annum.

18.10 Financial Aspects, Loans and Schemes Available

Capacity utilization is conserved to be 50% due to its operational costs and other limitations in the first year. Usually, during the first year 2–3 cycles are obtained whereas, in the second year around 5–6 cycles are carried out. Therefore, during the second year, 90% of capacity utilization is estimated. Revenue generated from vermicompost is Rs. 4500 per metric tonne, whereas the worms are sold at Rs. 200 per kg. During the second year, annual net income would drastically increase around Rs. 6,48,000.

According to the NABARD report, the initial payment is addressed to be 25% and bank loan up to 75%. Interest rates could vary among different banks, depending upon the RBI guidelines but more predominantly they range around 13–15%.

FPO- and FIG-mediated loans could be initiated among themselves, as such FPOs provide the raw materials and other required products to the FIGs and collect it back in the form of vermicompost products. Apart from this, government assistance is efficiently needed to cope up with the cost of infrastructure and capital costs.

The Government has also initiated many schemes to facilitate vermicomposting such as:

(i) Mission for Integrated Development of Horticulture (MIDH), which contributes to 50% of the capital cost. An approximate of Rs. 50,000 is provided for the unit size of 30 × 8 × 2.5 inches.
(ii) Paramparagat Krishi Vikas Yojana (PKVY), since 2015, has been contributing to the vermicompost production by providing finance around Rs. 5000 as an approach for purchasing of raw materials and worms, pit constructions, labour cost, etc. Rastriya Krishi Vikas Yojana (RKVY) acts more or less similar to PKVY.
(iii) Small Farmers Agribusiness Consortium (SFAC) Scheme provides loans under guarantee. Yet, certain norms need to be followed to attain the loan, such as: a proper balance sheet should be maintained at least for the recent year and also the organization should be minimum 2 years old. Moreover, it must possess a capital share of Rs. 3 lakhs.
(iv) NABKISAN provides loan completely based on merit and their strategy of business. Nascent entrepreneurs who do not possess knowledge on collaborations gain importance and they are funded up to Rs. 50 lakhs.
(v) National Mission for Sustainable Agriculture (NMSA) provides 50% of total cost,i.e. Rs. 5000/hectare and 10,000/unit to improve the quality of soil.
(vi) NABARD supports to 33% (Rs. 63 lakh) of the total cost in the construction of ceiling, biofertilizer units, size and production capacity.

18.11 Socioeconomic and Environmental Impacts

(i) Vermicomposting is a cost-effective method of processing organic waste and increasing the quality of soil, which leads to hygienic and litter-free environment especially in the rural areas.
(ii) Fields of agriculture and animal husbandry are enormously benefited by vermicomposting.
(iii) Surplus production of earthworm leads to protein-rich animal feed in poultry, fisheries and piggeries.
(iv) Enhanced opportunities among women and farmers to initiate an enterprise.
(v) It is interconnected with the economy of whole country as it's a platform for emerging entrepreneurs and notably it is in collaboration with National Programme for Organic Production (NPOP).

18.12 Conclusion

Managing all these business ideas and implementing it perfectly, along with the biological aspect of earthworm culturing and rearing, will cause a breakthrough in large-scale unit construction. Vermicomposting production is eco-friendly, inexpensive and accessible to the rural environment. Therefore, it is considered to be a sole source of enterprise origins. Its efficiency and profitability are clearly depicted so that there is a strong belief on starting up and enhancing large-scale unit constructions. Following the proper guidelines and analysing the characteristics of worms, vermicomposting will definitely be an advantage towards agriculture and animal husbandry.

References

Acharya SS, Agrawal NL (1999) Agricultural marketing in India. Oxford and IBH Publishing Co. Pvt. Ltd., New Delhi, India, p 402

Business model: Vermicomposting (2019) Capacity Building Toolkit Module 7, Deutsche Gesellschaft für Internationale Zusammenarbeit (GIZ) GmbH. https://birdlucknow.nabard.org/wp-content/uploads/2020/03/17-Vermicompost-BM.pdf

Card AB, Anderson JV and Davis JG (2004) Vermicomposting Horse Manure. Colorado State University Cooperative Extension no. 1.224

Darwin C (1892) The formation of vegetable mould, through the action of worms: with observations on their habits. J. Murray

Deepa Devkota SC, Dhakal D, Dhakal DD, Ojha RB (2014) Economics of production and Marketing of Vermicompost in Chitwan, Nepal. Int J Agric Soil Sci 2(7):112–117

Dheeraj Maan (n.d.), Project report on vermicompost, Agri Clinics & Agribusiness Centers (MANAGE)

Earles R, and Williams P (2005) Sustainable agriculture an introduction. ATTRA

Edwards CA, Lofty JR (1972) Biology of earthworms. Chapman and Hall Ltd., London, p 283

Eisenhauer, N., and Eisenhauer, E. (2020) The "intestines of the soil": the taxonomic and functional diversity of earthworms–a review for young ecologists

Elena S, Shetinina A, Potashova I (2019) Efficiency of vermicompost production and use in agriculture. E3S Web of Conferences 91:06006

Gaddie RE (Sr.) and Douglas DE (1975) Earthworms for ecology and profit, scientific earthworm farming. Bookworm Publishing Company, Cal., 1,180

Georg (2004) Feasibility of developing the organic and transitional farm market for processing municipal and farm organic wastes using large-scale vermicomposting. Good Earth Organic Resources Group, Halifax, Nova Scotia

Glenn Munroe (2007) Manual of On-Farm Vermicomposting and Vermiculture, Organic Agriculture Centre of Canada

ICAR report (n.d.) Vermicompost—production and practices, https://static.vikaspedia.in/media/files_en/agriculture/farm-based-enterprises/vermicompost-production-and-practices.pdf

NABARD project report (2019) Vermicomposting, Model scheme for vermi-composting units under agri-clinics, https://drive.google.com/file/d/1TGDYtBBAIN_T-ROD5yEdZkeGERdaylv5/view

Shivakumar C, Mahajanashetti SB, Murthy C, Basavaraja H, Hawaldar YN (2009) Production and marketing of vermicompost in Dharwad district: an economic analysis. Karnataka J Agric Sci 22:4

Siddharth K, Gauraha AK, Sharma S, Bhaskar SK (2021) An economic analysis of production and marketing of vermicompost in Balod District of Chhattisgarh. Int J Curr Microbiol App Sci 10(07):532–538

Srivastava RC, and Zamir Ahmed SK (2008) Micro business module in agriculture and allied fields for livelihood in A & N Islands, MBM-CARI-XIV, Vermicompost Production, 69–74

Tara Crescent (2003) Vermicomposting. Development Alternatives (DA) Sustainable Livelihoods

Chapter 19
Mass Production and Marketing of Microbial Compost

Swati Patel and Urjita Sheth

Abstract Microbial composting is a fusion of various metabolic activities for the degradation and conversion of complex material into available nutrients with a value added product. Diverse mesophilic and thermophilic microorganisms play a role in this recycling process. Composting process favours the environment with degradation of waste, reduction in greenhouse gas, toxicity and acts as a soil conditioner. Compost also serves as a fertilizer with biocontrol agents for organic farming. The addition of potent inoculum enhances the process of conversion. Commercial and mass production of compost is sustainable only when there are some proceeds for the producer. Marketing skills are indeed needed for viable compost production at the commercial scale.

Keywords Composting · Microbiota · Fertilizer · Biocontrol · Marketing · Value addition

19.1 Introduction

Compost is a preparation of fertilizer from degradable waste materials in a humid aerobic environment with a concentrated group of microorganisms. Composting is a natural self-heating biodegradation process that has the involvement of abundant microorganisms like bacteria, fungi and actinomycetes. It is an environment-friendly process for the transformation of different solid waste to organic fertilizer or soil conditioners. Composting process reduces greenhouse gases by 41% with the reduction in utilization of chemical fertilizer (Galgani et al. 2014). Anaerobic and

S. Patel (✉)
ASPEE SHAKILAM Biotechnology Institute, Navsari Agricultural University, Surat, Gujarat, India
e-mail: swati_20188@nau.in

U. Sheth
Department of Biotechnology, School of Science and Technology, Vanita Vishram Women's University, Surat, Gujarat, India

N. Amaresan et al. (eds.), *Agricultural Microbiology Based Entrepreneurship*, Microorganisms for Sustainability 39,
https://doi.org/10.1007/978-981-19-5747-5_19

aerobic digestion imparts economic benefits of 46,956 INR/day with the decline in eutrophication, acidification, and oxidation of phytochemicals and toxins (Singh and Basak 2018). As the world population rises, there is an addition of 70% of solid waste and it is expected to rise upto 3.40 billion tonns in 2050 (Kaza et al. 2018). There is an urgent need for organic compost to a sustainable environment as the survey indicates the presence of chemical remnants and leachate in soil due to the persistent application of chemical fertilizers. The global compost market is expected to rise to $9.2 billion by the year 2024 with the peak in the human population (Newswire 2019). Composting process involves potential microbiota for the complete degradation process. Diversity of microorganisms is according to the feedstock material such as agricultural waste, which is rich in cellulose; therefore cellulolytic microbes are sought in abundance for composting (Su et al. 2018). The dominant phyla for processing dairy waste and rice husk are *Actinobacteria Bacteroidetes, Firmicutes,* and *Chloroflexi* in the mesophilic phase and *Thermopolyspora* and *Thermomonospora* in the maturation phase (Meng et al. 2019). Composting process is influenced by various parameters like frequency of aeration, temperature, microbial diversity, and uptake of oxygen for degradation and humification (Zhang et al. 2021).

19.2 Stage of Microbial Composting

Composting process has been split into two stages:

1. Active composting.
   (a) Mesophilic stage.
   (b) Thermophilic stage.
2. Maturation or curing.

In the active phase, the highest activity of degradation has been observed by microorganisms. Initially, composting started with the mixing of waste material. It can be municipal solid waste, agricultural waste (wheat straw, rice husk, maize straw, vegetable waste, mushroom residues), river sediments, animal manure or food waste containing degradable, partially degradable, or non-degradable materials (Jha et al. 2003).

The active composting process starts with a short lag phase where microbes adapt to the environment and start to grow. Initially, the temperature is around 15–45 °C, therefore mesophilic microorganisms are found predominantly in abundance and they degrade readily digestible sugar, protein, and fat from the waste by the action of enzymes. It converts rigid material into humus or partially digested products. Mesophilic bacteria and actinomycetes are the most active microorganisms (Table 19.1), while some fungi are involved in the degradation of tough material that is used by bacteria. With the metabolic activity of microorganisms, there is a production of various organic acids such as lactic acid, citric acid, succinic acid,

Table 19.1 Biodiversity of bacteria during mesophilic and thermophilic stage during composting

Mesophilic stage		Thermophilic stage	
Bacterial speceis	Reference	Bacterial species	Reference
Bacillus subtilis, Bacillus flexus, Bacillus polymyxa, B. Pumilus, Bacillus badius, Bacillus cereus, B. Licheniformis	Li et al. (2019)	*Bacillus megaterium, Bacillus stearothermophilus, Bacillus subtilis, Bacillus pumilus, Bacillus benzoevorans, Bacillus coagulans, Bacillus flexus*	Partanen et al. (2010) Chandna et al. (2013a)
Paenibacillus	Ueda and Kurosawa (2015)	*Geobacillus thermodenitrificans, Geobacillus sp. Y4.1MC1, Geobacillus sp. WCH70,*	Ma et al. (2020)
Brevibacillus brevis	Chandna et al. (2013b)	*Brevibacillus brevis*	Hatayama et al. (2014)
Serratia marcescens	Chandna et al. (2013b)	*Amycolicicoccus subflavus*	Silva et al. (2021)
Staphylococcus aureus, staphylococcus xyloseus, Staphylococcus sciuri	Hefnawy et al. (2013)	*Rhodothermus marinus*	Braga et al. (2021)
Clostridium	Biyada et al. (2021)	*Clostridium thermocellum, Clostridium acidurici*	Palaniveloo et al. (2020)
Lysobacter	Thin Mar et al. (2018)	*Gemmatimonas aurantiaca*	Wright and Lima (2021)
Amycolicicoccus subflavus	Park and Park (2018)	*Amycolicicoccus subflavus*	Li et al. (2019)
Klebsiella pneumoniae	Parthasarathi et al. (2007)	*Klebsiella pneumoniae*	Parthasarathi et al. (2007)
Sphingomonas	Kobayashi et al. (2009)	*Solibacillus silvestris*	Zainudin et al. (2014)
Pseudomonas	Grgić et al. (2019)	*Pseudomonas putida, Pseudomonas mendocina*	Hefnawy et al. (2013)
Flavobacterium	Biyada et al. (2021)	*Paenibacillus mucilaginosus*	Antunes et al. (2016)
Nitrosomonas	Biyada et al. (2021)	*Thermaerobacter marianensis*	Antunes et al. (2016)
Cellulomonas cellulans	Ryckeboer et al. (2003)	*Thermobispora bispora*	Varma et al. (2018)
Caryophanon latum	Ivors et al. (2002)	*Thermobacillus composti*	Watanabe et al. (2007)
Stenotrophomonas maltophili	Andrews et al. (1994)	*Terribacillus halophilus*	Chandna et al. (2014)
Psychrobacter immobilis	Andrews et al. (1994)	*Thermobifida fusca*	Antunes et al. (2016)
Methylobacterium extorquens	Andrews et al. (1994)	*Thermosediminibacter oceani*	Bell et al. (2021)

oxalic acid, acetic acid, and formic acid (Wei et al. 2018), therefore there is a reduction in the pH (4.0–5.0) and a alacritous rise in temperature (Kausar et al. 2013). Now the compost has entered in succeeding phase of degradation.

The second phase in active composting is comprehended as the thermophilic stage as there is a rise in temperature in compost. All the microbial diversity in the mesophilic range has been decreased. There is a prompt degradation of composite substances like cellulose, hemicellulose, and lignin in this stage. The spores of bacteria and fungi, which can grow optimally at the relatively high temperature, are active in this phase. They do actively secrete the enzyme-like cellulase, hemicellulase, pectinase for additional degradation (Saini et al. 2015). Here the temperature ranges between 40 °C and 80 °C and only thermophilic microbes are at their optimum activity (Table 19.2). This higher temperature also favours the killing of pathogen, larvae, flies, and weed seeds. At this higher temperature, some of the microbes like the species of *Bacillus* are found to form spores for their survival. This phase lasts from 10 days to 60 days depending upon the type, configuration, and quantity of composting. At the end of this phase, there is a depletion of overall carbon sources and oxygen, therefore temperatures begin to decrease and it enters into the maturation phase.

The final stage of composting is known as maturation phase or curing. The remaining share of indigestible material starts to degrade as the temperature cools down around 25 °C due to a reduction in microbial metabolism (Diaz et al. 2020). This temperature becomes favourable for microbes rooming in the edges, and they start germinating from the spores. The leftover residues are finally degraded in this stage and converted into humus and a simpler form of nitrogen and phosphate. Suppression of disease is observed by the application of composting due to the metabolism of phytotoxic compounds chiefly by germinated spores of fungi. Plentiful bacteria and fungus are also observed in this last phase (Table 19.3). Curing is considered to be completed when after several turning and mixing, the compost retains temperature around 25 °C. The comprehensive time for this stage relies on end-users; it ranges from 1 month to 6 months (Table 19.4 and Fig. 19.1).

Composting can be followed by multifarious methods according to the quality and quantity of raw material and requirement of compost.

19.3 Composting Methods

1. Vermicomposting is a non-thermophilic primary technique used by farmers to alter a huge portion of agricultural waste into biofertilizers (Fig. 19.2). The microbes such as *Azosprillium, Azotobacter, Rhizobium, Bacillus, Pseudomonas* etc., are residing in the intestine of eartrhworms where they secrete the various enzymes like amylase, cellulase, protease, urease, chitinase, and lipase to digest all that agricultural waste and convert it into recapped compost (Nagavallemma et al. 2004). The plant growth-promoting bacteria (PGPR) are engulfed by earthworm and get activated in the gut; therefore vermicompost indirectly

Table 19.2 Biodiversity of fungus during mesophilic and thermophilic stage of composting

Mesophilic stage		Thermophilic stage	
Fungal speceis	Reference	Fungal species	Reference
Penicillium citrinum	Wahyuningsih (2019)	*Talaromyces thermophilus*	Guo et al. (2011)
Rhizopus nigricans	Hefnawy et al. (2013)	*Absidia corymbifera*	Van Heerden et al. (2002)
Fusarium moniliforme, Fusarium oxysporum	Hefnawy et al. (2013)	*Thermomyces sp., Thermomyces lanuginosu*	Langarica-Fuentes et al. (2014b)
Aspergillus niger, Aspergillus flavus, A. fumigatus	Saini et al. (2015) Van Heerden et al. (2002)	*Aspergillus fumigates var. Elpticus, Aspergillus fumigates, Aspergillus candidus*	Langarica-Fuentes et al. (2014b)
Streptomyces griseus, Streptomyces cinnaborinus, Streptomyces roseus, Streptomyces antibioticus	Hefnawy et al. (2013)	*Thermomyces lanuginosus, T. ibananensis*	Van Heerden et al. (2002)
Acremonium	Biyada et al. (2021)	*Pseudallescheria fimeti*	Langarica-Fuentes et al. (2014b)
Apiotrichum	Biyada et al. (2021)	*Rhizomucor miehei, R. pusillus*	Langarica-Fuentes et al. (2014b)
Mortierella	Biyada et al. (2021)	*Mortierella wolfii*	Langarica-Fuentes et al. (2014b)
Candida krusei	Bonito et al. (2010)	*Scytalidium thermophilum*	Langarica-Fuentes et al. (2014b)
Botrytis cinerea	Chamuris et al. (2000)	*Mycocladus corymbifer*	Langarica-Fuentes et al. (2014b)
Phoma sp.	Anastasi et al. (2005)	*Emericella nidulans*	Van Heerden et al. (2002)
Scopulariopsis sphaerospora	Anastasi et al. (2005)	*Paecilomyces variotii*	Van Heerden et al. (2002)

Table 19.3 Biodiversity of microorganisms during maturation or curing phase

Maturation/curing stage			
Bacterial species	Reference	Fungal species	Reference
Bacillus licheniformis, Bacillus composteris, Bacillus southcampusis, Bacillus circulans, Bacillus pumilus	Li et al. (2019)	*Scedosporium*	Anastasi et al. (2004)
Amycolicicoccus subflavus	Palaniveloo et al. (2020)	*Pseudallescheria boydii*	Langarica-Fuentes et al. (2014a)
Mycobacterium thermoresistibile, Mycobacterium xenopi	Li et al. (2019)	*Penicillium* roseopurpureum, *P. verrucosum var. verrucosum*	Anastasi et al. (2004)
Alcaligenes denitrificans	Van Heerden et al. (2002)	*Aspergillus aericomus, A.puniceus, A. caespitosus,*	Van Heerden et al. (2002)
Proteus vulgaris	Greff et al. (2021)	*Fusarium solani*	Van Heerden et al. (2002)
Pesudomonas aeruginosa	Van Heerden et al. (2002)	*Memmoniella echinata*	Van Heerden et al. (2002)
Paracoccus denitrificans	Wan et al. (2020)	*Paecilomyces lilacinus*	Van Heerden et al. (2002)

facilitates the growth of plants by solubilization of nutrients (Sinha et al. 2010). *Pseudomonas, Paenibacillus, Azoarcus, Burkholderia, Spiroplasm, Acaligenes,* and *Acidobacterium*, are the probable degraders, Firmicutes, viz., *Bacillus benzoevorans, B. cereus, B. licheniformis, B. megaterium, B. pumilus, B. subtilis, B. macroides;* Actinobacteria, namely, *Cellulosimicrobium cellulans, Microbacterium* spp., *M. oxydans; Proteobacteria* such as *Pseudomonas* spp., *P. libaniensis*; ungrouped genotypes *Sphingomonas* sp., *Kocuria palustris* and yeasts, namely, *Geotrichum* spp.; and *Williopsis californica* were documented from vermicompost (Vaz-Moreira et al. 2008). The microbial load present in the compost enhances the growth of plants by minimizing the disease infestation due to the control of pests and pathogens (Pathma and Sakthivel 2012).

2. Windrow Composting
   The type and construction of windrows depend on the climatic condition, type, and amount of the raw material. It has long narrow piles either in triangular or circular or another geometrical shape. It is 15–20 ft. wide, 6–10 ft. in height with a concave top to harbour moisture by collecting water (Fig. 19.3). During composting there is a generation of heat and these piles are continuously aerated. Aeration is a must in windrow and it can be done by frequent turning of the entire material, this turning process aids to lower the temperature, evenly mixing of raw material, and allowing the liberation of gases. This is heeded in the initial stages

Table 19.4 Biodiversity of microbes during vermicomposting

Species of earthworm	Bacterial species	Reference
Eisenia foetida	*Bacillus megaterium, B. megaterium, B. pumilus, Proteobacteria, Bacteroidetes, Firmicutes*	Vaz-Moreira et al. (2008) Yasir et al. (2009)
Lumbricus rubellus	*Pseudomonas putida, rhizobium japonicum*	Madsen and Alexander (1982)
L. terrestris	Filamentous *actinomycetes*, Fluorescent *pseudomonads*, *Bradyrhizobium japonicum*	Elmer (2009) Rouelle (1983)
Pheretima sp.	*Pseudomonas oxalaticus*	Khambata and Bhat (1953)
Eudrilus sp.	*Azotobacter, Azospirillum, Nitrobacter, Nitrosomonas*, fluorescent *pseudomonas*	Gopal et al. (2017)
Species of earthworm	**Fungal species**	**Reference**
Eisenia foetida	*Paecilomyces spp, Cephaliophora tropica,* *Trichoderma spp,* *Dactylaria biseptata*	Huang et al. (2013)
Eudrilus sp	*Mucor spp*	Nagavallemma et al. (2004)
Lumbricus rubellus	*Penicillium. piceum, P. brevicompactum*	Nováková and Nováková (2012)
Eudrilus eugeniae	*Aspergillus*	Manivannan et al. (2012)
Eudrilus eugeniae	*Haematonectria haematococca*	Anastasi et al. (2004)
Eisenia foetida	*Chrysosporium*	Gong et al. (2017)
Eudrilus eugeniae	*Myceliophthora*	Anastasi et al. (2004)
Eisenia foetida	*Scopulariopsis*	Martín-Gil et al. (2008)
Eudrilus eugeniae	*Cunninghamella*	Anastasi et al. (2004)
Lumbricus rubellus	*Talaromyces flavus* var. flavus	Anastasi et al. (2005)

of composting The frequency of turnings during composting changes from several weeks to months depending on the rise in the temperature in the pile, the steadiness of the manure, manual labour, availability of equipment, and the season during which the compost has been prepared. This method of composting is embraced in agriculture, cafeteria, and restaurant where large quantities of waste are generated. A large area of land is mandated for construction in this windrow type of composting method. Passively aerated type of windrow is employed for the raw material, which needs additional aeration. Perforated

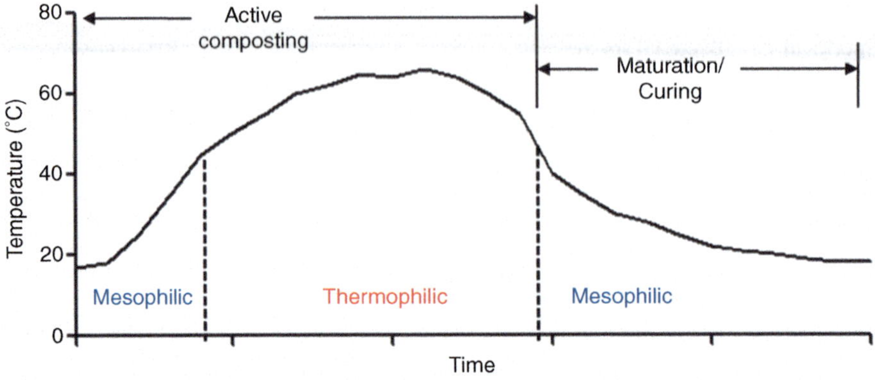

Fig. 19.1 Temperature profile during various stages of microbial composting

Fig. 19.2 (**a**) Organic waste (**b**) Waste converted into vermicompost by *Eisenia fetida* (Blouin et al. 2019)

Fig. 19.3 Windrow composting (FAO)

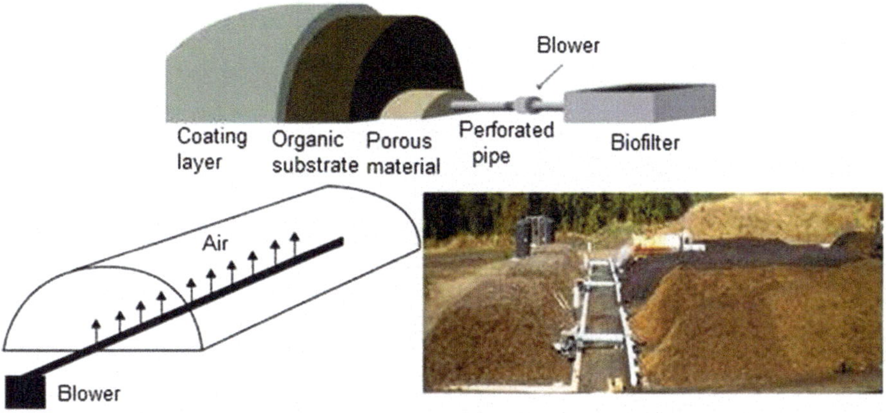

Fig. 19.4 Aerated static pile (Dincă et al. 2019)

pipes are inserted in the windrow for aeration. Here, top and basal layer should be made with a porous material like moss, finished compost, peat, or straw to pass air through those layers evenly.

3. Aerated Static Pile

   In this type of composting method, ambient air is replenished with either positive or negative pressure. Similar to the passively aerated windrow system, it requires a porous top and a basal layer of around 6 inches to continuously disseminate air in compost (Fig. 19.4). A carbon-rich bulking agent like sawdust, wheat straw, wood chips, sawdust, bio-char, shredded paper, or landscape waste is mandatory to be added in this. It requires diminutive space and a short time, around 3–5 months, for the preparation of compost. It is suitable for paper waste, food scrap, and yard trimmings; while it is not suitable for farm waste.

4. In-Vessel Composting

   This kind of composting is used at an industrial scale and requires an enormous amount of organic waste (Fig. 19.5). It can be constructed in different ways in multifarious types of containers.

   (a) Rectangular agitated bed: It has long rectangular beds with automated or manual turners for turning compost at recurring intervals. Sometimes it has blowers for aeration and it demands a prolonged curing period for compost.
   (b) Bin: This is constructed by unused vessel, wooden bin, or any suitable vessel with or without a covering. Air should be passed through bins by force or manually by turning to ensure an aerobic environment during composting.
   (c) Rotating tube: It has an arrangement of several baffles inside the rotary tube and compost is continuously passed through that for aeration. It is appropriate for a small quantity of waste.
   (d) Silo: It has an arrangement for loading of raw material from top and for collection of compost from the base. It also provides aeration in upward

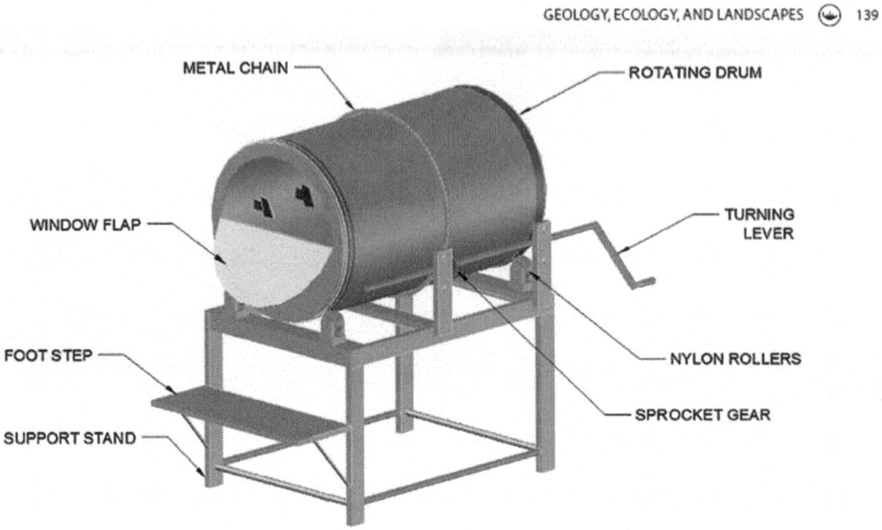

Fig. 19.5 In-Vessel—Rotary Drum Composter (Sharma and Yadav 2018)

directions. It too requires a lengthy curing period and biofilters for the treatment of odour released during composting.

19.4 Factors Affecting the Microbial Composting

19.4.1 C:N Ratio

This is the most critical factor for composting as carbon is the prime source of energy and with aerobic respiration, it is converted into CO_2 and another form is conjoined with nitrogen and further utilized for microbial growth. Similarly, nitrogen is present in the structure of essential macromolecules like DNA, RNA, proteins, amino acids, which are recycled continuously. The C: N ratio 25–50 indicates optimum composting activity (Petric et al. 2015). Higher C: N ratio reduces the rate of biodegradation and creates nutrient deficiency for microbes by piling of the substrate, whereas a lower C:N ratio indicates an increased amount of nitrogen, which will be lost in the process of leaching or volatilization in the form of ammonia, which in turn will not be used by microbiota for degradation (Rastogi et al. 2019).

19.4.2 Oxygen

As composting is an aerobic process, 15–25% percent of oxygen is required at all stages for optimum growth and composting by micobiota (Latifah et al. 2015). The level of oxygen is continiously maintained by turning the pile regularly and supplying sufficient oxygen at a lower level (<0.2 L min/kg/OM) (Cesaro et al. 2019). The optimum level of aeration influences the growth of microbes as well as controls the temperature, heat loss, and moisture content in regime.

19.4.3 pH

Acidity or alkanity of substrate is crucial for growth of bacteria, fungus, and actinomyces, which are predominant throughout all stages of composting. The initial pH of pile varies with the type of waste; afterwards there is a change in pH dependence on the stage. In the active phase of composting, with the increase in microbial activity and breakdown of carbon and notrogen materials, there is production of organic acid intermediates and organic acids, which results in acidic pH. This acidic pH is detrimental to bacterial species and it is again elevated by decomposition of organic acids. The ideal range of pH during composting is 6.7–9.0 as it supports the optimum growth of all microbes at several stages (Bernal et al. 2009). In all stages, there is no need to adjust the pH due to buffering capacity of composting piles and finally the pH stabilizes at the curing stages at 7.5–8.0.

19.4.4 MoistureContent

Moisture content or optimum humidity should be maintained in the degradation of waste. Microbiota needs 50–60% (w/v) moisture content adequate to type of waste raw regime. It has an inverse relationship with temperature; lower moisture level will increase the temperature and it is not favourable. Higher moisture content also creates the situation of logging in piles, cause leaching, and develops anerobic condition. Moisture content is proportional to free air space in piles (Jain et al. 2019). Therefore it is necessary to check the moisture level and retain moisture by wetting of piles at certain intervals. In degradation of some lignocellulytic waste material, which is rich in fiber, it is necessary to soak it in water to make it soften before actual degradation by microbiota.

19.4.5 Temperature

The stages of active degradation are divided according to temperature for the survival of mesophilic and thermophilic microorganisms. At the beginning of composting the temperature is 50 °C, where there is a dominancy of mesophilic bacteria, which starts decomposition. Further, with its metabolic activity, there is a rise in temperature up to 70 °C, and thermophilic bacteria and some strains of fungus continue with degradation. It is incredibly crucial to maintain the temperature at each stage of degradation for the existence of microbes. Ultimately, at the stabilization of compost during the curing stage, least microbial activity is noticed and temperature is lower than at the mesophilic range.

19.4.6 Porosity and Size of Particle

Porosity indicates air space in the regime and it provides an aerobic environment to the microbiota for its optimum activity. If there is higher water content then pores are filled with water and it restricts the growth and decomposition. Similarly, particle size in the compost also affects porosity, which regulates gas and water balance (Głąb et al. 2020). Shredding, chopping, and sieving are the ways to have a suitable particle size for the availability of a better surface area for rapid degradation.

19.5 Production and Marketing of Compost

19.5.1 Quality of Compost

Physical, chemical, and biological characterization is a must to check the quality and efficiency of the prepared compost. The physical characteristics comprise its colour, appearance, texture, size, porosity, odour, and debris or contaminants. The chemical characteristics include the richness of nutrients (available nitrogen, potassium, phosphorous, calcium, and magnesium), optimum moisture level (30–50%), pH (6.0–8.5), and soluble salts. Biologically, a compost should be rich in terms of biodiversity for soil enrichment. The final compost should have been stabilized in form of microbial activity so as to be non-competitive for plant nutrition uptake.

19.5.2 Cost of Compost

The ultimate cost of compost has been considered by the availability and price of raw material, compost method, land availability and operational cost, material handling,

Table 19.5 Total cost of compost production per tonne (Ali and Harper 2004)

Labour cost per tonne of final stabilized compost production	$12.00
Microbiome for composting, 3 kg per tonne of compost	$7.20
Final packing of baggaes—40 kg per tonne of compost	$5.00
Marketing cost	$6.00
Fixed cost	$4.00
Total cost of production	$34.20
Price for selling	$40.00
Net profit	$5.80

packaging, transportation, and marketing. The absolute price of compost should render revenue and profit to producers (Table 19.5). The price should be competitive with other producers and it should be reasonable for consumers. The price should be considered by the following equation:

$$\text{Price} = \text{Total cost of production} + \text{profit}.$$

19.5.3 Marketing of Compost

The popularity of organic farming has increased the competition between chemical fertilizer and compost. The ultimate success of compost production lies in sustainable soil fertilization capacity and the profit earned by farmers. A well-furnished production and marketing strategy has to be designated for several types of waste material degradation for ultimate benefit (Figs. 19.6 and 19.7). Production and marketing are critically affected by (a) the consistency of product and (b) the quality of compost. The target of the market is decided by:

$$\text{The volume of potential market} = \text{number of end} - \text{users in area} \times \text{Average use of customer} \times \text{Market share target}.$$

19.5.4 Market Segmentation

The present market and Potential market of a product can be separated by

1. Geographical segmentation: It is important to analyse the typical necessities of a particular compost in that particular area. The possibility of a customer choosing a product also affects business.

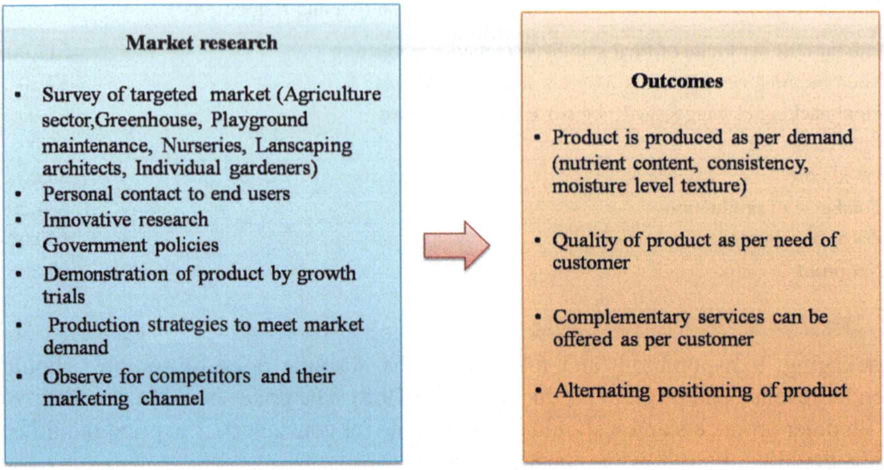

Fig. 19.6 Market research for compost and its outcomes

Fig. 19.7 Piles for succesful marketing of compost

2. Customer segmentation. A firm should know a customer who is most likely to purchase a product and what should be the ultimate price suitable for potent customers. The expectations of potent customers for product mix and their convenience should be considered for the final production of a product.
3. Crop type segmentation: The compost product is applied whether to edible or non-edible crops, accordingly quality has been optimized.
4. Segmentation on the basis of demand and frequency: The customers should be differentiated by their regular demand like agriculture sector has a massive demand but once or twice in a year according to season or crop and individual farmers or persons having private gardens have less requirement but it is required frequently in a year.

19.6 Demand and Production Relationship for Compost

There is demand and utilization of compost based on its market volume and market price for different end users (Fig. 19.8).

19.6.1 End Users of Compost

A large market volume is observed for agriculture, golf course and other playgrounds, topsoil preparation, and public gardens. Biofiltering, nurseries, greenhouses, construction site soil erosion, and contamination have intermediate market volume. The lowest market volume is for individuals doing home or rooftop gardens and retail marketing. Production of compost depends on factors like quality of compost, its performance, availability of the product as per demand, transportation economics to reach to market, standards and specifications, and various challenges of compost applications (Zurbrügg et al. 2005). It is a challenge to keep a balance between the production of compost and its existing demand. It should reach to the end users by certain ways of marketing strategy (Fig. 19.9).

19.6.2 Labelling of Compost

Compost is prepared and marketed by considering its end user. Agriculture is a major sector that has a choice and the highest demand for compost. Farmers have organic waste material from which they can prepare compost and reduce the cost of

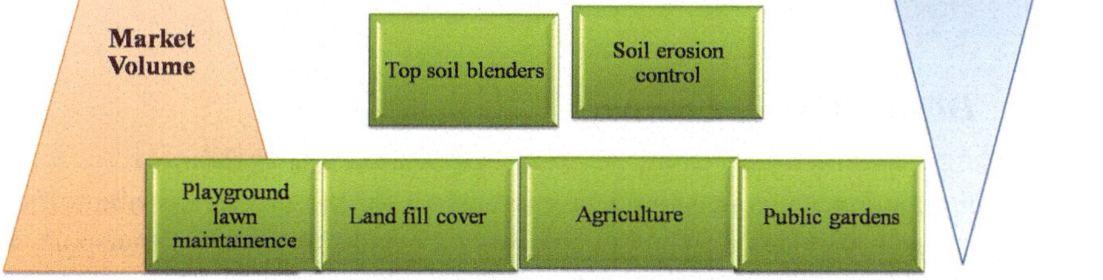

Fig. 19.8 Relationship of market volume and market price for compost for various consumers

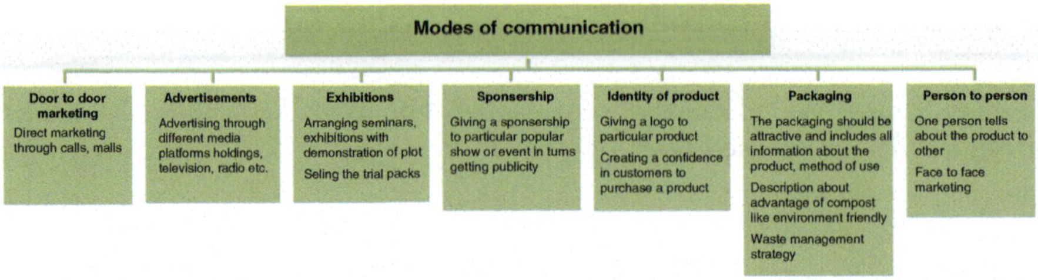

Fig. 19.9 Communication modes for marketing of compost

other chemical fertilizers as well as manage agricultural waste. Globally, one-third of the food being used for feeding people is wasted, which could be utilized for compost preparation. For the selling of a product, customers should know the basic features of the compost product and it should be labelled on the packaging.

1. Maturity index.
2. Content of nutrients.
3. pH of compost,
4. Methods and duration of application for different soil and crops.
5. Contaminating seeds and other inert contaminants.
6. Soluble salt content.
7. Brief method of production.
8. Level of heavy metal.
9. Potential toxic elements.
10. Any value-added product.

19.6.3 Value Addition for Compost

The additives such as charcoal, minerals, biochar, rock phosphate, zeolite, phosphoric acid, corn stalk, and ash can enhance nutrient uptake and metal mobility and also decrease the duration of composting (Barthod et al. 2018). It is more than compost or enrichment of compost with the addition of fungicide, antibacterial agent, anti-nematodes, or viricide, which serves as a biocontrol agent in addition to fertilization and mineralization (Ayilara et al. 2020).

19.6.4 Distribution of Compost

A strong distribution network should be established for the proper distribution of compost in time. It includes a chain of people and the paths or means of transport. A distributor is a person who is stocking a product and when demand is there, they will distribute it to other agents (dealers) or direct customers. Choose the best or most efficient route of distribution in terms of customer satisfaction and profit margin.

19.7 Conclusion

Microbial composting is a cost-effective method as waste management strategy and for sustainable environment. This can be done by various methods according to the type of feedstock material used. Industrially meaningful microorganisms that can produce cellulase, hemicellulase, pectinase, and lipase are enriched in compost and are present in different stages of compost. In a developing or developed country, the marketing of compost is indispensable for gaining benefit. Market research and survey is helping to sell and improve the product with time. In the future, more approaches with potent degraders will help in obtaining optimized compost for customers.

19.8 Future Perspectives

Future and sustainability of microbial composting has numerous challenges to develop a cost-effctive system. Some of the following perspectives should be considered for successful composting from waste.

1. There is strong need of optimization of composting process in terms of waste material with the respective microbial population, which will augment the efficiency of degradation in a short span.
2. The microbial inocula should be maintained on cheap or low cost material so that it can be marketed and made available to all small- and large-scale producers for reduction of production cost and to maintain its consistency.
3. A number of Governmant policies are helpful in marketing of compost to the farmers and other individuals.

Conflict of Interest The authors declare that there is no conflict of interest as regards the publication of this chapter.

References

Ali M, Harper M (2004) Sustainable composting. Water, Engineering and Development Center (WEDC), Loughborough University, UK

Anastasi A, Varese GC, Voyron S, Scannerini S, Marchisio VF (2004) Characterization of fungal biodiversity in compost and vermicompost. Comp Sci Util 12(2):185–191

Anastasi A, Varese GC, Filipello Marchisio V (2005) Isolation and identification of fungal communities in compost and vermicompost. Mycologia 97(1):33–44

Andrews SA, Lee H, Trevors JT (1994) Bacterial species in raw and cured compost from a large-scale urban composter. J Ind Microbiol Biotechnol 13(3):177–182

Antunes LP, Martins LF, Pereira RV, Thomas AM, Barbosa D, Lemos LN et al (2016) Microbial community structure and dynamics in thermophilic composting viewed through metagenomics and metatranscriptomics. Sci Rep 6(1):1–13

Ayilara MS, Olanrewaju OS, Babalola OO, Odeyemi O (2020) Waste management through composting: Challenges and potentials. Sustainability 12(11):4456

Barthod J, Rumpel C, Dignac M-F (2018) Composting with additives to improve organic amendments. A review. Agron Sustain Dev 38(2):1–23

Bell E, Rattray JE, Sloan K, Sherry A, Pilloni G, & Hubert CRJ (2021) Extremely thermophilic endospores germinate and metabolise organic carbon in sediments heated to above 80° C. *bioRxiv*

Bernal MP, Alburquerque JA, Moral R (2009) Composting of animal manures and chemical criteria for compost maturity assessment. A review. Bioresour Technol 100(22):5444–5453

Biyada S, Merzouki M, Dėmčėnko T, Vasiliauskienė D, Ivanec-Goranina R, Urbonavičius J et al (2021) Microbial community dynamics in the mesophilic and thermophilic phases of textile waste composting identified through next-generation sequencing. Sci Rep 11(1):1–11

Blouin M, Barrere J, Meyer N, Lartigue S, Barot S, Mathieu J (eds) (2019) Vermicompost significantly affects plant growth. A meta-analysis, vol 39. Springer, pp 1–15

Bonito G, Isikhuemhen OS, Vilgalys R (2010) Identification of fungi associated with municipal compost using DNA-based techniques. Bioresour Technol 101(3):1021–1027

Braga LPP, Pereira RV, Martins LF, Moura LMS, Sanchez FB, Patané JSL et al (2021) Genome-resolved metagenome and metatranscriptome analyses of thermophilic composting reveal key bacterial players and their metabolic interactions. BMC Genomics 22(1):1–19

Cesaro A, Conte A, Belgiorno V, Siciliano A, Guida M (2019) The evolution of compost stability and maturity during the full-scale treatment of the organic fraction of municipal solid waste. J Environ Manag 232:264–270

Chamuris GP, Koziol-Kotch S, Brouse TM (2000) Screening fungi isolated from woody compost for lignin-degrading potential. Compost Science & Utilization 8(1):6–11

Chandna P, Mallik S, Kuhad RC (2013a) Assessment of bacterial diversity in agricultural by-product compost by sequencing of cultivated isolates and amplified rDNA restriction analysis. Appl Microbiol Biotechnol 97(15):6991–7003

Chandna P, Nain L, Singh S, Kuhad RC (2013b) Assessment of bacterial diversity during composting of agricultural byproducts. BMC Microbiol 13(1):1–14

Chandna P, Gupta S, Rajam MV, Kuhad RC (2014) Molecular identification and in vitro screening of antagonistic bacteria from agricultural byproduct compost: effect of compost on development and photosynthetic efficiency of tomato plant. Ann Microbiol 64(2):571–580

Diaz LF, Savage GM, Eggerth LL, Golueke CG (2020) Composting and recycling: municipal solid waste. CRC Press

Dincă MN, Ferdeş M, Paraschiv G, Ungureanu N, Zăbavă B, Ionescu M, Moiceanu G (2019) Recovery of organic waste through composting process. Acta Tech Corviniensis Bull Eng 12(1):119–122

Elmer WH (2009) Influence of earthworm activity on soil microbes and soilborne diseases of vegetables. Plant Dis 93(2):175–179

Galgani P, van der Voet E, Korevaar G (2014) Composting, anaerobic digestion and biochar production in Ghana. Environmental–economic assessment in the context of voluntary carbon markets. Waste Manag 34(12):2454–2465

Głąb T, Żabiński A, Sadowska U, Gondek K, Kopeć M, Mierzwa-Hersztek M, Stanek-Tarkowska J (2020) Fertilization effects of compost produced from maize, sewage sludge and biochar on soil water retention and chemical properties. Soil Tillage Res 197:104493

Gong X, Wei L, Yu X, Li S, Sun X, Wang X (2017) Effects of rhamnolipid and microbial inoculants on the vermicomposting of green waste with Eisenia fetida. PLoS One 12(1):e0170820

Gopal M, Bhute SS, Gupta A, Prabhu SR, Thomas GV, Whitman WB, Jangid K (2017) Changes in structure and function of bacterial communities during coconut leaf vermicomposting. Antonie Van Leeuwenhoek 110(10):1339–1355

Greff B, Lakatos E, Szigeti J, Varga L (2021) Co-composting with herbal wastes: potential effects of essential oil residues on microbial pathogens during composting. Crit Rev Environ Sci Technol 51(5):457–511

Grgić DK, Domanovac MV, Domanovac T, Šabić M, Cvetnić M, Bulatović VO (2019) Influence of Bacillus subtilis and Pseudomonas aeruginosa BSW and Clinoptilolite addition on the biowaste composting process. Arab J Sci Eng 44(6):5399–5409

Guo J-P, Tan J-L, Wang Y-L, Wu H-Y, Zhang C-P, Niu X-M et al (2011) Isolation of talathermophilins from the thermophilic fungus Talaromyces thermophilus YM3-4. J Nat Prod 74(10):2278–2281

Hatayama K, Shoun H, Ueda Y, Nakamura A (2014) Brevibacillus fulvus sp. nov., isolated from a compost pile. Int J Syst Evol Microbiol *64*(Pt_2):506–512

Hefnawy M, Gharieb M, Nagdi OM (2013) Microbial diversity during composting cycles of rice straw. Int J Adv Biol Biomed Res 7(1):30–43

Huang K, Li F, Wei Y, Chen X, Fu X (2013) Changes of bacterial and fungal community compositions during vermicomposting of vegetable wastes by Eisenia foetida. Bioresour Technol 150:235–241

Ivors K, Beyer D, Wuest P, Kang S (2002) Survey of fungal diversity in mushroom compost using sequences of PCR-amplified genes encoding 18S ribosomal RNA. In: Microbiology of composting. Springer, Berlin, pp 17–24

Jain MS, Daga M, Kalamdhad AS (2019) Variation in the key indicators during composting of municipal solid organic wastes. Sustain Environ Res 29(1):1–8

Jha MK, Sondhi OAK, Pansare M (2003) Solid waste management-a case study. Indian J Environ Prot 23:1153–1160

Kausar H, Ismail MR, Saud HM, Othman R, Habib S (2013) Use of lignocellulolytic microbial consortium and pH amendment on composting efficacy of rice straw. Compost Sci Util 21(3-4):121–133

Kaza S, Yao L, Bhada-Tata P, Van Woerden F (2018) What a waste 2.0: a global snapshot of solid waste management to 2050. World Bank Publications

Khambata SR, Bhat JV (1953) Studies on a new oxalate-decomposing bacterium, pseudomonas oxalaticus. J Bacteriol 66(5):505–507

Kobayashi T, Murai Y, Tatsumi K, Iimura Y (2009) Biodegradation of polycyclic aromatic hydrocarbons by Sphingomonas sp. enhanced by water-extractable organic matter from manure compost. Sci Total Environ 407(22):5805–5810

Langarica-Fuentes A, Handley PS, Houlden A, Fox G, Robson GD (2014a) An investigation of the biodiversity of thermophilic and thermotolerant fungal species in composts using culture-based and molecular techniques. Fungal Ecol 11:132–144

Langarica-Fuentes A, Zafar U, Heyworth A, Brown T, Fox G, Robson GD (2014b) Fungal succession in an in-vessel composting system characterized using 454 pyrosequencing. FEMS Microbiol Ecol 88(2):296–308

Latifah O, Ahmed OH, Susilawati K, Majid NM (2015) Compost maturity and nitrogen availability by co-composting of paddy husk and chicken manure amended with clinoptilolite zeolite. Waste Manag Res 33(4):322–331

Li Z, Yang Y, Xia Y, Wu T, Zhu J, Wang Z, Yang J. (2019) The succession pattern of bacterial diversity in compost using pig manure mixed with wood chips analyzed by 16S rRNA gene analysis. *BioRxiv*, 674069

Ma L, Zhao Y, Meng L, Wang X, Yi Y, Shan Y et al (2020) Isolation of thermostable lignocellulosic bacteria from chicken manure compost and a M42 family endocellulase cloning from Geobacillus thermodenitrificans Y7. Front Microbiol 11:281

Madsen EL, Alexander M (1982) Transport of Rhizobium and Pseudomonas through soil. Soil Sci Soc Am J 46(3):557–560

Manivannan N, Thajuddin N, Daniel T, Gunasekaran M (2012) Effect of Aspergillus sp. as a bioinoculant in vermicompost with special reference to phosphate enrichment. Am Eurasian J Sustain Agric 6(4):205–208

Martín-Gil J, Gómez-Sobrino E, Correa-Guimaraes A, Hernández-Navarro S, Sánchez-Báscones M, del Carmen Ramos-Sánchez M (2008) Composting and vermicomposting experiences in the treatment and bioconversion of asphaltens from the Prestige oil spill. Bioresour Technol 99(6):1821–1829

Meng Q, Xu X, Zhang W, Men M, Xu B, Deng L et al (2019) Bacterial community succession in dairy manure composting with a static composting technique. Canadian J Microbiol 65(6):436–449

Nagavallemma KP, Wani SP, Lacroix S, Padmaja VV, Vineela C, Rao MB, Sahrawat KL (2004) Vermicomposting: recycling wastes into valuable organic fertilizer. Global Theme on Agroecosystems Report no. 8

Newswire PR (2019) New York. NY,(May 23, 2002), 2

Nováková A, Nováková A (2012) Collection of microscopic fungi ISB: catalogue of strains. Institute of Soil Biology, Biology Centre AS CR, České Budějovice

Palaniveloo K, Amran MA, Norhashim NA, Mohamad-Fauzi N, Peng-Hui F, Hui-Wen L et al (2020) Food waste composting and microbial community structure profiling. PRO 8(6):723

Park C, Park W (2018) Survival and energy producing strategies of alkane degraders under extreme conditions and their biotechnological potential. Front Microbiol 9:1081

Partanen P, Hultman J, Paulin L, Auvinen P, Romantschuk M (2010) Bacterial diversity at different stages of the composting process. BMC Microbiol 10(1):1–11

Parthasarathi K, Ranganathan LS, Anandi V, Zeyer J (2007) Diversity of microflora in the gut and casts of tropical composting earthworms reared on different substrates. J Environ Biol 28(1): 87–97

Pathma J, Sakthivel N (2012) Microbial diversity of vermicompost bacteria that exhibit useful agricultural traits and waste management potential. SpringerPlus 1(1):1–19

Petric I, Avdihodžić E, Ibrić N (2015) Numerical simulation of composting process for mixture of organic fraction of municipal solid waste and poultry manure. Ecological engineering 75:242–249

Rastogi M, Nandal M, Nain L (2019) Additive effect of cow dung slurry and cellulolytic bacterial inoculation on humic fractions during composting of municipal solid waste. Int J Recycling Organic Waste Agric 8(3):325–332

Rouelle J (1983) Introduction of amoebae and Rhizobium japonicum into the gut of Eisenia fetida (Sav.) and Lumbricus terrestris L. In: Earthworm ecology. Springer, Dordrecht, pp 375–381

Ryckeboer J, Mergaert J, Coosemans J, Deprins K, Swings J (2003) Microbiological aspects of biowaste during composting in a monitored compost bin. J Appl Microbiol 94(1):127–137

Saini A, Aggarwal NK, Sharma A, Yadav A (2015) Actinomycetes: a source of lignocellulolytic enzymes. Enzyme Res 2015:279381

Sharma D, Yadav KD (2018) Application of rotary in-vessel composting and analytical hierarchy process for the selection of a suitable combination of flower waste. Geol Ecol Landscapes 2(2): 137–147

Silva JP, Ticona ARP, Hamann PRV, Quirino BF, Noronha EF (2021) Deconstruction of lignin: from enzymes to microorganisms. Molecules 26(8):2299

Singh A, Basak P (2018) Economic and environmental evaluation of municipal solid waste management system using industrial ecology approach: Evidence from India. J Clean Prod 195:10–20

Sinha RK, Herat S, Bharambe G, Brahambhatt A (2010) Vermistabilization of sewage sludge (biosolids) by earthworms: converting a potential biohazard destined for landfill disposal into a pathogen-free, nutritive and safe biofertilizer for farms. Waste Manag Res 28(10):872–881

Su X, Zhang S, Mei R, Zhang Y, Hashmi MZ, Liu J et al (2018) Resuscitation of viable but nonculturable bacteria to enhance the cellulose-degrading capability of bacterial community in composting. Microbial Biotechnol 11(3):527–536

Thin Mar T, Kyaw EP, Lynn TM, San Yu S (2018) The effects of compost based biofertilizer on eggplant (Solanum melongena L.) growth. Int J Plant Biol Res

Ueda J, Kurosawa N (2015) Characterization of an extracellular thermophilic chitinase from Paenibacillus thermoaerophilus strain TC22-2b isolated from compost. World J Microbiol Biotechnol 31(1):135–143

Van Heerden I, Cronjé C, Swart SH, Kotzé JM (2002) Microbial, chemical and physical aspects of citrus waste composting. Bioresour Technol 81(1):71–76

Varma VS, Dhamodharan K, Kalamdhad AS (2018) Characterization of bacterial community structure during in-vessel composting of agricultural waste by 16S rRNA sequencing. 3 Biotech 8(7):1–8

Vaz-Moreira I, Silva ME, Manaia CM, Nunes OC (2008) Diversity of bacterial isolates from commercial and homemade composts. Microb Ecol 55(4):714–722

Wahyuningsih S (2019) Biodegradation of pulp sludge by Phanerochaete chrysosporium, Penicillium oxalicum and Penicillium citrinum after six months incubation. Indonesian J Forest Res 6(2):95–105

Wan L, Wang X, Cong C, Li J, Xu Y, Li X et al (2020) Effect of inoculating microorganisms in chicken manure composting with maize straw. Bioresour Technol 301:122730

Watanabe K, Nagao N, Yamamoto S, Toda T, Kurosawa N (2007) Thermobacillus composti sp. nov., a moderately thermophilic bacterium isolated from a composting reactor. Int J Syst Evol Microbiol 57(7):1473–1477

Wei Y, Zhao Y, Shi M, Cao Z, Lu Q, Yang T et al (2018) Effect of organic acids production and bacterial community on the possible mechanism of phosphorus solubilization during composting with enriched phosphate-solubilizing bacteria inoculation. Bioresour Technol 247:190–199

Wright MS, Lima IM (2021) Identification of microbial populations in blends of worm castings or sugarcane filter mud compost with biochar. Agronomy 11(8):1671

Yasir M, Aslam Z, Kim SW, Lee S-W, Jeon CO, Chung YR (2009) Bacterial community composition and chitinase gene diversity of vermicompost with antifungal activity. Bioresour Technol 100(19):4396–4403

Zainudin MHM, Hassan MA, Shah UKM, Abdullah N, Tokura M, Yasueda H et al (2014) Bacterial community structure and biochemical changes associated with composting of lignocellulosic oil palm empty fruit bunch. Bioresources 9(1):316–335

Zhang X, Zhan Y, Zhang H, Wang R, Tao X, Zhang L, Li J (2021) Inoculation of phosphate-solubilizing bacteria (Bacillus) regulates microbial interaction to improve phosphorus fractions mobilization during kitchen waste composting. Bioresour Technol 340:125714

Zurbrügg C, Drescher S, Rytz I, Md Maqsood Sinha AH, Enayetullah I (2005) Decentralised composting in Bangladesh, a win-win situation for all stakeholders. Resour Conserv Recycl 43(3):281–292

Chapter 20
Cost-Benefit Assessment of Solid-Based Biofertilizer Production

R. Kannan and T. Seenivasa Moorthy

Abstract Due to an ever-increasing human population and a significant demand for agricultural food security in the public domain, biofertilizer is one of the best technologies for creating a sustainable ecosystem in the globe. Solid-based biofertilizer manufacturing uses a variety of low-cost substrates as a carrier for commercially produced bacterium, fungus, and algae-based biofertilizer. This chapter discusses the cost-effectiveness of solid-based biofertilizer production and the impact of soil health on agriculture productivity, as well as the results obtained by using microbial-based biofertilizers to increase soil microbial population, soil health, and toxic-free agriculture products while reducing the use of synthetic fertilizers. Solid-based biofertilizers are a cost-effective product that saves money while also benefiting the environment and increasing productivity.

Keywords Cost effective · Biofertilizer · Ash · Eco-friendly · Agriculture · Soil health

20.1 Introduction

Biofertilizers are living creatures that symbiotically or asymbiotically supplement plant nutrition supply. Azotobacter and, Azospirillum, among the asymbiotic, nitrogen-fixing bacteria, lead to a significant increase in crop output of 15–20% while minimizing soil nutrient depletion (Motsara et al. 1995). In addition to these advantages, biofertilizers can save at least 20–30 kg/ha of inorganic nitrogen fertilizers because they have a high capacity for nitrogen fixation (Tilak 1991). Continuous usage of inorganic fertilizers resulted in nutritional inadequacy, physicochemical imbalances in the soil, and unsustainable crop output (Tandon 1987). With the rising expense of inorganic fertilizers, small and marginal farmers are finding it difficult to apply the recommended dose. As a result, for supplementing

R. Kannan (✉) · T. S. Moorthy
Kalasalingam School of Agriculture and Horticulture (KSAH), Kalasalingam Academy of Research and Education (KARE), Krishnankoil, Tamil Nadu, India

N. Amaresan et al. (eds.), *Agricultural Microbiology Based Entrepreneurship*, Microorganisms for Sustainability 39,
https://doi.org/10.1007/978-981-19-5747-5_20

and complementing chemical fertilizers, renewable and low-cost sources of plant nutrients should be substituted, which can be afforded by the majority of the farming population (Gupta et al. 1999).

Biofertilizers are living or latent cells of effective strains of microorganisms, usually immobilized on a carrier that assist crops take up nutrients through their interactions in the rhizosphere, resulting in increased production. Chemical fertilizers have a number of advantages over biofertilizers. These include their eco-friendliness and, in the case of developing countries, their suitability for small- and medium-scale production using locally accessible materials. The International Federation of Organic Agriculture Movements recognizes biofertilizers as organic agriculture inputs (IFOAM 2005). Seed inoculation, in which the inoculant (bacteria-carrier mixture) is mixed with water to make a slurry form and then mixed with seeds, or "soil inoculation," in which the inoculant (bacteria-carrier mixture) is spread over the field during cultivation, are two methods of supplying biofertilizers to the soil. The carrier in the event of seed inoculation must be a fine powder. Biofertilizers are typically made up of carrier-based inoculants containing viable microorganisms. The incorporation of microorganisms into carrier materials allows for easy handling, long-term storage and has excellent biofertilizer efficacy. Bacterial inoculants are a type of biofertilizer that comprises rhizobia, nitrogen-fixing rhizobacteria, plant growth-promoting rhizobacteria, phosphate-solubilizing bacteria, and others. Basically, a standard process can be used to make a carrier-based inoculant containing these microorganisms.

Small and marginal farmers can benefit from biofertilizer as a cost-effective means of achieving their ultimate aim of increased productivity. To augment chemical fertilizers, biofertilizers are a low-cost, effective, and sustainable source of plant nutrients. Bacteria, fungus, and blue green algae are examples of microorganisms that can be employed as biofertilizers. These organisms are placed in the plant's rhizosphere to increase their activity in the soil (Jambhhekar 2002).

Punjab during the 2013–2014 and 2014–2015 Rabi seasons. The experimental farm is situated in India's Trans-Gangetic agro-climatic zone, within Punjab's central plain region. The experimental farm's soil is loamy sand in texture, with low organic carbon (0.32%) and available nitrogen (119.8 kg ha^{-1}) but medium available P (13.6 kg ha^{-1}) and available K. (161 kg ha^{-1}). The experiment included four phosphorus levels (0, 20, 30, and 40 kg P2O5 ha^{-1}) and two/four biofertilizer treatments (Uninoculated control and Rhizobium + plant growth promoting rhizobacteria (PGPR) in 2013–2014) replicated three times in Randomized Complete Block Design (Uninoculated control, Rhizobium, PGPR, and Rhizobium + PGPR in 2014–15) (RCBD). On November 11, 2013 and November 6, 2014, lentil cultivar "LL 699" was sown (Singh et al. 2017).

Phosphorus was applied at planting using a single super-phosphate (16% P2O5), according to the treatments. Prior to sowing, seeds were inoculated with Rhizobium (LLR 12) and PGPR (RB 2) as indicated in the treatments. According to the treatments, Rhizobium and PGPR were utilized as single inoculations or as a combined inoculation (Rhizobium + PGPR). Before sowing, the inoculated seeds were dried in the shade. At the time of planting, a consistent basal dose of N fertilizer

Table 20.1 Level of input of biofertilizers and phosphorus use for different treatments for lentil cultivation

Type of input	Cost (Rs. ha^{-1})
Rhizobium	50
PGPR	50
Rhizobium + PGPR	100
20 kg P2O5ha^{-1}	2862
30 kg P2O5ha^{-1}	4292
40 kg P2O5ha^{-1}	5725
Labor charges for *rhizobium* inoculation	50
Labor charges for PGPR inoculation	50
Labor charges for *rhizobium* + PGPR inoculation	50
General cost of cultivation (except the treatment costs)	20,500 (in 2013–14) 22,500 (in 2014–15)

Table 20.2 Effect of phosphorus and biofertilizers on economic returns of lentil in 2013–2014

Treatment	Cost of cultivation (Rs. ha^{-1})	Returns (Rs. ha^{-1})		B:C[a]
		Gross	Net	
0 kg P2O5ha^{-1}(unfertilized control)	20,500	35,400	14,900	1.73
20 kg P2O5ha^{-1}	23,362	40,644	17,282	1.74
30 kg P2O5ha^{-1}	24,792	43,704	18,912	1.76
40 kg P2O5ha^{-1}	26,225	45,015	18,790	1.72
Rhizobium + PGPR	20,650	41,081	20,431	1.99
Rhizobium + PGPR +20 kg P2O5ha^{-1}	23,512	44,141	20,629	1.88
Rhizobium + PGPR +30 kg P2O5ha^{-1}	24,942	44,578	19,636	1.79
Rhizobium + PGPR +40 kg P2O5ha^{-1}	26,375	45,889	19,514	1.74
CD ($p = 0.05$)	5281	NS		NS

[a]*B:C* Benefit–Cost Ratio

of 12.5 kg ha^{-1} through urea (46% N) was broadcast. In 2014 and 2015, the crop was harvested on April 8 and 14, respectively. The crop was grown in accordance with the guidelines (PAU 2013). On November 11, 2013 and November 6, 2014, lentil cultivar "LL 699" was sown (Singh et al. 2017).

Singh et al. (2017) used the grain yield multiplied by the minimum support price (MSP), which was Rs. 2950 quintal-1. Human labor, mechanical energy, seed, fertilizer, biofertilizers, pesticide, irrigation, and other variable expenses are included in the cost of agriculture (i.e., total variable costs). Table 20.1 shows the details of the expenses associated with various inputs. Total variable costs were subtracted from gross profits to arrive at net returns. By dividing the gross returns by the entire cost of cultivation, the benefit cost ratio (B:C) was computed. The gross and net returns were calculated in rupees per hectare.

Gross Returns: The gross returns rose as the phosphorus dose was raised from 0 to 40 kg P2O5 ha^{-1} (Tables 20.2 and 20.3). The control treatment, which did not

Table 20.3 Effect of phosphorus and biofertilizers on economic returns of lentil in 2014–2015

Treatment	Cost of cultivation (Rs. ha^{-1})	Returns (Rs. ha^{-1}) Gross	ha^{-1}	ha^{-1}
P2O5(kg ha^{-1})				
0	22,587	47,140	24,553	2.08
20	25,449	53,534	28,085	2.10
30	26,879	56,402	29,523	2.09
40	28,312	57,377	29,065	2.02
CD (p = 0.05)	1607	1607		NS
Biofertilizers				
Uninoculated	25,719	50,454	24,735	1.96
Rhizobium	25,819	53,845	28,026	2.08
PGPR	25,820	52,838	27,018	2.04
Rhizobium + PGPR	25,869	57,316	31,447	2.21
CD (p = 0.05)	1607	1607		0.06

[a]*B:C* Benefit–Cost Ratio

use phosphate or biofertilizers, yielded the lowest gross returns. The coinoculated treatment (Rhizobium + PGPR) produced greater gross returns than the uninoculated control or solitary inoculations of Rhizobium and PGPR among biofertilizer treatments. The largest net returns from inoculation, according to Jain et al. (2006), may be related to maximum grain and straw yield.

The combination treatment of Rhizobium + PGPR +40 kg P2O5ha^{-1}(Rs. 45,902) yielded the highest gross returns (Table 20.2), which could be related to increased grain yields. The combined use of phosphorus and biofertilizers (Rhizobium, PGPR, and Rhizobium + PGPR) boosted gross yields when compared to the solitary application of 20, 30, and 40 kg P2O5 ha^{-1}. If the value of the increase in crop output due to the quantity of fertilizer applied is greater than the cost of fertilizer used, then applying a unit fertilizer is cost-effective. Even after a steady increase in production, if a unit of fertilizer does not raise yield enough to pay for its cost, its application will be inefficient and will not generate profit (Singh 2004).

The application of essential plant nutrients in optimum quantity and right proportion is the key to increase the profit.

Net Returns: Among the various phosphorus levels tested, 30 kg P2O5 ha^{-1} yielded the best net yields (Tables 20.2 and 20.3). The coinoculated treatment had larger net returns than the uninoculated control (Table 20.2), as well as the uninoculated control and single Rhizobium and PGPR inoculations (Table 20.3). In chickpea, Jain et al. found higher net returns (Rs. 11,312 ha^{-1}) with the application of Rhizobium + phosphorus solubilizing bacteria compared to the uninoculated control (Rs. 8282 ha^{-1}) and single Rhizobium (Rs. 9883 ha^{-1}) and PGPR (Rs. 9697 ha^{-1}) inoculations (2006).

The integrated application of Rhizobium + PGPR +20 kg P2O5ha^{-1}(Rs. 20,620) yielded the highest net returns (Table 20.2), which could be related to the high grain production and low cultivation cost. Previously, Jain et al. (2006) observed that higher gross returns (Rs. 30,538 ha^{-1}) in chickpea resulted in higher net returns (Rs. 22,067 ha^{-1}). The application of 20 kg P2O5 ha^{-1} with consortium (Rhizobium + PGPR) was more profitable than 40 kg P2O5ha^{-1}+ Rhizobium + PGPR, which could be attributable to the lower cost of single superphosphate and biofertilizers compared to the increased grain production (Kanwar et al. 2013). Thus, using Rhizobium + PGPR+ 20 kg P2O5ha^{-1} instead of Rhizobium + PGPR +40 kg P2O5ha^{-1} or 40 kg P2O5ha^{-1} alone resulted in a net saving of 20 kg P2O5ha^{-1} without losing the yield economic returns.

Benefit-Cost Ratio: Tables 20.2 and 20.3 show that 20 and 30 kg P2O5ha^{-1} produced higher B:C than 0 and 40 kg P2O5ha^{-1}, while the differences were not statistically significant in 2014–2015. Inoculation treatment resulted in considerably greater B:C than uninoculated control (Table 20.2) and uninoculated control as well as single Rhizobium or PGPR inoculation.

Except for Rhizobium + PGPR +20 kg P2O5ha^{-1} (1.88), the combined usage of Rhizobium+ PGPR +20 kg P2O5ha^{-1} (1.88) resulted in a higher B:C than all other treatments (Table 20.2). These findings are comparable to those of Jain et al. (2006), who found that higher grain yield was related to improved nutrient uptake (N and P). Maximum benefits can be obtained by reducing the phosphorus dose and, as a result, lowering the cost per unit of output through increased yield. Phosphorus or biofertilizers alone were unable to produce a superior B:C than the combined application of both. It demonstrates the importance of biofertilizers as well as fertilizers. In chickpea, the low cost of biofertilizers is responsible for improving the B:C ratio in PGPR (4.33) above the uninoculated control (3.54) when compared to phosphorus fertilizers (Tanwar et al. 2010).

20.2 Carrier Material

As carrier materials, peat soil, lignite, vermiculite, charcoal, press mud, farmyard manure, wood ash, and soil mixture can be employed. It has been discovered that neutralized peat soil/lignite are better carrier materials for biofertilizer manufacturing.

In such diets, wood ash has been used to replace limestone as a source of calcium (Van Ryssen et al. 2014). Limestone, on the other hand, is a relatively inexpensive and readily available source of calcium. This reduces the demand for additional calcium sources. In their study, Van Ryssen et al. (2014) discovered that fine wood ash tended to separate from other dietary items, except when macadamia oil cake meal, which had relatively high levels of oil, was included. Apart from Ca, another feature of wood ash and wood bark ash is the range of other mineral elements it contains (Van Ryssen and Ndlovu 2018).

Given the predominant presence of Ca in wood ash and the low temperatures of wood burning in homestead fires, it may be predicted that a substantial proportion of trace elements in wood ash would be present as carbonates and bicarbonates, as suggested by Van Ryssen and Ndlovu (2018). When plant material, such as wood, is burned, the inorganic residue left is ash, which is commonly available in subsistence farming groups. Although it has the potential to be utilized in agriculture, ash is considered a waste product. Wood ash is a source of calcium that can be employed in resource-constrained farming circumstances (Van Ryssen and Ndlovu 2018).

In reality, when applied to forestry and agricultural soils, wood ash has been discovered to be an effective fertilizer and may be used as a source of plant mineral nutrients (Naylor and Schmidt 1986; Etiegni and Campbell 1991; Olanders and Steenari 1995).

Naylor and Schmidt (1986), reported wood ash has low fertilizing properties, it can be used as a substitute for lime and limestone to neutralize acidic soils through supplementing Ca, K, P, Mg, and replacing microelements that would have been depleted from the soil during plant growth and harvesting. Liming with wood ash may reduce the toxic effects of Al and Mn on plants grown in acidic soils. An indirect benefit of using wood ash as a fertilizer would be the rise in the pH level of acidic soils that would increase the bioavailability of elements such as P, Mg, Mo, and Se in plants (Reid and Horvath 1980), which in turn would be beneficial to animals consuming those plants. Wood ash can also contain compounds detrimental to agricultural crops and the entire chain of production.

Farmers use those biofertilizers or bio-inoculants as a substitute for NPK (nitrogen-phosphorous-potash) fertilizer in paddy and a number of crops to increase yield. If the farmers were to use bio-inoculants, they would spend only a couple of hundred rupees. This is not the only savings, though. The use of wood ash will result in increase in yield by 10% to 15%. This will be an additional income for the farmers. The larger picture is that the State will have to spend less on fertilizers and also fertilizer subsidy.

Various types of materials are used as carrier for seed or soil inoculation. For preparation of seed inoculant, the carrier material is milled to fine powder with particle size of 10–40 μm. According to the "Handbook for Rhizobia" (Somasegaran and Hoben 1994), the properties of a good carrier material for seed inoculation are:

1. Nontoxic to inoculant bacterial strain.
2. Good moisture absorption capacity.
3. Easy to process and free of lump-forming materials.
4. Easy to sterilize by autoclaving or gamma-irradiation.
5. Available in adequate amounts.
6. Inexpensive.
7. Good adhesion to seeds.
8. Good pH buffering capacity. Needless to say.
9. Nontoxic to plant, is another important property.

Peat is the most frequently used carrier material for seed inoculation. Peat-based rhizobial inoculant is already used in many countries and a number of information is available on the properties and effect of the inoculant. For soil inoculation, carrier material with granular form (0.5–1.5 mm) is generally used. Granular forms of peat, perlite, charcoal or soil aggregates are suitable for soil inoculation. Various types of materials are used or tested as carrier for bacterial inoculant (mostly Rhizobia). Other essential criteria for carrier selection relating to survival of the inoculant bacteria should be considered.

1. Survival of the inoculant bacteria on seed. Seeds are not always sown immediately after seed coating with the inoculant bacteria. The bacteria have to survive on seed surface against drying condition until placed into soil.
2. Survival of the inoculant bacteria during the storage period.
3. Survival of the inoculant bacteria in soil.

After being introduced into the soil, the inoculant bacteria have to compete with native soil microorganisms for the nutrient and habitable niche, and have to survive against grazing protozoa. Such carrier materials that offer the available nutrient and/or habitable micro-pore to the inoculant bacteria will be desirable. In this sense, materials with micro-porous structure, such as soil aggregate and charcoal, will be good carriers for soil inoculant.

20.3 Carrier-Based Biofertilizers

At present, biofertilizers are supplied as carrier-based microbial inoculants, which are added to the soil to enrich the soil fertility. The carrier is a medium that can carry the microorganisms in sufficient quantities and keep them viable under specified conditions, easy to supply to the farmers. The use of ideal carrier material is necessary in the production of good quality biofertilizer. A good carrier should have the following qualities:

- Highly absorptive (water-holding capacity) and easy to process.
- Nontoxic to microorganisms.
- Easy to sterilize effectively.
- Available in adequate amounts and low-cost.
- Provide good adhesion to seeds.
- Has good buffering capacity.
- High organic matter content and water-holding capacity of more than 50%.

Other essential criteria for carrier selection relating to the survival of the inoculant bacteria should be considered.

- Survival of the inoculant bacteria on seeds. Seeds are not always sown immediately after seed coating with the inoculant bacteria. The bacteria have to survive on seed surface against drying condition until placed into soil.
- Survival of the inoculant bacteria during the storage period.
- Survival of the inoculant bacteria in soil. After being introduced into the soil, the inoculant bacteria have to compete with native soil microorganisms for the nutrient and habitable niche, and have to survive against grazing protozoa. Such carrier materials that offer the available nutrient and/or habitable micropores to the inoculant bacteria will be desirable. In this sense, materials with

micro-porous structure, such as soil aggregate and charcoal, will be good carriers for soil inoculants.

20.4 Liquid Biofertilizers

The strength of biofertilizers is determined by two basic parameters: number of cells and efficiency of the microorganisms to fix nitrogen or solubilize phosphates.

Liquid biofertilizers are liquid formulations containing the dormant form of desired microorganisms and their nutrients along with the substances that encourage formation of resting spores or cysts for longer shelf life and tolerance to adverse conditions. The dormant forms, on reaching the soil, germinate to produce a fresh batch of active cells. These cells grow and multiply by utilizing the carbon source in the soil or from root exudates.

As an alternative to conventional carrier-based biofertilizers, liquid formulation technology, which has more advantages than the carrier-based inoculants, has been developed in the Department of Agricultural Microbiology, TNAU, Coimbatore. The advantages of liquid biofertilizers over conventional carrier-based biofertilizers are as follows:

- Longer shelf life, 12–24 months.
- No contamination.
- No loss of properties due to storage up to 45 ° C.
- Greater potential to fight with native populations.
- High populations can be maintained at more than 10^9 cells/ml up to 12–24 months.
- Easy identification by typical fermented smell.
- Cost saving on carrier material, pulverization, neutralization, sterilization, packing and transport.
- Quality control protocols are easy and quick.
- Better survival on seeds and soil.
- No need of running biofertilizer production units throughout the year.
- Very much easy to use by the farmer.
- Dosages are ten times less than those of carrier-based powder biofertilizers.
- High commercial revenues.
- High export potential.
- Very high enzymatic activity, since contamination is nil.

Among different techniques to produce biofertilizer, the concept of effective microorganisms (EM), which are available in liquid form, was introduced in 1991 by Dr.TeruoHiga of Japan. The major groups of microorganisms contained in the EM include filamentous fungi, yeast, lactic acid bacteria, and other soil bacteria. The application of EM aims to function as inoculum of microorganisms to the soil in which it will help to establish or reestablish soil ecosystems. EM is commercially available in concentrated form that needs to be processed before the application.

According to the procedure suggested by the EM manufacturer, the concentrated EM (EM Bokashi) can be used directly by mixing with molasses and water. However, the common method is to use EM Bokashi as a starter to ferment the raw materials and produce either liquid or solid biofertilizer. The common raw materials include leftover plant or animal materials in the farms. The fermentation period was suggested to be at least 7 days and the product is recommended to be used within 3 months. Today, the production of ready-to-use liquid biofertilizer from EM is becoming available in the market due to the convenience for small-scale farming or domestic application in which the users do not have space and raw materials available for fermentation.

20.5 Mode of Application

There are three ways of using biofertilizers.

20.5.1 Seed Treatment

Seed treatment is the most common method adopted for all types of inoculants. The seed treatment is effective and economic. For a small quantity of seeds (up to 5 kg), the coating can be done in a plastic bag. For this purpose, a plastic bag sized 21″ × 10″ or larger can be used. The bag should be filled with 2 kg of seeds or more. The bag should be closed in such a way so as to trap the air as much as possible. The bag should be squeezed for 2 min or more until all the seeds are uniformly wetted. Then the bag is opened, inflated again, and shaken gently. The shaking should stop after each seed gets a uniform layer of culture coating. The bag is opened and the seeds are shade-dried for 20–30 min. For large amounts of seeds, coating can be done in a bucket and the inoculant can be mixed directly by hand. Seed treatment with *Rhizobium, Azotobacter, Azospirillum*, along with PSM can be done. The seed treatment can be done with any of two or more bacteria. There is no side (antagonistic) effect. The important things that have to be kept in mind are that the seeds must be coated first with *Rhizobium, Azotobacter*, or *Azospirillum*. When each seed gets a layer of these bacteria, the PSM inoculant has to be coated as an outer layer. This method will provide maximum cell counts of all bacteria required for better results. Treatments of seeds with any two bacteria will not provide a maximum number of bacteria on individual seeds.

20.5.2 Root Dipping

This method is used for application of *Azospirillum*/PSM on paddy transplating/vegetable crops. The required quantity of *Azospirillum*/PSM has to be mixed with 5–10 liters of water at one corner of the field and the roots of seedlings has to be dipped for a minimum of half an hour before transplantation.

20.5.3 Soil Application

Use 200 ml of PSM per acre. Mix PSM with 400–600 kg of cow dung farmyard manure along with ½ bag of rock phosphate, if available. The mixture of PSM, cow dung, and rock phosphate has to be kept under any tree or in the shade overnight and 50% moisture should be maintained. The mixture is used for soil application in rows or during leveling of soil.

References

Etiegni L, Campbell AG (1991) Physical and chemical characteristics of wood ash. Bioresour Technol 37:173–178. https://doi.org/10.1016/0960-8524(91)90207-Z

Gupta RP, Sharma VP, Singh DK, Srivastava KJ (1999) Effect of organic manures and inorganic fertilizers on growth yield and quality of onion variety. Agrifound dark red. Natl Hortic Res Dev Found News Lett 19(2/3):7–11

IFOAM (2005) Norms of IFOAM basic standards for organic production and processing. IFOAM, Germany. http://www.ifoam.org. Accessed 8 Nov 2011

Jain LK, Singh P, Balyan JK (2006) Productivity and profitability of chickpea (*Cicer arietinum* L.) cultivation as influenced by biofertilizers and phosphorus fertilization. Indian J Dryland Agric Res Dev 21:201–203

Jambhhekar H (2002) Vermiculture in India online training material. Agricultural Bioteks, Pune

Kanwar P, Singh P, Singh P, Singh P (2013) Effect of Rhizobium, PSB and phosphorus on yield and economics of mungbean. Ann Plant Soil Res 15:164–166

Motsara MR, Bhattacharya P, Srivastaba B (1995) Biofertiliser, technology, marketing and usage. New Delhi, Fertilizer Development Consultation Organization, p 183

Naylor LM, Schmidt EJ (1986) Agricultural use of wood ash as a fertilizer and liming material. Tappi J 69(10):114–119

Olanders B, Steenari B-M (1995) Characteristics of ashes from wood and straw. Biomass Bioenergy 8:105–115

PAU (2013) Packages of practices for Rabi crops of Pun-jab. Punjab Agricultural University, Ludhiana, India

Reid RL, Horvath DJ (1980) Soil chemistry and mineral problems in farm livestock. A review. Anim Feed Sci Technol 5:95–167

Singh SS (2004) Soil fertility and nutrient management, 2nd edn. Kalyani Publishers, New Delhi

Singh N, Singh G, Aggarwal N (2017) Economic analysis of application of phosphorus, single and dual inoculation of Rhizobium and plant growth promoting rhizobacteria in lentil (Lens *culinaris medikus*). J Appl Nat Sci 9(2):1008–1011

Somasegaran P, Hoben HJ (1994) Handbook for rhizobia. Methods in Legume-Rhizobium technology. Springer-Verlag, New York, pp 332–341

Tandon HLS (1987) Fertilizer recommendation for horticultural crops. A guide book. Fertilizer Development and Consultation Organization, New Delhi, pp 10–14

Tanwar SPS, Rokadia P, Singh AK (2010) Productivity, nutrient balance and economics of *kabuli* chickpea (*Cicerkabulium*) as influenced by integrated nutrient management. Indian J Agron 55: 33–37

Tilak KVBR (1991) Bacterial fertilizers. Publication of Indian Council of Agricultural Research, New Delhi, p 65

Van Ryssen JBJ, Ndlovu H (2018) Wood ash in livestock nutrition: 1. Factors affecting the mineral composition of wood ash. Appl Anim Husb Rural Dev 11:53–61

Van Ryssen JBJ et al (2014) Different levels of macadamia oil cake meal, and wood ash vs. feed lime as dietary sources of calcium on bone characteristics of slow-growing chickens. South Afr J Anim Sci 44(1):71–79

Printed by Printforce, United Kingdom